Edited by Pramod Rastogi and Erwin Hack

Optical Methods for Solid Mechanics

Related Titles

Osten, W., Reingand, N. (eds.)

Optical Imaging and Metrology
Selected Topics

2012
ISBN: 978-3-527-41064-4

Kaufmann, G. H. (ed.)

Advances in Speckle Metrology and Related Techniques

2011
ISBN: 978-3-527-40957-0

Dörband, B., Müller, H., Gross, H.

Handbook of Optical Systems
Volume 5: Metrology of Optical Components and Systems

2012
ISBN: 978-3-527-40381-3

Gåsvik, K. J.

Optical Metrology

2002
ISBN: 978-0-470-84300-0

Edited by Pramod Rastogi and Erwin Hack

Optical Methods for Solid Mechanics

A Full-Field Approach

WILEY-VCH Verlag GmbH & Co. KGaA

The Editors

Prof. Pramod Rastogi
Swiss Federal Institute of Technology
Lausanne (EPFL)
Applied Computing and
Mechanics Laboratory
Station 18
CH-1015 Lausanne
Switzerland

Dr. Erwin Hack
Swiss Federal Laboratories for
Materials Science and Technology (EMPA)
Electronics/Metrology/Reliability
Ueberlandstr. 129
CH-8600 Duebendorf
Switzerland

Cover
courtesy K. Ramesh (adapted)

All books published by **Wiley-VCH** are carefully produced. Nevertheless, authors, editors, and publisher do not warrant the information contained in these books, including this book, to be free of errors. Readers are advised to keep in mind that statements, data, illustrations, procedural details or other items may inadvertently be inaccurate.

Library of Congress Card No.: applied for

British Library Cataloguing-in-Publication Data
A catalogue record for this book is available from the British Library.

Bibliographic information published by the Deutsche Nationalbibliothek
The Deutsche Nationalbibliothek lists this publication in the Deutsche Nationalbibliografie; detailed bibliographic data are available on the Internet at <http://dnb.d-nb.de>.

© 2012 Wiley-VCH Verlag & Co. KGaA, Boschstr. 12, 69469 Weinheim, Germany

All rights reserved (including those of translation into other languages). No part of this book may be reproduced in any form – by photoprinting, microfilm, or any other means – nor transmitted or translated into a machine language without written permission from the publishers. Registered names, trademarks, etc. used in this book, even when not specifically marked as such, are not to be considered unprotected by law.

Print ISBN: 978-3-527-41111-5

Cover Design   Adam-Design, Weinheim, Germany
Typesetting   Laserwords Private Limited, Chennai, India
Printing and Binding   Markono Print Media Pte Ltd, Singapore

Contents

Preface *XIII*
List of Contributors *XV*

1	**Basic Optics** *1*	
	Krishna Thyagarajan and Ajoy Ghatak	
1.1	Introduction *1*	
1.2	Light as an Electromagnetic Wave *1*	
1.2.1	Reflection and Refraction of Light Waves at a Dielectric Interface *7*	
1.3	Rays of Light *9*	
1.4	Imaging through Optical Systems *11*	
1.4.1	Thin Lens *13*	
1.4.2	Thick Lens *13*	
1.4.3	Principal Points of a Lens *14*	
1.5	Aberrations of Optical Systems *17*	
1.5.1	Monochromatic Aberrations *17*	
1.5.2	Spherical Aberration *17*	
1.5.3	Coma *18*	
1.5.4	Astigmatism and Curvature of Field *18*	
1.5.5	Distortion *18*	
1.6	Interference of Light *19*	
1.6.1	Young's Double-Slit Arrangement *19*	
1.7	Coherence *25*	
1.8	Diffraction of Light *27*	
1.8.1	Resolution of Optical Instruments *30*	
1.9	Anisotropic Media *33*	
1.10	Jones Calculus *35*	
1.11	Lasers *38*	
1.11.1	Principle *39*	
1.11.2	Coherence Properties of the Laser *41*	
1.12	Optical Fibers *42*	
1.13	Summary *44*

2	**Electronic Image Sensing and Processing** 47
	Thomas Baechler
2.1	Introduction 47
2.2	Image Formation 49
2.2.1	Geometrical Optics and Imaging Concepts 50
2.2.2	Optical Distortion and Other Limiting Factors 54
2.3	Image Sensing: From Photons to Electrons 56
2.3.1	Image Formation Related to Image Sensing 58
2.3.2	Operating Modes of State-of-the-Art Electronic Image Sensors 62
2.3.3	Noise Sources in State-of-the-Art Electronic Image Sensors 64
2.3.4	From Electrons to Digital Image Data 67
2.4	Image Processing 70
2.4.1	Image Histograms 71
2.4.2	Linear Point Operations 73
2.4.3	Multi-image Operations for Noise Reduction 76
2.4.4	Morphological Image Operations 77
2.4.5	Introduction to Feature and Motion Detection 78
2.5	Conclusions 79

3	**Phase Decoding and Reconstruction** 83
	Jan Burke
3.1	Introduction 83
3.2	Basic Concepts 85
3.2.1	Deriving a Generic Phase-Shifting Formula 91
3.2.2	Phase Shifting with Three Steps 94
3.2.3	General N-Step Method 96
3.2.4	Symmetrical N + 1 Step Formulae 97
3.2.5	Extended Averaging 99
3.2.6	Summary 100
3.3	Methods of Phase Shifting 101
3.3.1	Temporal Phase Shifting (TPS) 102
3.3.2	Spatial Phase Shifting (SPS) 107
3.3.3	Spatiotemporal Phase Shifting (STPS) 112
3.4	Designing and Analyzing Phase-Shift Methods with the Complex-Polynomial Method 115
3.5	Sources and Removal of Errors 124
3.5.1	Phase-Shift Miscalibration and Calibration 124
3.5.2	Signal Nonlinearity and Harmonics 126
3.5.3	Vibrations 127
3.5.4	Random Noise 127
3.5.5	Digitization Noise 128
3.6	Phase Unwrapping 129
3.6.1	Spatial Phase Unwrapping 129
3.6.2	Temporal Phase Unwrapping 133
3.6.3	Multiwavelength Unwrapping Techniques 134

4	**Experimental Stress Analysis – An Overview** *141*	
	Krishnamurthi Ramesh	
4.1	Introduction *141*	
4.2	Concept of Stress and Strain *141*	
4.3	Stress-Strain Relations *146*	
4.4	Rudiments of a Tension Test *147*	
4.5	Principal Stress and Strain *148*	
4.6	Concept of Stress Concentration *151*	
4.7	Birth of Fracture Mechanics *153*	
4.8	Peculiarities of Experimental Approach *154*	
4.9	Information Directly Obtainable from Various Experimental Techniques and Their Typical Applications *155*	
4.9.1	Photoelasticity *155*	
4.9.2	Holography *155*	
4.9.3	Grid Methods *156*	
4.9.4	Geometric Moiré *157*	
4.9.5	Moiré Interferometry *157*	
4.9.6	Speckle Interferometry *157*	
4.9.7	Digital Image Correlation *158*	
4.9.8	Thermoelastic Stress Analysis *159*	
4.9.9	Brittle Coating *159*	
4.9.10	Strain Gauge *159*	
4.9.11	Caustics *160*	
4.9.12	Coherent Gradient Sensor *161*	
4.10	Selection of an Experimental Technique *161*	
4.11	Case Studies *165*	
4.12	Experimental Study on Investigation of Random Failure of Chain Plates *165*	
4.12.1	Possible Investigation Methodologies *167*	
4.12.1.1	Analysis of Combined Stress Fields *167*	
4.12.1.2	Decoupled Analysis of Assembly and Application Stress Fields *168*	
4.12.2	Analysis of Assembly Stress due to Interference Fit between Bush and Inner Plate Using Transmission Photoelasticity *168*	
4.12.3	Analysis of a Chain Plate and Bush Assembly Using Reflection Photoelasticity *170*	
4.12.4	Stress Concentration Factor due to Applied Load *171*	
4.13	Comprehensive Experimental Study on a MEMS Pressure Sensor *172*	
4.13.1	Microscale Speckle Interferometry and Shearography *173*	
4.13.2	Deflection Measurement of the Pressure Sensor *175*	
4.14	Conclusions *178*	
	Acknowledgements *179*	

5	**Digital Image Correlation** *183*
	François Hild and Stéphane Roux
5.1	Introduction *183*
5.2	Correlation Principles *185*
5.2.1	Determination of the Optical Flow *185*
5.2.2	Local DIC *187*
5.2.3	Global DIC *188*
5.2.4	Gray-Level Interpolation *189*
5.2.5	Relaxation of the Gray Level Conservation *191*
5.3	2-D Digital Image Correlation *199*
5.3.1	Multiscale Analyses and Sequences of Pictures *199*
5.3.2	Local Approach *200*
5.3.2.1	FFT-Based DIC *200*
5.3.2.2	Measurement Uncertainties *203*
5.3.3	Global Approach *208*
5.3.3.1	General Formulation *208*
5.3.3.2	Q4-DIC: A Global Approach Using Q4 Elements *209*
5.4	3-D Digital Image Correlation *216*
5.4.1	Basic Principles *216*
5.4.2	Camera Calibration *216*
5.5	Digital Volume Correlation *220*
5.5.1	3-D Extensions of 2-D Approaches *220*
5.5.2	Resolution Analysis *221*
5.6	Summary *225*
5.7	Problems *225*
5.7.1	Mean Strain Extractor *225*
5.7.2	On the Use of a Global Approach When Analyzing Experiments on Beams *226*
5.7.3	Propagation of Uncertainties *227*
5.7.4	Measuring Displacement Fields with a Global Approach in the Presence of Cracks *227*
5.7.5	DIC Coupled with Finite-Element Analyses *227*
5.7.6	Application of Integrated DIC to a Brazilian Test *228*
6	**Rough Surface Interferometry** *229*
	Kay Gastinger, Pierre Slangen, Pascal Picart, and Peter Somers
6.1	Introduction *229*
6.2	Speckle *230*
6.2.1	Speckle in Monochromatic Light *231*
6.2.2	Speckle in Low-Coherent Light *232*
6.3	Electronic Speckle Pattern Interferometry–ESPI *235*
6.3.1	Introduction *235*
6.3.2	General Principle *235*
6.3.2.1	Interference Equation *236*
6.3.3	Evaluation *237*

6.3.3.1	Subtraction Mode	*237*
6.3.3.2	Time Averaged Method	*237*
6.3.4	Configuration	*239*
6.3.4.1	Sensitivity Vectors	*240*
6.3.5	Characteristics	*240*
6.3.5.1	Optimization of the Interference Signal	*240*
6.3.6	Applications	*241*
6.3.6.1	Compact Tension Notch Sample	*242*
6.3.6.2	Tensile Test Sample	*244*
6.4	Low-Coherence Speckle Interferometry–LCSI	*246*
6.4.1	Introduction	*246*
6.4.2	General Principle	*247*
6.4.2.1	Measurement Principle	*248*
6.4.3	Interferometer Setup	*251*
6.4.4	Evaluation	*252*
6.4.4.1	Full-Field OCT Mode	*252*
6.4.4.2	LCSI Mode	*254*
6.4.5	Characteristics	*256*
6.4.6	Applications	*257*
6.4.6.1	NDT of Interfacial Instabilities of Adhesive Bonded Joints	*257*
6.4.6.2	Membrane Deformation of a MEMS Pressure Sensor	*259*
6.5	Speckle Pattern Shearing Interferometry – Shearography	*262*
6.5.1	Introduction	*262*
6.5.2	General Principle	*262*
6.5.3	Evaluation	*264*
6.5.4	Configurations	*266*
6.5.5	Characteristics	*268*
6.5.6	Applications	*269*
6.5.6.1	Excitation or Loading of the Object	*269*
6.5.6.2	Static Applications	*270*
6.5.6.3	Dynamic Applications	*274*
6.6	Digital Holography	*278*
6.6.1	Introduction	*278*
6.6.2	General Principle	*279*
6.6.3	Evaluation	*282*
6.6.3.1	Discrete Fresnel Transform	*282*
6.6.3.2	Convolution Algorithm	*282*
6.6.4	Applications	*284*
6.6.4.1	2D Deformation Measurement Using a Two-Color Digital Holographic Interferometer	*284*
6.6.4.2	Real-Time Three-Sensitivity Measurement	*287*
6.6.4.3	Vibration Analysis with Digital Fresnel Holography	*292*
6.6.4.4	Time-Averaging Mode	*293*
6.6.4.5	Stroboscopic Regime	*295*
6.7	Summary	*298*

7	**Fringe Projection Profilometry** *303*
	Jan Buytaert and Joris Dirckx
7.1	General Introduction *303*
7.1.1	Non-Optical Topography *303*
7.1.2	Optical Full-Field Profilometry *304*
7.1.2.1	Coherence-Based Techniques *304*
7.1.2.2	Triangulation-Based Techniques *304*
7.2	Grid Projection Profilometry: the Basics *310*
7.3	Fourier Transform Profilometry *311*
7.3.1	Theory *311*
7.3.2	Extensions *314*
7.3.3	Simulation Example *316*
7.4	Moiré profilometry *316*
7.4.1	Shadow Versus Projection Moiré *316*
7.4.2	Theory of Projection Moiré *321*
7.4.2.1	Basic Principles *321*
7.4.2.2	Optical Geometric Interference *322*
7.4.2.3	Grid Noise Removal *323*
7.4.2.4	Digital Geometric Interference *325*
7.4.2.5	Phase-Shifting Algorithms *326*
7.4.2.6	Nonlinearity and Fringe Plane Distance *329*
7.4.3	Practical Considerations *331*
7.4.4	Practical Implementation *333*
7.4.5	Demonstration Measurements *337*
7.5	Noncontinuous Surfaces *337*
7.6	Summary *342*
8	**Thermoelastic Stress Analysis** *345*
	Janice M. Dulieu-Barton
8.1	Introduction *345*
8.2	The Thermoelastic Effect *345*
8.3	Infrared Thermography *348*
8.4	Obtaining Thermoelastic Measurements from an Infrared System *352*
8.5	Temperature Dependence of Thermoelastic Response *354*
8.6	Derivation of the Thermoelastic Constant *354*
8.7	Nonadiabatic Conditions *356*
8.8	Paint Coatings *358*
8.9	Temperature Dependence of the Material Elastic Properties *359*
8.10	Progress, Applications, and Prospects *363*
	Acknowledgements *365*
9	**Photoelasticity** *367*
	Eann A. Patterson
9.1	Introduction *367*

9.2	Polariscope Theory and Design	*369*
9.3	Isoclinic and Isochromatic Fringes	*373*
9.4	Fractional Fringe Analysis Using Compensation Techniques	*376*
9.5	Digital Fringe Analysis	*379*
9.6	Material and Load Selection	*383*
9.7	Stress Analysis	*386*
9.8	Conclusions	*390*

Color Plates *393*

References *405*

Abbreviations and Notations *421*

Index *427*

Preface

Solid mechanics is a widespread field that draws contributions from diverse disciplines such as material science, design theory, physics and mathematics, and their implementation in numerical analyses. The complexities arising from the interactions of these disciplines pose great challenges for both the engineer who simulates a structure and the one who actually builds it, specifically in relation to the structure's behavior under operating conditions. Strain and stress are the key quantities to describe the behavior of a structure and its limitations, and it would be of paramount importance to be able to know them. The last two decades have seen a substantial growth in optical methods dedicated to the quantitative measurement of deformations and strain on an entire object surface, without necessarily invoking a need for *a priori* information, and which by far overcome the limitations that point-based measurement techniques are known to be saddled with.

The ability to measure in real time and in a noninvasive manner the variations in certain important physical quantities such as displacements and strains at each point on a structure's surface, and to compute the results with speed and accuracy are some of the hallmark features that whole-field optical methods have brought into the reach of solid mechanics for addressing some of its measurement needs. Rapidly advancing technology in computing and electronic image sensing and the increasing availability of digital signal and image processing tools have benefited the optical methods by furthering their abilities for measurements to explore new opportunities in solid mechanics.

While giving lectures in the area of whole-field optical methods and their application to deformation measurements we became aware of a lack of a comprehensive resource on the subject written in a tutorial style. In this context and in order to fill up the lacuna, we approached some of the top specialists in the field with requests to write chapters on their respective areas of expertise. This book is the outcome of this endeavor. It is addressed to students of engineering with an interest in testing their designs and structures, and who are keen to learn about the possibilities offered by modern digital methods in optical measurements.

The book written in a tutorial style is meant to provide a self-contained treatment of the subject and aims to familiarize the reader with the essentials of imaging and full-field optical measurement techniques in solid mechanics with emphasis on their digital avatars wherever and whenever possible, to help him or her identify the appropriate techniques to meet the measurement requirements, and to assist him or her in assessing measurement systems. Solved problems are integrated in the text to inculcate a better understanding of the subject.

The book is structured such that the basic principles common to all optical techniques covered by the book are included in its first part composed of Chapters 1 to 3. Chapter 4 dedicated to strain and stress analysis guides the reader to the appropriate full-field technique, depending on the problem at hand, the most important of which are then described in the second part of the book made up of Chapters 5 to 9. All chapters include examples of applications, tutorial exercises interspersed in the text, and unsolved problems. The solutions to the latter will be offered on-line. Boxes are used to touch upon topics that could easily be left aside on a first read, but which nevertheless introduce important insights into the subject for the interested reader.

We expect the book to be of relevance to graduate and advanced undergraduate level students, with the material presented in a form that should make it useful as a source of reference, for individual study, and for developing short courses in this exciting field.

February 2012

Pramod Rastogi
Erwin Hack

List of Contributors

Thomas Baechler
CSEM – Centre Suisse
d'Electronique et de
Microtechnique
Switzerland

Jan Burke
Australian Centre for
Precision Optics
CSIRO Industrial Physics
PO Box 218
Lindfield
2070 NSW
Australia

Jan Buytaert
University of Antwerp
(Groenenborgercampus)
Laboratory of Biomedical Physics
Groenenborgerlaan 171
B-2020 Antwerp
Belgium

Joris Dirckx
University of Antwerp
(Groenenborgercampus)
Laboratory of Biomedical Physics
Groenenborgerlaan 171
B-2020 Antwerp
Belgium

Janice M. Dulieu-Barton
University of Southampton
Engineering and the
Environment
Southampton
Hampshire SO17 1BJ
UK

Kay Gastinger
NTNU
Faculty of Natural Sciences and
Technology
NO-7491 Trondheim
Norway

Ajoy Ghatak
Department of Physics
Indian Institute of Technology
Delhi
New Delhi 110016
India

François Hild
Laboratoire de Mécanique et
Technologie (LMT-Cachan)
ENS Cachan/CNRS/UPMC/
PRES UniverSud Paris
61 avenue du Président Wilson
F-94235 Cachan Cedex
France

Eann A. Patterson
School of Engineering
University of Liverpool
The Quadrangle
Brownlow Hill
Liverpool, L69 3GH
UK

Pascal Picart
ENSIM – École Nationale
Supérieure d'Ingénieurs du Mans
LAUM – Laboratoire
d'Acoustique de l'Université du
Maine
rue Aristote
72085 LE MANS cedex 9
France

Krishnamurthi Ramesh
Department of Applied
Mechanics
Indian Institute of Technology
Madras
Chennai
Tamil Nadu 600 036
India

Stéphane Roux
Laboratoire de Mécanique et
Technologie (LMT-Cachan)
ENS Cachan/CNRS/UPMC/
PRES UniverSud Paris
61 avenue du Président Wilson
F-94235 Cachan Cedex
France

Pierre Slangen
Ecole des Mines d'Alés
LGEI-ISR
Equipe Risques Industriels et
Naturels
6 Av. de Clavieres
F-30319 ALES Cedex
France

Peter Somers
Delft University of Technology
Dept. of Imaging Science &
Technology
Optics Group
Lorentzweg 1
2628 CJ Delft
The Netherlands

Krishna Thyagarajan
Department of Physics
Indian Institute of Technology
Delhi
New Delhi 110016
India

1
Basic Optics

Krishna Thyagarajan and Ajoy Ghatak

1.1
Introduction

This chapter on optics provides the reader with the basic understanding of light rays and light waves, image formation and aberrations, interference and diffraction effects, and resolution limits that one encounters because of diffraction. Laser sources are one of the primary sources used in various applications such as interferometry, thermography, photoelasticity, and so on, and Section 1.10 provides the basics of lasers with their special characteristics. Also included is a short section on optical fibers since optical fibers are used in various applications such as holography, and so on. The treatment given here is condensed and short; more detailed analyses of optical phenomena can be found in many detailed texts on optics [1–6].

1.2
Light as an Electromagnetic Wave

Light is a transverse electromagnetic wave and is characterized by electric and magnetic fields which satisfy Maxwell's equations [1, 2]. Using these equations in free space, Maxwell showed that each of the Cartesian components of the electric and magnetic field satisfies the following equation:

$$\nabla^2 \Psi = \varepsilon_0 \mu_0 \frac{\partial^2 \Psi}{\partial t^2} \tag{1.1}$$

where ε_0 and μ_0 represent the dielectric permittivity and magnetic permeability of free space. After deriving the wave equation, Maxwell could predict the existence of electromagnetic waves whose velocity (in free space) is given by

$$c = \frac{1}{\sqrt{\varepsilon_0 \mu_0}} \tag{1.2}$$

Since

$$\varepsilon_0 = 8.854 \times 10^{-12} \, C^2 \, N^{-1} \, m^{-2} \text{ and } \mu_0 = 4\pi \times 10^{-7} \, Ns^2 \, C^{-2} \tag{1.3}$$

Optical Methods for Solid Mechanics: A Full-Field Approach, First Edition. Edited by Pramod Rastogi and Erwin Hack.
© 2012 Wiley-VCH Verlag GmbH & Co. KGaA. Published 2012 by Wiley-VCH Verlag GmbH & Co. KGaA.

we obtain the velocity of light waves in free space

$$c = \frac{1}{\sqrt{\varepsilon_0 \mu_0}} = 2.99794 \times 10^8 \text{ ms}^{-1} \tag{1.4}$$

In a linear, homogeneous and isotropic medium, the velocity of light is given by

$$v = \frac{1}{\sqrt{\varepsilon \mu}} \tag{1.5}$$

where ε and μ represent the dielectric permittivity and magnetic permeability of the medium. The refractive index n of the medium is given by the ratio of the velocity of light in free space to that in the medium

$$n = \sqrt{\frac{\varepsilon \mu}{\varepsilon_0 \mu_0}} \tag{1.6}$$

In most optical media, the magnetic permeability is very close to μ_0 and hence we can write Eq. (1.6) as

$$n \approx \sqrt{\frac{\varepsilon}{\varepsilon_0}} = \sqrt{K} \tag{1.7}$$

where K represents the relative permittivity of the medium, also referred to as the *dielectric constant*.

The most basic light wave is a plane wave described by the following electric and magnetic field variations:

$$\mathbf{E} = \hat{x} E_0 e^{i(\omega t - kz)}$$
$$\mathbf{H} = \hat{y} H_0 e^{i(\omega t - kz)} \tag{1.8}$$

Here, we have assumed the direction of propagation to be along the $+z$-direction and the electric field to be oriented along the x-direction. Note that the electric and magnetic fields are oscillating in phase.

The wave given by Eq. (1.8) represents a plane wave since the surface of constant phase is a plane perpendicular to the z-axis. It is a monochromatic wave since it is described by a single angular frequency ω. It has its electric field along the x-direction and hence is a linearly polarized wave, in this case an x-polarized wave. The electric field amplitude is described by E_0. The amplitudes of the electric and magnetic fields are related through the following equation:

$$H_0 = \frac{k}{\omega \mu} E_0 = \sqrt{\frac{\varepsilon}{\mu}} E_0 \tag{1.9}$$

and ω and k are related through

$$k^2 = \varepsilon \mu \omega^2 = \frac{\omega^2}{v^2} \tag{1.10}$$

The propagation constant of the wave represented by k is related to the wavelength through the following equation:

$$k = \frac{2\pi}{\lambda_0} n \tag{1.11}$$

where λ_0 is the free space wavelength and n represents the refractive index of the medium in which the plane wave is propagating.

The intensity or irradiance of a light wave is the amount of energy crossing a unit area perpendicular to the propagation direction per unit time and is given by the time average of the Poynting vector **S**, which is defined by

$$\langle S \rangle = \frac{1}{2} \text{Re}\left(\boldsymbol{E} \times \boldsymbol{H}^* \right) \tag{1.12}$$

where the * in the superscript represents complex conjugate and angular brackets represent time average. Thus for the wave described by Eq. (1.8), the intensity is given by

$$\langle S \rangle = \frac{k}{2\omega\mu} E_0^2 \hat{z} \tag{1.13}$$

Thus, the intensity of the wave is proportional to the square of the electric field amplitude. If E_0 is a complex quantity, then in Eq. (1.13) E_0^2 gets replaced by $|E_0|^2$. Equation (1.13) can be written in terms of refractive index of the medium by using Eq. (1.11) as

$$I = \frac{n}{2c\mu_0} |E_0|^2 \tag{1.14}$$

where we have assumed the magnetic permeability of the medium to be μ_0.

The following represents an x-polarized wave propagating along the $-z$-direction:

$$\boldsymbol{E} = \hat{x} E_0 e^{i(\omega t + kz)} \tag{1.15}$$

A y-polarized light wave propagating along the z-direction is described by the following expression:

$$\boldsymbol{E} = \hat{y} E_0 e^{i(\omega t - kz)} \tag{1.16}$$

and is orthogonal to the x-polarized wave. The x- and y-polarized waves form two independent polarization states and any plane wave propagating along the z-direction can be expressed as a linear combination of the two components with different amplitudes and phases.

Thus the following two combinations represent right circularly polarized (RCP) and left circularly polarized (LCP) waves, respectively:

$$\boldsymbol{E} = (\hat{x} - i\hat{y}) E_0 e^{i(\omega t - kz)} \tag{1.17}$$

and

$$\boldsymbol{E} = (\hat{x} + i\hat{y}) E_0 e^{i(\omega t - kz)} \tag{1.18}$$

They are represented as superpositions of the x- and y-polarized waves with equal amplitudes and phase differences of $+\pi/2$ or $-\pi/2$. If the amplitudes of the two components are unequal then they would represent elliptically polarized waves, which are the most general polarization states.

The general expression for the electric field of a plane wave propagating along the z-direction is given by

$$\boldsymbol{E} = \hat{x} E_1 e^{i(\omega t - kz)} + \hat{y} E_2 e^{i(\omega t - kz)} \tag{1.19}$$

where E_1 and E_2 are complex quantities and represent the components of the electric field vector along the x- and y-directions, respectively. Linear, circular, and elliptical polarization states are special cases of the expression given in Eq. (1.19).

In the above discussion, we have written a circularly polarized wave as a linear combination of two orthogonally linearly polarized components. In fact, instead of choosing two orthogonally linearly polarized waves as a basis, we can as well choose any pair of orthogonal polarization states as basis states. Thus if we choose right circular and left circular polarization states as the basis states, then we can write a linearly polarized light wave as a superposition of a right and a left circularly polarized wave as follows:

$$\mathbf{E} = \hat{\mathbf{x}} E_0 e^{i(\omega t - kz)} = \frac{1}{2}(\hat{\mathbf{x}} - i\hat{\mathbf{y}}) E_0 e^{i(\omega t - kz)} + \frac{1}{2}(\hat{\mathbf{x}} + i\hat{\mathbf{y}}) E_0 e^{i(\omega t - kz)} \quad (1.20)$$

The first term on the rightmost equation represents an RCP wave and the second term represents an LCP wave.

The choice of the basis set depends on the problem. Thus, for anisotropic media (which is discussed in Section 1.9), it is appropriate to choose the basis set as linearly polarized waves, since the eigen modes are linearly polarized components. On the other hand, in the case of Faraday effect, it is appropriate to choose circularly polarized wave for the basis set, because when a magnetic field is applied to a medium along the propagation of a linearly polarized light wave, the plane of polarization rotates as the wave propagates.

In Section 1.10, we shall discuss Jones vector representation for the description of polarized light and the effect of various polarization components on the state of polarization of the light wave.

A plane wave propagating along a general direction can be written as

$$\mathbf{E} = \hat{\mathbf{n}} E_0 e^{i(\omega t - \mathbf{k} \cdot \mathbf{r})} \quad (1.21)$$

where $\mathbf{k}$ represents the direction of propagation of the wave and its magnitude is given by Eq. (1.11). The state of polarization of the wave is contained in the unit vector n. Since light waves are transverse waves we have

$$\hat{\mathbf{n}} \cdot \hat{\mathbf{k}} = 0 \quad (1.22)$$

that is, the propagation vector is perpendicular to the electric field vector of the wave. The vectors $\mathbf{E}$, $\mathbf{H}$, and $\mathbf{k}$ form a right-handed coordinate system (Figure 1.1). The vector $\mathbf{k}$ gives the direction of propagation of the wavefronts.

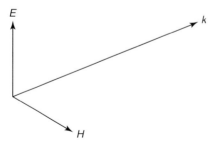

Figure 1.1 A plane wave propagating along a direction specified by k. The electric and magnetic fields associated with the wave are at right angles to the direction of propagation.

Tutorial Exercise 1.1

Consider the superposition of an x-polarized and a y-polarized wave with unequal amplitudes E_1 and E_2 but with the same phase. Write the resulting wave, discuss its nature, and obtain the intensity of this wave.

Solution:

Combining Eqs. (1.1) and (1.4) with amplitudes E_1 and E_2 instead of E_0, we have

$$\mathbf{E} = \hat{x} E_1 e^{i(\omega t - kz)} + \hat{y} E_2 e^{i(\omega t - kz)} = (\hat{x} E_1 + \hat{y} E_2) e^{i(\omega t - kz)}$$

The above equation represents another linearly polarized wave with its electric field vector $\mathbf{E}$ making an angle of $\tan^{-1}(E_2/E_1)$ with the x-axis. Its intensity is given by

$$I = \frac{n}{2c\mu_0} |E|^2 = \frac{n}{2c\mu_0} (\hat{x} E_1 + \hat{y} E_2) e^{i(\omega t - kz)} \cdot (\hat{x} E_1^* + \hat{y} E_2^*) e^{-i(\omega t - kz)}$$

$$= \frac{n}{2c\mu_0} (|E_1|^2 + |E_2|^2) = I_1 + I_2$$

Tutorial Exercise 1.2

Consider a laser emitting a power of 1 mW and having a beam diameter of 2 mm. Calculate the intensity of the laser beam and its field amplitude in air.

Solution:

Intensity is power per unit area; thus

$$I = \frac{10^{-3} \text{ W}}{\pi \times (10^{-3} \text{ m})^2} = \frac{10^3}{\pi} \frac{\text{W}}{\text{m}^2}$$

From Eq. (1.15), the field amplitude is given by $|E_0| = \sqrt{(2c\mu_0 I)}$. This corresponds to an electric field amplitude of approximately 500 V m^{-1}.

Absorbing media (such as metals) can be described by a complex refractive index:

$$n = n_\text{r} - i n_\text{i} \tag{1.23}$$

where n_r and n_i represent the real and imaginary parts of the refractive index. The propagation constant also becomes complex and is given by

$$k = k_\text{r} - i k_\text{i} \tag{1.24}$$

In such media, a plane wave propagating along the z-direction has the following variation of electric field

$$\mathbf{E} = \hat{x} E_0 e^{i(\omega t - \{k_\text{r} - i k_\text{i}\} z)} = \hat{x} E_0 e^{-k_\text{i} z} e^{i(\omega t - k_\text{r} z)} \tag{1.25}$$

which shows that the wave attenuates exponentially as it propagates. The attenuation constant is given by k_i.

If we consider a point source of light, then the waves originating from the source would be spherical waves and the electric field of a spherical wave would be described by

$$E = \frac{E_0}{r} e^{i(\omega t - kr)} \tag{1.26}$$

Here r is the distance from the point source and is the radial coordinate of the spherical polar coordinate system. The amplitude of the electric field decreases as $1/r$ so that the intensity, which is proportional to the square of the amplitude, decreases as $1/r^2$ in order to satisfy energy conservation. The surfaces of constant phase are given by $r =$ constant and thus represent spheres. Thus the wavefronts are spherical. The wave given by Eq. (1.26) represents a diverging spherical wave. A converging spherical wave would be given by

$$E = \frac{E_0}{r} e^{i(\omega t + kr)} \tag{1.27}$$

Light emanating from incoherent sources such as incandescent lamps or sodium lamp are randomly polarized. In many experiments, it is desired to have linearly polarized waves and this can be achieved by passing the randomly polarized light through an optical element called a *polarizer*. The polarizer could be an element that absorbs light polarized along one orientation while passing that along the perpendicular orientation. A Polaroid sheet is such an element and consists of long-chain polymer molecules that contain atoms (such as iodine) that provide high conductivity along the length of the chain. These long-chain molecules are aligned so that they are almost parallel to each other. When a light beam is incident on such a Polaroid, the molecules (aligned parallel to each other) absorb the component of electric field that is parallel to the direction of alignment because of the high conductivity provided by the iodine atoms; the component perpendicular to it passes through. Thus, linearly polarized light waves are produced.

When a randomly polarized laser light is incident on a Polaroid, then the Polaroid transmits only half of the incident light intensity (assuming that there are no other losses and that the Polaroid passes entirely the component parallel to its pass axis). On the other hand, if an x-polarized beam is passed through a Polaroid whose pass axis makes an angle θ with the x-axis, then the intensity of the emerging beam is given by

$$I = I_0 \cos^2 \theta \tag{1.28}$$

where I_0 represents the intensity of the emergent beam when the pass axis of the polarizer is also along the x-axis (i.e., when $\theta = 0$). Equation (1.28) represents Malus' law.

Linearly polarized light can also be produced by the simple process of reflection. If a randomly polarized plane wave is incident at an interface separating media of refractive indices n_1 and n_2 at an angle of incidence (θ) such that

$$\theta_1 = \theta_B = \tan^{-1}\left(\frac{n_2}{n_1}\right) \tag{1.29}$$

then the reflected beam will be linearly polarized with its electric vector perpendicular to the plane of incidence (Tutorial Exercise 1.4). The above equation is known as *Brewster's law* and the angle θ_B is known as the *polarizing angle* (or *Brewster angle*).

Using anisotropic media, it is possible to change the state of polarization of an input light into another desired state of polarization. Anisotropic media are briefly discussed in Section 1.9.

1.2.1
Reflection and Refraction of Light Waves at a Dielectric Interface

Electric and magnetic fields need to satisfy certain boundary conditions at an interface separating two media. When a light wave is incident on an interface separating two dielectrics, in general, it will generate a reflected wave and a transmitted wave (Figure 1.2). The angles of reflection and transmission as well as the amplitudes of the reflected and transmitted waves can be deduced by applying the boundary conditions at the interface.

It follows that the angle of reflection is equal to the angle of incidence, while the angle of refraction and the angle of incidence are related through Snell's law:

$$n_1 \sin \theta_1 = n_2 \sin \theta_2 \tag{1.30}$$

Thus, as the wave propagates from a rarer medium to a denser medium ($n_1 < n_2$), the wave bends toward the normal of the interface as it gets refracted. On the other hand, when the wave propagates from a denser medium to a rarer medium ($n_1 > n_2$), the wave bends away from the normal. In fact for a certain angle of incidence, the angle of refraction would become equal to $90°$ and this angle is referred to as the *critical angle*. The critical angle θ_c is given by

$$\sin \theta_c = \frac{n_2}{n_1} \tag{1.31}$$

For an interface between glass of refractive index 1.5 and air, the critical angle is approximately given by $41.8°$. Hence for angles of incidence greater than the critical angle, there would be no refracted wave, and such a phenomenon is referred to as *total internal reflection*.

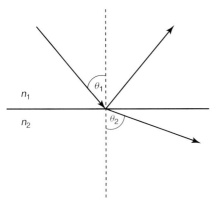

Figure 1.2 A plane wave incident on a dielectric interface generates a reflected and a transmitted wave.

For a light wave polarized in the plane of incidence, the amplitude reflection coefficient, which is the ratio of the amplitude of the electric field of the reflected wave and the amplitude of the electric field of the incident wave, is given by [1]

$$r_p = \frac{n_1 \cos\theta_2 - n_2 \cos\theta_1}{n_1 \cos\theta_2 + n_2 \cos\theta_1} \tag{1.32}$$

where n_1 and n_2 represent the refractive indices of the two media and θ_1 and θ_2 represent the angles of incidence and refraction (Figure 1.2).

Similarly, for a light wave polarized perpendicular to the plane of incidence, the amplitude reflection coefficient is given by

$$r_s = \frac{n_1 \cos\theta_1 - n_2 \cos\theta_2}{n_1 \cos\theta_1 + n_2 \cos\theta_2} \tag{1.33}$$

The corresponding energy reflection coefficients are given by $R_p = |r_p|^2$ and $R_s = |r_s|^2$, respectively.

Tutorial Exercise 1.3

A light wave is normally incident on an air–glass interface; the refractive index of glass is 1.5. Calculate the amplitude reflection coefficient and the corresponding energy reflection coefficient.

Solution:

For normal incidence, the incident angle is $\theta_1 = 0$; hence the refraction angle, from Eq. (1.30), is also $0°$. Using Eq. (1.32) then yields an amplitude reflection coefficient of -0.2 and the energy reflection coefficient is 0.04. The negative sign in the amplitude reflection coefficient signifies a phase change of π on reflection. If the wave is incident on the glass–air interface, then the amplitude reflection coefficient would be positive and there is no phase change on reflection.

Tutorial Exercise 1.4

The reflection coefficient for parallel polarization, r_p, in Eq. (1.32) can become zero. Calculate the corresponding angle. What is the reflection coefficient, r_s for this case? What will happen, if the incoming light is randomly polarized?

Solution:

From Eq. (1.32), it follows that the reflection coefficient will become zero if

$$n_1 \cos\theta_2 = n_2 \cos\theta_1$$

Using Snell's law (Eq. (1.30)), we can simplify the above equation to obtain the condition $\theta_1 + \theta_2 = \pi/2$; this gives us the following angle of incidence:

$$\theta_B = \tan^{-1}\left(\frac{n_2}{n_1}\right)$$

which is referred to as the *Brewster angle*. Inserting $\cos\theta_2 = n_2/n_1 \cos\theta_1$ into Eq. (1.33) yields

$$r_s = \frac{n_1^2 - n_2^2}{n_1^2 + n_2^2}$$

When unpolarized light is incident at the Brewster angle, the reflected light will show s-polarization only, because $r_p = 0$. Hence, unpolarized light will be polarized under reflection at the Brewster angle.

When the light wave undergoes total internal reflection, the angle of refraction satisfies the following inequality derived from Snell's law

$$\sin\theta_2 = \frac{n_1}{n_2}\sin\theta_1 > 1 \qquad (1.34)$$

Since $\sin\theta_2 > 1$, $\cos\theta_2$ becomes purely imaginary. Thus the amplitude reflection coefficient (for example, for the case when the incident light is polarized perpendicular to the plane of incidence) becomes

$$r_s = \frac{n_1\cos\theta_1 + i\alpha}{n_1\cos\theta_1 - i\alpha} \qquad (1.35)$$

where

$$\alpha = \sqrt{n_1^2\sin^2\theta_1 - n_2^2} \qquad (1.36)$$

is a real quantity. Thus, r_s will be a complex quantity with unit magnitude and can be written as

$$r_s = e^{-i\Phi} \qquad (1.37)$$

where

$$\Phi = -2\tan^{-1}\left(\frac{\sqrt{n_1^2\sin^2\theta_1 - n_2^2}}{n_1\cos\theta_1}\right) \qquad (1.38)$$

Thus, under total internal reflection, the energy reflection coefficient R_s is unity, that is, the intensity of the reflected wave and the incident wave are equal. The reflected wave undergoes a phase shift on reflection and this phase shift is a function of the angle of incidence of the wave. It can be shown that under total internal reflection, there is still a wave in the rarer medium; the amplitude of this wave decreases exponentially as we move away from the interface and it is a wave that is propagating parallel to the interface. This wave is referred to as an *evanescent wave*.

1.3
Rays of Light

Like any wave, light waves also undergo diffraction as they propagate (Section 1.8). However, we can neglect diffraction effects whenever the dimensions of the object with which light interacts are very large compared to the wavelength or when we do not look closely at points such as the focus of a lens or a caustic. In such a case, we can describe light propagation in terms of light rays. Light rays are directed lines perpendicular to the wavefront and represent the direction of propagation of energy. Thus the light rays corresponding to a plane wave would be just parallel straight arrows; for a spherical wave it would correspond to arrows emerging from the point source and for a more complex wavefront, light rays would be represented by arrows perpendicular to the wavefront at every point. The field of optics dealing with rays is referred to as *geometrical optics* since simple geometry can be used to construct the position of images and their magnification formed by optical instruments. In Section 1.4, we will discuss image formation by optical systems using the concept of light rays.

However, they cannot be used to estimate, for example, the ultimate resolution of the instruments since this is determined by diffraction effects. In Section 1.8, we shall discuss the diffraction phenomenon and how it ultimately limits the resolution of optical instruments such as microscopes, cameras, and so on.

Box 1.1: Rays in an Inhomogeneous Medium

The path of rays in a medium with a general refractive index variation given by $n(x, y, z)$ is described by the following ray equation [1]:

$$\frac{d}{ds}\left(n\frac{d\mathbf{r}}{ds}\right) = \nabla n$$

where ds is the arc length along the ray and is given by

$$ds = dz\sqrt{1 + \left(\frac{dx}{dz}\right)^2 + \left(\frac{dy}{dz}\right)^2}$$

For a given refractive index distribution, the solution of the ray equation will give us the path of rays in that medium.

Example: Consider a medium with a parabolic index variation given by

$$n^2(x) = n_1^2\left(1 - 2\Delta\left(\frac{x}{a}\right)^2\right)$$

Substituting the value of $n^2(x)$ in the ray equation, we obtain

$$\frac{d^2x}{dz^2} + \Gamma^2 x = 0$$

where

$$\Gamma = \frac{n_1\sqrt{2\Delta}}{\tilde{\beta}a}$$

The solution of the ray equation gives us the ray paths as

$$x(z) = A\sin\Gamma z + B\cos\Gamma z$$

showing that the rays in such a medium follow sinusoidal paths. The constants A and B are determined by the initial launching conditions on the ray.

In homogeneous media, the refractive index n is constant and light rays travel along straight lines. However, in graded index media, in which n depends on the spatial coordinates, light rays propagate along curved paths (Box 1.1). For example, in a medium with a refractive index varying with only x, we can assume the plane of propagation of the ray to be the x–z plane and the ray equation becomes

$$\frac{d^2 x}{dz^2} = \frac{1}{2\tilde{\beta}^2} \frac{dn^2(x)}{dx} \tag{1.39}$$

where

$$\tilde{\beta} = n(x) \cos\theta(x) \tag{1.40}$$

with $\theta(x)$ representing the x-dependent angle made by the ray with the z-axis. The quantity $\tilde{\beta}$ is a constant of motion for a given ray, and as the ray propagates this quantity remains constant.

Tutorial Exercise 1.5

Consider a medium with the following refractive index variation:

$$n^2(x) = n_1^2(1 + \alpha x)$$

over the region $0 < x < x_0$. This represents a linear variation of refractive index. Calculate the ray path.

Solution:

In such a medium, the ray equation can be integrated easily and the ray path in the region $0 < x < x_0$ is given by

$$x(z) = \left(\frac{\alpha n_1^2}{4\tilde{\beta}^2}\right) z^2 + C_1 z + C_2$$

where C_1 and C_2 are constants determined by initial launch conditions of the ray. The ray paths are thus parabolic.

The ray analysis given above is used in obtaining ray paths and imaging characteristics of optical systems containing homogeneous lenses or graded index (GRIN) lenses, in understanding light propagation through multimode optical fibers, and in many other applications.

1.4 Imaging through Optical Systems

Since the wavelength of the light is negligible in comparison to the dimensions of optical devices such as lenses and mirrors, we can approximate the propagation of light through geometrical optics.

The propagation of rays through such systems is primarily based on the laws of refraction at the boundary between media of different refractive indices and reflection at mirror surfaces. In the paraxial approximation, we assume that the angle made by the rays with the axis of the optical system is small and also that they propagate close to the axis. Using such an approximation, we find that optical systems can form perfect images and that we can obtain the basic properties such as position of the images, their magnifications, and so on, using ray optics. In this section, we shall use the paraxial approximation to study formation of images by lenses and then in Section 1.5, we shall discuss the various aberrations suffered by the images. More details can be found in Ref. [7].

When an optical system consists of many components then in the paraxial approximation, it is easy to formulate the properties of the optical system in terms of matrices which describe the changes in the height and the angle made by the ray with the axis of the system as the ray undergoes refraction at different interfaces or propagates through different thicknesses of the media. These matrices are obtained by applying Snell's law at the interfaces and also using the fact that rays propagate along straight lines in homogeneous media.

We represent a ray by a column matrix as follows:

$$\begin{pmatrix} x \\ \alpha \end{pmatrix} \tag{1.41}$$

where x represents the height or radial distance of the ray from the axis and α represents the angle made by the ray with the axis. Since we are using the paraxial approximation, we assume $\sin\alpha \approx \tan\alpha \approx \alpha$. We assume that rays propagate from left to right and that convex surfaces have positive radii of curvature while concave surfaces have negative radii of curvatures. We also define that rays pointing upwards have a positive value of α while rays pointing downwards have a negative value of α.

Figure 1.3(a) shows a ray refracting at a spherical interface of radius of curvature R between media of refractive indices n_1 and n_2 and Fig. 1.3(b) propagation through a homogeneous medium of thickness d. The effect of refraction at the interface

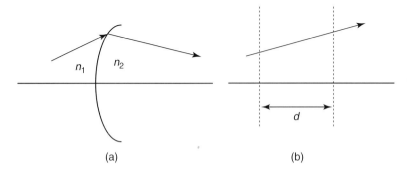

Figure 1.3 (a) Shows a ray refracting at an interface between two media of refractive indices n_1 and n_2. (b) A ray propagating through homogeneous medium.

and propagation through the medium are represented by square matrices.

$$\text{Refraction:} \begin{pmatrix} 1 & 0 \\ \frac{(n_1-n_2)}{n_2 R} & \frac{n_1}{n_2} \end{pmatrix}$$

$$\text{Propagation:} \begin{pmatrix} 1 & d \\ 0 & 1 \end{pmatrix}$$

1.4.1
Thin Lens

A thin lens consists of two refracting surfaces and since it is thin we neglect the effect of the propagation of the ray between the two interfaces within the lens. The effect of the two interfaces is a product of the matrices corresponding to refraction at the first interface of radius of curvature R_1 between media of refractive indices n_1 and n_2 and the matrix corresponding to refraction at the second interface of radius of curvature R_2 between media of refractive indices n_2 and n_1. Thus the effect of a thin lens is given by matrix corresponding to a thin lens:

$$\begin{pmatrix} 1 & 0 \\ \frac{(n_2-n_1)}{n_1 R_2} & \frac{n_2}{n_1} \end{pmatrix} \begin{pmatrix} 1 & 0 \\ \frac{(n_1-n_2)}{n_2 R_1} & \frac{n_1}{n_2} \end{pmatrix} = \begin{pmatrix} 1 & 0 \\ -\frac{1}{f} & 1 \end{pmatrix} \tag{1.42}$$

where

$$\frac{1}{f} = \frac{(n_2 - n_1)}{n_1}\left(\frac{1}{R_1} - \frac{1}{R_2}\right) \tag{1.43}$$

f is the focal length of the lens.

1.4.2
Thick Lens

In case we cannot neglect the finite thickness of the lens, then we also need to consider the effect of propagation of the ray between the two refracting surfaces. Thus if the thickness of the lens is d (the distance between the points of intersection of the lens surfaces with the axis), then the effect of a thick lens is given by

$$\begin{pmatrix} 1 & 0 \\ \frac{(n_2-n_1)}{n_1 R_2} & \frac{n_2}{n_1} \end{pmatrix} \begin{pmatrix} 1 & d \\ 0 & 1 \end{pmatrix} \begin{pmatrix} 1 & 0 \\ \frac{(n_1-n_2)}{n_2 R_1} & \frac{n_1}{n_2} \end{pmatrix} = \begin{pmatrix} A & B \\ C & D \end{pmatrix} \tag{1.44}$$

where

$$A = 1 - \frac{(n_2 - n_1)}{n_2 R_1}d \tag{1.45}$$

$$B = \frac{n_1}{n_2}d \tag{1.46}$$

$$C = \frac{(n_2 - n_1)}{n_1}\left(\frac{1}{R_1} - \frac{1}{R_2}\right) - \frac{(n_2 - n_1)^2}{n_1 n_2 R_1 R_2}d \tag{1.47}$$

$$D = 1 + \frac{(n_2 - n_1)}{n_2 R_2} d \qquad (1.48)$$

Note that for thin lenses, we can assume $d = 0$ and we get back the matrix for a thin lens, as shown in Eq. (1.42).

We can use the formalism given above to study a combination of lenses. For example, if we have two lenses of focal lengths f_1 and f_2 separated by a distance d, then the overall matrix representing the combination would be given by the product of the matrices corresponding to the first lens, propagation through free space, and the second lens:

$$\begin{pmatrix} 1 & 0 \\ -\frac{1}{f_2} & 1 \end{pmatrix} \begin{pmatrix} 1 & d \\ 0 & 1 \end{pmatrix} \begin{pmatrix} 1 & 0 \\ -\frac{1}{f_1} & 1 \end{pmatrix} = \begin{pmatrix} 1 - \frac{d}{f_1} & d \\ -\frac{1}{f_1} - \frac{1}{f_2} + \frac{d}{f_1 f_2} & 1 - \frac{d}{f_2} \end{pmatrix}$$

The first element in the second row contains the focal length of the lens combination and hence the combination is equivalent to a lens of focal length

$$\frac{1}{f} = \frac{1}{f_1} + \frac{1}{f_2} - \frac{d}{f_1 f_2} \qquad (1.49)$$

1.4.3
Principal Points of a Lens

Let us consider a pair of planes enclosing an optical system, which could be a thick lens, a combination of thin or thick lenses, and so on (Figure 1.4). Knowing the components and their spacing, we can use the above formalism to get the matrix connecting the rays between the two planes. Let us assume a ray starts from a height x_o making an angle of θ_o from the input plane and let us assume that the coordinates of the ray in the final plane are x_f and θ_f. Hence we have

$$\begin{pmatrix} x_f \\ \theta_f \end{pmatrix} = \begin{pmatrix} A & B \\ C & D \end{pmatrix} \begin{pmatrix} x_o \\ \theta_o \end{pmatrix} \qquad (1.50)$$

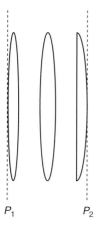

Figure 1.4 An optical system formed by three lenses. The system matrix describes the matrix of propagation of the ray from plane P_1 to plane P_2.

This matrix equation is equivalent to the two vector component equations:

$$x_f = Ax_o + B\theta_o$$
$$\theta_f = Cx_o + D\theta_o \tag{1.51}$$

If we choose an input plane such that θ_f is independent of θ_o, then this would imply the condition

$$D = 0 \tag{1.52}$$

This input plane must be the front focal plane because all rays from a point on the front focal plane entering the optical system under different angles emerge parallel from the final plane. Similarly, if x_f is independent of x_o, then this implies that rays coming into the optical system at a particular angle converge to one point x_i and the corresponding plane should be the back focal plane. Hence for the back focal plane

$$A = 0 \tag{1.53}$$

If the two planes are such that $B = 0$, then this implies that the output plane is the image plane and A represents the magnification. Similarly if $C = 0$, then a parallel bundle of rays will emerge as a parallel beam and D represents the angular magnification.

Tutorial Exercise 1.6

As an example, we consider a combination of two thin lenses shown in Figure 1.5. The first lens is assumed to be a converging lens of focal length 20 cm and the second a diverging lens of focal length 8 cm. We assume that they are separated by a distance of 13 cm. Calculate the ABCD matrix connecting the plane S_1 and S_2 and the focal length of this lens system:

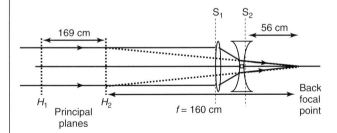

Figure 1.5 An optical system consisting of a double convex and a double concave lens. H_1 and H_2 are the principal planes.

Solution:

Inserting the values (in centimeters) into the corresponding matrices we obtain

$$\begin{pmatrix} A & B \\ C & D \end{pmatrix} = \begin{pmatrix} 1 & 0 \\ \frac{1}{8} & 1 \end{pmatrix} \begin{pmatrix} 1 & 13 \\ 0 & 1 \end{pmatrix} \begin{pmatrix} 1 & 0 \\ -\frac{1}{20} & 1 \end{pmatrix} = \begin{pmatrix} 0.35 & 13 \\ -0.00625 & 2.625 \end{pmatrix}$$

The focal length of the combination is given by $1/0.00625 = 160$ cm.

1 Basic Optics

In the example discussed above, we obtained the focal length of the lens combination. However, we need to know from where this distance needs to be measured. For a single thin lens, we measure distances from the center of the lens; however, for combination of lenses, we need to determine the planes from where we must measure the distances. To understand this let us consider a plane at a distance u in front of the first lens and a plane at a distance v from the second lens. The matrix connecting the rays at these two planes is

$$\begin{pmatrix} 1 & v \\ 0 & 1 \end{pmatrix} \begin{pmatrix} A & B \\ C & D \end{pmatrix} \begin{pmatrix} 1 & u \\ 0 & 1 \end{pmatrix} = \begin{pmatrix} A + Cv & B - Au + v(D - Cu) \\ C & D - Cu \end{pmatrix} \tag{1.54}$$

The image plane is determined by the condition

$$B - Au + v(D - Cu) = 0$$

which can be simplified to

$$-\frac{1}{(u - u_p)} + \frac{1}{(v - v_p)} = \frac{1}{f} \tag{1.55}$$

where

$$u_p = \frac{(D - 1)}{C}; \quad v_p = \frac{(1 - A)}{C}; \quad f = -\frac{1}{C} \tag{1.56}$$

Thus imaging by the lens combination can be described by the same formula as for a simple lens provided we measure all distances from appropriate planes. The object distance is measured from a point with a coordinate u_p with respect to the front plane S_1 of the lens combination and the image distance is measured from the point with the coordinate v_p with respect to the back plane S_2 of the lens combination. Positive values of these quantities imply that they are on the right of the corresponding planes and negative values imply that they are on the left of the corresponding planes. The planes perpendicular to the axis and passing through these points are referred to as the *principal planes* of the imaging system. The back focal length of the system is f and measured from the second principal plane. Hence, it is at a distance $-A/C$ from the back plane S_2. Similarly, the front focal distance measured from the plane S_1 is $-D/C$.

Tutorial Exercise 1.7

Obtain the positions of the front and back principal planes of Tutorial Exercise 1.6.

Solution:

Substituting the values of the various quantities, we obtain the back focal distance from the second lens as $-0.35/(-0.000625) = 56$ cm and the front focal distance from the first lens as $2.625/0.00625 = 420$ cm. For the given system $u_p = 260$ cm and $v_p = -104$ cm. The first and second principal planes are situated as shown in Figure 1.3.

The advantage of using matrices to describe the optical system is that it is easily amenable to programing on a computer and complex systems can be analyzed easily. Of course, the analysis is based on paraxial approximation. To obtain the quality of the image in terms of aberrations, and so on, one would need to carry out more precise ray tracing through the optical system.

1.5
Aberrations of Optical Systems

The paraxial analysis used in Section 1.4 assumes that the rays do not make large angles with the axis and also lie close to the axis. Principally in this analysis $\sin\theta$ is replaced by θ, where θ is the angle made by the ray with the axis of the optical system. In such a situation, perfect images can be formed by optical systems. In actual practice, not all rays forming images are paraxial and this leads to imperfect images or aberrated images. Thus, the approximation of replacing $\sin\theta$ by θ fails and higher order terms in the expansion have to be considered. The first additional term that would appear would be of third power in θ and hence the aberrations are termed as *third-order aberrations*. These aberrations do not depend on the wavelength of the light and are termed as *monochromatic aberrations*. When the illumination is not monochromatic, we have additional contribution coming from chromatic aberration because of the dispersion of the optical material used in the lenses.

1.5.1
Monochromatic Aberrations

There are five primary monochromatic aberrations: spherical aberration, coma, astigmatism, curvature of field, and distortion.

1.5.2
Spherical Aberration

According to paraxial optics, all rays parallel to the axis of a converging system focus at one point behind the lens. However, if we trace rays through the system then we find that rays farther from the axis intersect the axis at a different point compared to rays closer to the axis. This is termed as *longitudinal spherical aberration*. Thus, if we place a plane corresponding to the paraxial focal plane, then all rays would not converge at one point and rays farther from the axis will intersect the plane at different heights. The transverse distance measured on the focal plane is termed *transverse spherical aberration*. This implies that if we begin with a very small aperture in front of the lens, then the image would almost be perfect. However, as we increase the size of the aperture, the size of the focused spot on the focal plane will increase, leading to a drop in quality of the image.

It is possible to minimize spherical aberration of a single converging lens by ensuring that both the surfaces contribute equally to the focusing of the incident light. By using lens combinations, it is possible to eliminate spherical aberration by using aspherical lenses or combination of lenses. Thus a plano-convex lens with the spherical surface facing the incident light would have lower aberration than the same lens with the plane surface facing the incident light. In fact, for far off objects the plano-convex lens is close to having the smallest spherical aberration and hence is often used in optical systems.

For points lying on the axis of the optical system, the aberration of the image is only due to spherical aberration.

1.5.3
Coma

For off-axis points, each circular zone of the lens has a different magnification and forms a circular image and the circular images together form a coma-type image shape. By adjusting the radii of the surfaces of the lens, keeping the focal length the same, it is possible to minimize coma similar to spherical aberration. An optical system free of spherical aberration and coma is referred to as an *aplanatic lens*.

1.5.4
Astigmatism and Curvature of Field

Consider rays emerging from an off-axis point and passing through an optical system that is free of spherical aberration and coma. In this case, rays in the tangential plane and the sagittal planes come to focus at single points but they do not coincide. Thus, on the planes where these individually focus the images are in the form of lines brought about by the other set of rays. This aberration is termed as *astigmatism*. At a point intermediate between these two line images, the image will become circular and this is termed as the *circle of least confusion*.

Now for different off-axis points, the surface where the tangential rays form the image is not perpendicular to the axis but is in the form of a paraboloidal surface. Similarly, the sagittal rays form an image on a different paraboloidal surface. Elimination of astigmatism would imply that these two surfaces coincide but they would still not be a plane perpendicular to the axis. This effect is called *curvature* of field.

1.5.5
Distortion

Even if all the above aberrations are eliminated, the lateral magnification could be different for points at different distances from the axis. If the magnification increases with the distance then it leads to pincushion distortion while if it decreases with distance from the axis, then it is referred to as *barrel distortion*. In either of the two cases, the image will be sharp but will be distorted.

Aberration minimization is very important in optical system design. Nowadays, lens design programs are available that can trace different sets of rays through the system without making any approximations and by plotting points where they intersect an image plane, one can form a spot diagram. By changing the parameters of the optical system, one can see changes in the spot diagram, and some optimization routine is used to finally get an optimized optical system.

1.6
Interference of Light

Light waves follow the superposition principle and hence when two or more light waves superpose at any point in space then the total electric field is a superposition of the electric fields of the two waves at that point and depending on their phase difference, they may interfere constructively or destructively. This phenomenon of interference leads to very interesting effects and has wide applications in nondestructive testing, vibration analysis, holography, and so on.

In order that the interfering waves form an observable interference pattern, they must originate from coherent sources. Section 1.7 discusses the concept of coherence.

1.6.1
Young's Double-Slit Arrangement

Figure 1.6 shows Young's double-slit experiment in which light emerging from a slit S with infinitesimal width illuminates a pair of infinitesimal slits S_1 and S_2 separated by distance d from each other. On the other side of the double slit, there is a screen placed at a distance D from the slits. If we consider a point P at a distance x from the axis of the setup, then assuming $D \gg d$, the path difference between the waves arriving at the point P from S_1 and S_2 would be approximately given by

$$\Delta = d\frac{x}{D} \tag{1.57}$$

When the path difference is an even multiple of $\lambda_0/2$, then the waves from S_1 and S_2 interfere constructively, leading to a maximum in intensity. On the other hand, if the path difference is an odd multiple of $\lambda_0/2$, then the waves will interfere destructively, leading to a minimum in intensity. Thus the positions of constructive interference in the screen are given by

$$x_{\max} = m\frac{\lambda_0 D}{d}; \quad m = 0, \pm 1, \pm 2, \ldots \tag{1.58}$$

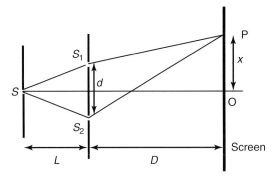

Figure 1.6 Young's double-slit experiment. Light from a source S illuminates the pair of slits S_1 and S_2. Light waves emanating from S_1 and S_2 interfere on the screen to produce an interference pattern.

The separation β between two adjacent maxima, which is referred to as the *fringe width* is

$$\beta = \frac{\lambda_0 D}{d} \tag{1.59}$$

The intensity pattern on the screen along the x-direction would be given by

$$I = 4I_0 \cos^2\left(\frac{\pi x d}{\lambda_0 D}\right) \tag{1.60}$$

where I_0 is the intensity produced by the source S_1 or S_2 in the absence of the other source.

Box 1.2: Antireflective Coating and Fabry–Perot Filters

If we consider a thin film of refractive index n_f and thickness d coated on a medium of refractive index n_s placed in air, then light waves at a wavelength λ_0 incident normally on the film will undergo reflection at both the upper and lower interfaces. If the reflectivities at the interfaces are not large, then we can neglect multiple reflections of the light waves. The path difference between the two reflected waves (one reflected from the upper surface and one from the lower surface) would be

$$\Delta = 2n_f d$$

Since in this case each of the two reflections occurs from a denser medium, it undergoes a phase shift of π on reflection; hence the only phase difference between the two waves is because of the extra path length that the reflected wave from the lower surface takes. If we assume that $1 < n_f < n_s$, then we would have constructive interference when

$$\Delta = m\lambda_0; \quad m = 1, 2, 3, \ldots$$

and for

$$\Delta = \left(m + \frac{1}{2}\right)\lambda_0; \quad m = 1, 2, 3, \ldots$$

we would have destructive interference. If the refractive index of the film is the geometric mean of the refractive indices of the two media surrounding the film [2] then the amplitudes of the two interfering beams are equal and we would have complete destructive interference. This is the principle behind antireflection coatings. The required minimum film thickness is

$$d = \frac{\lambda_0}{4n_f}$$

Example: Let us calculate the refractive index and thickness of a film to reduce the reflection at a wavelength of 600 nm from a medium of refractive index 2 placed in air. The required refractive index is $\sqrt{2} \approx 1.414$ and thickness is 0.106 μm.

If the reflectivities of the two surfaces become large, then we cannot neglect the presence of multiple reflections. If we assume that the reflectivities of the two surfaces are equal and is represented by R, then the overall intensity transmittance of the film would be given by

$$T = \frac{1}{1 + F \sin^2\left(\frac{\delta}{2}\right)}$$

where

$$F = \frac{4R}{(1-R)^2}$$

is called the *coefficient of finesse* and δ represents the phase difference accumulated during one back and forth propagation of the wave through the film. For normal incidence, it is just $(2\pi/\lambda_0)2n_f d$. If the angle made by the waves inside the film is θ_f, then

$$\delta = \frac{2\pi}{\lambda_0} 2n_f d \cos\theta_f$$

Note that when δ is an integral multiple of 2π, then all the incident light gets transmitted. If R is close to unity, then the transmittance drops very quickly as δ changes. Figure 1.7 shows a typical transmission spectrum of a film with $R = 0.9$. The changes in δ could be brought about by changes in the wavelength of the incident radiation, the thickness or the refractive index of the medium between the two highly reflecting surfaces, or by changes in the angle of illumination. Thus in transmission, this produces very sharp interference effects. This interference phenomenon is referred to as *multiple beam interference* and the Fabry–Perot interferometer and the Fabry–Perot etalon are based on this principle. By scanning the distance between the two highly reflecting surfaces, it is possible to measure very precisely the spectrum of the incident radiation and this is widely used in spectroscopy. Fabry–Perot etalons are also widely used inside laser cavities for selecting a single frequency of oscillation.

The resolving power of a Fabry–Perot interferometer is given by

$$R = \frac{\lambda_0}{\Delta\lambda} = \frac{\pi d\sqrt{F}}{\lambda}$$

where it is assumed that the Fabry–Perot interferometer operates at normal incidence.

In almost all interferometers light from the given source is split into two parts by for example using beam splitters, and made to interfere after propagating through two different paths by optical components such as mirrors. This ensures that the interfering beams are coherent, leading to formation of good contrast interference. Creating any difference in the propagation paths of the two interfering beams changes the interference condition, leading to changes in intensity. This is used in instrumentation for measuring spectra of light sources, for very accurate displacement measurement, for surface evaluation, and for many other applications.

Figure 1.8 shows a Michelson interferometer arrangement. S represents an extended near monochromatic source, G represents a 50% beam splitter, and M_1 and M_2 are two plane mirrors. The mirror M_2 is fixed while the mirror M_1 can be moved either toward or away from G. Light from the source S is incident on

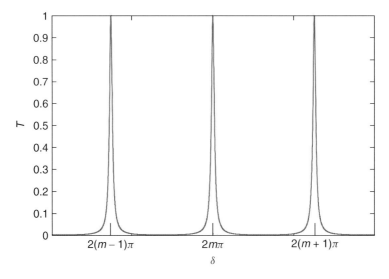

Figure 1.7 The transmittance of a Fabry–Perot interferometer. The maxima of transmission correspond to integral multiples of 2π. The higher the reflectivity of the mirrors the sharper would be the peaks.

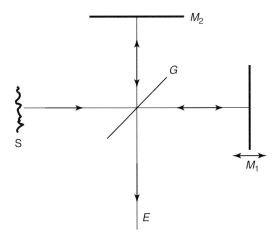

Figure 1.8 Michelson interferometer arrangement.

G and is divided into two equal amplitude portions; one part travels toward M_1 and is reflected back to the same beam splitter and the other part is reflected back from M_2 to the beam splitter. At the beam splitter, both beams partially undergo reflection and transmission and interfere and produce interference fringes that are visible from the direction E. If the two mirrors are perpendicular to each other and the beam splitter is at 45° to the incident beam, then the system is equivalent to interference fringes formed by a parallel plate illuminated by a near monochromatic extended source, and we obtain circular fringes of equal inclination. If the mirror M_1 is moved toward or away from the beam splitter, depending on its distance from the beam splitter vis-à-vis mirror M_2, the circular fringes either contract toward the center or expand away from the center. Each fringe collapsing at the center of the pattern corresponds to a movement of $\lambda_0/2$ of the mirror M_1. Thus, movements can be measured very precisely using this interferometer. When illuminated by broadband sources, the fringes will appear only when the path lengths to both mirrors are almost equal. (In this case, one has to use an additional glass plate identical to the beam splitter to ensure that the optical path lengths of the two arms are the same for all wavelengths of the incident light beam.) A measurement of variation of contrast with the displacement of mirror M_1 can be used to measure the spectrum of the source and this is referred to as *Fourier transform spectrometry*.

Instead of incoherent sources, lasers are often used with the interferometers and in this case even large path differences achieve good contrast interference. Laser interferometers are used for high precision measurements, measuring and controlling displacements from a few nanometers to about 100 m, used for measuring angles very precisely, to measure flatness of surfaces, velocity of moving objects, vibrations of objects, and so on. It is also possible to use interference among two lasers with slightly different frequencies leading to beating of the two laser beams. This is termed *heterodyne interferometry*.

1 Basic Optics

In holography (Chapter 6) interference between a reference wave and the wave scattered from the object is recorded. This is used in many applications such as for measuring tiny displacements of objects in real time or to identify defects using interference between the object wave emanating from the object under two different strain conditions. Vibration interferometry can be used to identify the state of vibration of the object using this principle. In holographic interferometry, the object need not be specularly reflecting, and thus this is a very powerful technique for nondestructive testing of various objects.

When objects with a rough surface are viewed under laser illumination, one sees a granular pattern, which is referred to as the *speckle pattern*. In this, the light waves get scattered from different points and propagate in different directions. Thus the light reaching any point on a screen consists of these various scattered waves. Owing to the nature of the surface the phases of the various waves reaching the given point on the screen may lie anywhere between 0 and 2π. With an illumination from a coherent source such as a laser, when scattered waves with these random phases are added, the resultant could lie anywhere between a maximum and a minimum value. At a nearby point, the waves may add to generate a completely different intensity value. In such a circumstance what we observe on the screen is a speckle pattern. The mean speckle diameter is approximately given by

$$s \approx 1.22 \frac{\lambda_0 L}{d} \tag{1.61}$$

where λ_0 is the wavelength of illumination, L is the distance between the screen and the rough surface, and d is the diameter of the region of illumination of the object. If instead of allowing the light to fall on a screen, an imaging system is used to form an image, then again we see a speckle pattern due to interference effects. In this case, the mean speckle diameter s is approximately given by the following relation:

$$s \approx 1.22 \lambda_0 F(1 + M) \tag{1.62}$$

where F is the F-number of the lens (focal length divided by diameter of the lens) and M is the magnification.

Tutorial Exercise 1.8

Obtain the speckle size for the following situations:

1) Reflections from a laser spot of $d = 2$ cm and a wavelength of $\lambda = 500$ nm on a screen in a distance of $L = 1$ m.
2) Let us assume that we observe from a distance of 100 cm with the naked eye a rough surface illuminated by a laser with a wavelength of 600 nm. Assume a pupil diameter of 4 mm and an eye length of 24 mm.

Solution:
1) Substituting the values into Eq. (1.61) we obtain $s \sim 30$ μm.

2) From the pupil diameter and the length of the eye we obtain an *F*-number of 6. Using Eq. (1.61), we obtain for the approximate size of the speckle as formed on the retina $s \sim 4.4\,\mu m$.

Speckle contrast measurement has proved to be a powerful tool for the non-destructive testing of small surface roughness within the light wavelength. More details of speckle techniques are given in Chapter 6.

1.7 Coherence

Light sources are never perfectly monochromatic; they emit over a range of wavelengths. If the spectral bandwidth of emission is very small the source is termed *quasi-monochromatic*. When the source emits more than one wavelength, then in an interference setup each wavelength would form its own interference pattern and what one observes is a superposition of the interference patterns from different wavelengths. Thus, if we consider a source emitting two wavelengths (λ_0 and $\lambda_0 - \Delta\lambda$), then when we start from zero path difference between the two interfering waves, the maxima and minima of the two wavelengths would coincide and we will observe very good contrast fringes. Since the fringe width depends on the wavelength, with increasing path difference between the two waves, the maxima and minima of each wavelength would occur at different positions and thus the contrast would begin to fall. In the case of two wavelengths, the contrast will become zero when the maxima of one wavelength fall on the minima of the other wavelength and vice versa. This will happen for a path difference l_c given by

$$l_c = m\lambda_0 = \left(m + \frac{1}{2}\right)(\lambda_0 - \Delta\lambda) \tag{1.63}$$

Eliminating *m* from the two equations, we obtain

$$l_c = \frac{\lambda_0^2}{2\Delta\lambda} \tag{1.64}$$

For a source emitting a continuous range of wavelengths from λ_0 to $(\lambda_0 - \Delta\lambda)$, the expression for coherence length becomes

$$l_c = \frac{\lambda_0^2}{\Delta\lambda} = \frac{c}{\Delta\nu} \tag{1.65}$$

where $\Delta\nu$ represents the spectral width in frequency.

Coherence length is a very important property of a source as it defines the maximum path difference permitted between the interfering waves so that the interference pattern formed has good contrast. For good contrast fringes, the path difference must be much smaller than the coherence length and as the path difference approaches the coherence length and exceeds it, the contrast in the fringes would decrease steadily, and finally no interference pattern will be visible.

Tutorial Exercise 1.9

Consider two lasers operating at 800 nm with spectral widths of (i) 1 nm and (ii) 0.001 nm and calculate the coherence lengths. Which one would you prefer to measure displacements of a few centimeters?

Solution:

Using Eq. (1.65) the coherence length of the two lasers would be (i) 0.64 mm and (ii) 64 cm, respectively. The latter laser has a much larger coherence length and setting up interference experiments with that laser would be much easier. Using the first laser, one would have to ensure that the maximum path difference is much less than 0.64 mm for good contrast interference.

A laser beam has a well-defined phase front and all points across the wavefront are coherent with respect to each other. Thus if we illuminate a double slit with the laser beam, then the waves emerging from the two slits will exhibit a good contrast interference pattern. When extended incoherent sources such as sodium lamps are used, then the different points across the source are not coherent with respect to each other. In order to form good contrast interference, we would also need to ensure spatial coherence of the source.

Consider Young's double-slit experiment (Figure 1.6); assume we had two sources, one on the axis, S, as shown in the figure, and another source S' displaced from the axis by a distance l. Then at point O, the source S will produce a maximum of intensity, but the intensity produced by S' would depend on l. In fact, if L represents the distance from the plane of the source S to the double-slit arrangement, and if l is such that the path difference $(S'S_2 - S'S_1) = \lambda_0/2$, then the source S' would produce a minimum of intensity at O and the contrast in the interference pattern at O will be nearly zero. This will happen when the following condition is satisfied (assuming $L \gg l, d$):

$$S'S_2 - S'S_1 = d\frac{l}{L} = \frac{\lambda_0}{2} \tag{1.66}$$

If we represent by $\theta (\sim l/L)$ the angle subtended by the pair of sources S and S' at the plane of the slits, then Eq. (1.66) gives

$$d = l_w = \frac{\lambda_0}{2\theta} \tag{1.67}$$

where l_w is referred to as the *lateral coherence width*. The above discussion pertains to a pair of sources; if the source is an extended source and subtends an angle θ at the plane of the slits, then the lateral coherence width is given by

$$l_w = \frac{\lambda_0}{\theta} \tag{1.68}$$

Tutorial Exercise 1.10

Consider a double-slit experiment to be conducted using a sodium lamp ($\lambda_0 = 589$ nm) with a pinhole having a diameter of 2 mm placed in front of it.

The distance of the slits is assumed to be 0.3 mm. What is the minimum distance to the lamp needed to obtain good contrast fringes?

Solution:

The lateral coherence width of the source at a distance of 1 m would be $l_w \sim 0.3$ mm. Thus for forming good contrast interference pattern, the distance to the lamp must be larger than 1 m. The lateral coherence width can be increased either by decreasing the size of the source or by increasing the distance between the source and the double slit.

1.8
Diffraction of Light

Plane waves described by Eq. (1.9) have infinite extent in the transverse direction; for such a wave, the transverse amplitude distribution remains the same as the wave propagates; only the phase changes because of propagation. However, any wave that has an amplitude or phase depending on the transverse coordinate will undergo changes in the transverse field distribution. This phenomenon is referred to as *diffraction*.

If $f(x,y)$ is the transverse complex amplitude distribution of a wave on a plane $z = 0$, then the amplitude distribution on any plane z is given by the following equation [1]:

$$f(x,y,z) = \frac{i}{\lambda z} e^{-ikz} \int_{-\infty}^{\infty}\int_{-\infty}^{\infty} f(\xi,\eta) \exp\left[-i\frac{k}{2z}\left\{(x-\xi)^2 + (y-\eta)^2\right\}\right]d\xi\,d\eta \quad (1.69)$$

Thus the transverse amplitude distribution changes, in general, as the field propagates and this is referred to as *Fresnel diffraction*. Note that Fresnel diffraction is a two-dimensional convolution operation as given below:

$$f(x,y,z) = \frac{i}{\lambda z} e^{-ikz} f(x,y) \otimes \exp\left[-i\frac{k}{2z}(x^2 + y^2)\right] \quad (1.70)$$

If the distance of the observation plane z satisfies the following condition

$$z \gg \frac{a^2}{\lambda} \quad (1.71)$$

where a is the transverse dimension of the field pattern on the plane $z = 0$, then the Fresnel diffraction pattern reduces to the Fraunhofer diffraction pattern. In this case, we can approximate the field distribution also referred to as *far field distribution* by the following equation:

$$f(x,y,z) = C\int_{-\infty}^{\infty}\int_{-\infty}^{\infty} f(\xi,\eta) e^{2\pi i(u\xi + v\eta)} d\xi\,d\eta \quad (1.72)$$

where

$$u = \frac{x}{\lambda z}; \quad v = \frac{y}{\lambda z} \qquad (1.73)$$

represent spatial frequencies along the coordinate directions x and y respectively, and C is a constant. It can be seen that Eq. (1.72) is nothing but a two-dimensional Fourier transform. Thus the Fraunhofer diffraction is the Fourier transform of the field distribution on the initial plane.

When we observe the field distribution on the back focal plane of a lens, we obtain the Fraunhofer diffraction pattern. This can be seen from the fact that a parallel beam of light falling on the lens focuses on a point on the back focal plane. Thus, each point on the back focal plane corresponds to a certain direction of propagation in space and the pattern on the back focal plane represents the Fraunhofer diffraction pattern.

If we consider a plane wave falling on a converging lens, then the wave focuses to the focal point of the lens. In this case, the transverse dimension of the lens acts as an aperture and thus on the back focal plane of the lens we would observe the Fraunhofer diffraction pattern corresponding to a circular aperture of dimension equal to that of the lens aperture. Thus, although geometrical optics predicts focusing on a point image (in the absence of aberrations), the focused image will have a finite size because of diffraction effects.

The Fraunhofer diffraction through a circular aperture is evaluated in many texts and is given by Ghatak [2]

$$I = I_0 \left[\frac{2 J_1(\beta)}{\beta} \right]^2 \qquad (1.74)$$

where I_0 is the intensity on the axis of the lens and

$$\beta = ka \sin \theta \qquad (1.75)$$

where θ is defined in Figure 1.9 and J_1 is the Bessel function of first kind of order 1. The distribution given by Eq. (1.74) is referred to as the *Airy pattern* (Figure 1.9b). The focal pattern consists of a bright central spot surrounded by rings of smaller and smaller intensities. The radius of the first dark ring corresponds to the first zero of J_1 and occurs at $\beta \sim 3.83$, which corresponds to

$$\sin \theta \approx \frac{1.22 \lambda}{2a} \qquad (1.76)$$

and is referred to as the *radius* of the Airy pattern.

For small θ, that is, paraxial condition, we can approximate $\sin \theta$ by θ; in such a case, the transverse dimension of the Airy spot is approximately given by

$$\Delta r \approx \frac{1.22 \lambda f}{2a} = 1.22 \lambda f^{\#} \qquad (1.77)$$

where f is the focal length of the lens and $f^{\#}$, the ratio of focal length to the diameter of the lens is called the *f-number* of the lens. The smaller the *f*-number, the smaller is the spot size and the sharper is the image. For a given focal length, the larger the diameter of the lens, the smaller is the focused spot. The quantity $1.22 \lambda / 2a$

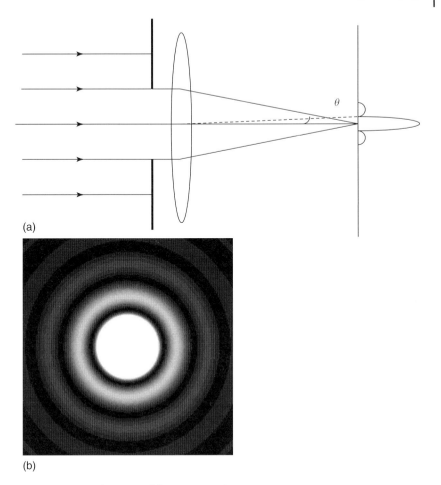

(a)

(b)

Figure 1.9 (a) A plane wave falling on a circular aperture and forming an Airy pattern on the focal plane of the lens. (b) Airy pattern that will be observed on the back focal plane of the lens.

corresponds to the angle made by the first dark ring of the Airy spot with the center of the lens.

Box 1.3: Diffraction Gratings

A diffraction grating is a very important optical component that is used in many instruments such as spectrometers for dispersing different wavelengths present in the illumination. It consists of a number of equally spaced identical long narrow slits placed parallel to each other. Each of the slits produces its own diffraction and the interference among the diffracted light waves from the various apertures produces the diffraction pattern of the grating given by

$$I(\theta) = I_0 \left(\frac{\sin\beta}{\beta}\right)^2 \left(\frac{\sin N\gamma}{\sin\gamma}\right)^2$$

where

$$\beta = \frac{\pi b \sin\theta}{\lambda_0}$$

$$\gamma = \frac{\pi d \sin\theta}{\lambda_0}$$

Here b represents the width of each slit and d is the spacing between the slits. The first term within the brackets represents the diffraction by individual slits and the second term represents the interference between the various diffraction patterns.

Whenever $\gamma = m\pi$, $m = 0, 1, 2, \ldots$, the second bracketed term in the first equation becomes N^2 and these correspond to the principal maxima of the diffraction grating. This gives us the angles of the principal maxima as

$$d \sin\theta = m\lambda_0$$

where m represents the order of the diffraction. Since the angles of the principal maxima depend on the wavelength, the diffraction grating disperses the incident wavelengths along different directions, thus forming a spectrum.

The resolving power of the diffraction grating is given by

$$RP = \frac{\lambda_0}{\Delta\lambda} = mN$$

where N is the total number of slits in the grating that is illuminated by the incident light and $\Delta\lambda$ is the minimum resolvable wavelength difference. If we consider a diffraction grating with 5000 lines per centimeter and illuminate 2 cm width of the grating, then in the first order the resolving power would be 10 000. Thus at a wavelength of 500 nm, the grating can resolve two lines spaced by 0.05 nm.

1.8.1
Resolution of Optical Instruments

According to ray optics, an aberration-less optical system should form point images of point objects and thus have infinite resolution. Since light has a finite wavelength, diffraction effects limit the size of the focused spot for a point object and thus diffraction would decide the resolving power of optical instruments.

When a telescope is used to image far-off objects such as stars, then each star produces an Airy pattern and if the Airy patterns of two stars are too close to each other then it would not be possible for us to resolve the two stars. Thus diffraction effects will ultimately limit the resolving ability of telescopes. If the aperture of the objective of the telescope is d and the wavelength is λ_0, then the angular resolution of the telescope is

$$\Delta\theta_{\min} = \frac{1.22\lambda_0}{d} \quad (1.78)$$

The larger the diameter of the objective lens, the better is the angular resolution. Terrestrial-based telescopes do not actually operate at this limit because of atmospheric turbulence and hence practical values may be higher than the value predicted by Eq. (1.78). The diameter of the objective of the Keck telescope is 10 m. With this size of the objective, the expected resolution is about 0.01 arc sec.

In the case of a microscope in which the object is illuminated by incoherent illumination, the spatial resolution is given by

$$\Delta x = \frac{0.61\lambda_0}{\text{NA}} \quad (1.79)$$

where NA is the numerical aperture of the microscope objective and is given by $n \sin \theta_m$ where n is the refractive index in which the object is placed (for example, in oil immersion objectives, n can be larger than 1) and θ_m is the limiting angular aperture. Typically a 40× objective with an NA of 0.65 will have a resolution of about 0.42 μm at 550 nm wavelength. As demand for increased resolution of microscopes increases, many techniques to overcome this limitation are being developed and microscopes with resolutions of better than 100 nm are now available [8].

Tutorial Exercise 1.11

Calculate the resolution for the following optical systems for a wavelength of 550 nm, which corresponds to the center of the visible spectrum.

1) An $f/16$ lens, that is, a lens with an F-number of 16 is used to focus a laser beam. What is the diameter of the focused spot? How does this diameter change when the pupil of the lens is changed to $f/1.4$?
2) The average pupil size of the human eye is about 4 mm and the focal length is about 17 mm. What is the diameter of a far away object such as a star on the retina?
3) Compare the resolution of the 200 in. Mt Palomar telescope with the human eye. What is its spatial resolution when observing the moon surface?
4) Let us consider a microscope with an NA of 0.8. What is the minimum spatial resolution?

Solution:

1) Using Eq. (1.79), the diameter of the focused image would be about 21 μm, but only 1.9 μm for the $f/1.4$ lens. The $f/1.4$ lens has obviously better resolution than the $f/16$ lens. In camera lenses you would see the f-number written on the lens and that gives you an indication of its resolution capabilities.

2) The diameter of the first dark ring of the Airy pattern would be about 3 μm.
3) For the telescope with an objective diameter of 200 in., the minimum angular resolution is given by 0.13×10^{-6} rad. This telescope can resolve spatial details of 50 m on the moon (assuming the distance between earth and moon to be 384 000 km).
4) The minimum spatial resolution is given by 0.41 μm.

Box 1.4: Diffraction of a Gaussian Beam

The beam of a laser has a Gaussian distribution in the transverse plane. It is described by the following equation:

$$f(x, y) = A \exp\left[-\frac{(x^2 + y^2)}{w_0^2}\right]$$

where w_0 represents the spot size of the Gaussian beam. We can use Eq. (1.69) to study the propagation of such a Gaussian beam. Substituting in Eq. (1.69) and integrating, we obtain the field distribution on any plane z as

$$u(x, y, z) \approx \frac{a}{(1 - i\gamma)} \exp\left[-\frac{x^2 + y^2}{w^2(z)}\right] e^{-i\Phi}$$

where

$$\gamma = \frac{\lambda z}{\pi w_0^2},$$

$$w(z) = w_0 \sqrt{1 + \gamma^2},$$

$$\Phi = kz + \frac{k}{2R(z)}(x^2 + y^2)$$

$$R(z) \equiv z\left(1 + \frac{1}{\gamma^2}\right)$$

Thus, the intensity distribution $|u|^2$ at any z is given by

$$I(x, y, z) = \frac{I_0}{1 + \gamma^2} \exp\left[-\frac{2(x^2 + y^2)}{w^2(z)}\right]$$

which again represents a Gaussian distribution. Thus, as a Gaussian beam propagates in a homogeneous medium, the transverse distribution remains Gaussian with its spot size changing with z as given by $w(z)$.

On the plane $z = 0$, the phase is independent of the transverse coordinates x and y and hence on this plane the Gaussian beam has a plane wavefront. As the beam propagates, the phase distribution changes and is given by the function Φ. Now, we note that the transverse phase distribution at any value of z is given by

$$\exp\left[-\mathrm{i}\frac{k}{2R(z)}(x^2+y^2)\right]$$

The above phase distribution corresponds to the paraxial approximation of the phase distribution of a spherical wave and thus as the Gaussian beam propagates, its phase front becomes spherical with a radius of curvature given by $R(z)$. The radius of curvature of the phase front is infinite at $z=0$ representing a plane phase front and as z tends to infinity it again tends to infinity.

The equations representing the variation of the spot size and radius of curvature of the phase front of the Gaussian beam are valid for all values of z (positive or negative) and z is measured from the plane where the beam has a plane phase front. This plane is called the *waist* of the Gaussian beam, and as can be seen from $w(z)$ the beam has the minimum value of spot size at the waist.

For values of z satisfying

$$z \gg \pi w_0^2/\lambda$$

that is, in the far field, the spot size increases linearly with z:

$$w(z) \approx \frac{\lambda z}{\pi w_0}$$

which is similar to the case of a circular aperture except for a different factor. In fact, for a beam having a transverse dimension w, the angle of diffraction is approximately given by λ/w.

1.9 Anisotropic Media

The electric displacement vector $\mathbf{D}$ inside a material and the applied electric field $\mathbf{E}$ are related through the following equation:

$$\mathbf{D} = \varepsilon \mathbf{E} \tag{1.80}$$

where ε is the electric permittivity of the medium. In isotropic media, the displacement and the electric field are parallel to each other and ε is a scalar quantity. In anisotropic media, the two vectors are not parallel to each other and ε is a tensor and we write Eq. (1.80) in the form

$$\mathbf{D} = \bar{\varepsilon}\mathbf{E} \tag{1.81}$$

where we have put a bar on ε to indicate that it is not a scalar.

In the principal axis system of the medium, $\bar{\varepsilon}$ can be represented by a diagonal matrix:

$$\bar{\varepsilon} = \begin{pmatrix} \varepsilon_{xx} & 0 & 0 \\ 0 & \varepsilon_{yy} & 0 \\ 0 & 0 & \varepsilon_{zz} \end{pmatrix} \tag{1.82}$$

And the three diagonal terms give the principal dielectric permittivities of the medium.

- For isotropic media

$$\varepsilon_{xx} = \varepsilon_{yy} = \varepsilon_{zz} = \varepsilon \tag{1.83}$$

- For uniaxial media

$$\varepsilon_{xx} = \varepsilon_{yy} \neq \varepsilon_{zz} \tag{1.84}$$

- and for biaxial media

$$\varepsilon_{xx} \neq \varepsilon_{yy} \neq \varepsilon_{zz} \tag{1.85}$$

The principal dielectric constants and the principal refractive indices of the anisotropic medium are and defined through the following equations:

$$K_{ij} = \frac{\varepsilon_{ij}}{\varepsilon_0}; \quad n_{ij}^2 = \sqrt{K_{ij}} \tag{1.86}$$

Since in the principal axis system ε is diagonal, the principal refractive indices are also sometimes referred to as n_x, n_y, and n_z. For isotropic media $n_x = n_y = n_z$; for uniaxial media $n_x = n_y \neq n_z$, while for biaxial media all the three principal refractive indices are different.

In isotropic media, the speed of wave propagation is independent of the state of polarization of the light beam and is also independent of the direction of propagation. On the other hand, in anisotropic media, along any given direction of propagation there are two linearly polarized eigenmodes which propagate, in general, with different phase velocities. Thus any incident light beam can be broken into the two eigenmodes of propagation and since their velocities are in general different, the phase difference between the two components changes as the light beam propagates. This results in a change of the state of polarization of the light beam as it propagates along the medium.

In uniaxial media, there is one direction of propagation along which the two velocities are equal and this direction is referred to as the *optic axis*. There is only one such direction in uniaxial media and hence the name. In biaxial media, there are two such optic axes.

When a plane wave propagates in a uniaxial medium, one of the polarization components travels always at the same velocity, which is independent of the direction of propagation. This wave is referred to as the *ordinary wave* and its velocity is given by c/n_o. On the other hand, the orthogonal polarization, which is referred to as the *extraordinary wave*, has a velocity of propagation that depends on the direction of propagation with respect to the optic axis. If the direction of the propagation vector of the extraordinary wave makes an angle of ψ with respect to the optic axis, then the velocity of the extraordinary wave is given by $c/n_e(\psi)$ where

$$\frac{1}{n_e^2(\psi)} = \frac{\cos^2 \psi}{n_o^2} + \frac{\sin^2 \psi}{n_e^2} \tag{1.87}$$

The velocity of the extraordinary wave varies between c/n_o and c/n_e.

For the o-wave, the displacement vector **D** and the electric field vector **E** are perpendicular to the plane containing the propagation vector **k** and the optic axis. For the e-wave, the displacement vector **D** lies in the plane containing the propagation vector **k** and the optic axis and is normal to **k**.

If we consider a plane wave propagating perpendicular to the optic axis of a uniaxial crystal, then the two eigen polarizations have refractive indices n_o and n_e. If the free space wavelength of the wave is λ_0, then in propagating through a distance l, the phase change due to propagation would be

$$\Delta\phi = \frac{2\pi}{\lambda_0}(|n_o - n_e|)l \tag{1.88}$$

If a linearly polarized wave is incident on such a medium with the polarization direction making an angle θ with the optic axis and propagating perpendicular to the optic axis, then since in the linearly polarized wave the two components are in phase, as the wave propagates through the uniaxial medium, it will develop a phase difference given by Eq. (1.88). If the phase difference is π, then the polarization state would be still linear but now oriented at an angle $-\theta$ with the optic axis. Such a device is called a half wave plate (HWP) and the required thickness is

$$l_h = \frac{\lambda_0}{2|n_o - n_e|} \tag{1.89}$$

Using a HWP it is possible to change the orientation of the linearly polarized wave by appropriately orienting the optic axis of the HWP.

If the thickness of the uniaxial medium is half of that given by Eq. (1.89), then the phase difference introduced is $\pi/2$ and such a plate is called a quarter wave plate (QWP). If the angle θ is chosen to be $\pi/4$, then the output of a QWP will have two equal components of linear polarizations but with a phase difference of $\pi/2$; such a wave would correspond to a circularly polarized wave. Thus a QWP can be used to convert a linearly polarized wave into a circularly polarized wave and vice versa. If the angle θ is not $\pi/4$, then the output would be an elliptically polarized wave.

HWP and QWPs are very useful components in experiments involving polarization (Chapter 9).

1.10
Jones Calculus

Jones calculus is a very convenient method to determine the changes in the state of polarization of a light wave as it traverses various polarization components. In this calculus, the state of polarization is represented as a (2 × 1) column vector and the polarization components are represented by (2 × 2) matrices. The propagation through each polarization component is represented by multiplying the column vector by the Jones matrix corresponding to the polarization component.

We first discuss the Jones vector representation of the state of polarization of a plane wave. Equation (1.19) gives a general expression for an elliptically polarized plane light wave propagating along the z-direction. The complex amplitudes of the electric field vector can be expressed as the components of the Jones vector $\mathbf{J}$ where

$$\mathbf{J} = \begin{pmatrix} E_x \\ E_y \end{pmatrix} \tag{1.90}$$

If we write

$$|x\rangle = \begin{pmatrix} 1 \\ 0 \end{pmatrix} \tag{1.91}$$

and

$$|y\rangle = \begin{pmatrix} 0 \\ 1 \end{pmatrix} \tag{1.92}$$

then the Jones vector can also be written as

$$\mathbf{J} = E_x \begin{pmatrix} 1 \\ 0 \end{pmatrix} + E_y \begin{pmatrix} 0 \\ 1 \end{pmatrix} = E_x |x\rangle + E_y |y\rangle \tag{1.93}$$

The vectors $|x\rangle$ and $|y\rangle$ represent normalized Jones vectors for an x-polarized and a y-polarized wave, respectively. This is an equivalent representation of the fact that a general polarized wave can be represented as a superposition of a linearly polarized wave along x and a linearly polarized wave along y with appropriate amplitude and phase.

The normalized Jones vector representing a linearly polarized wave making an angle ϕ with the x-axis is given by

$$|\phi\rangle = \begin{pmatrix} \cos\phi \\ \sin\phi \end{pmatrix} = \cos\phi \, |x\rangle + \sin\phi \, |y\rangle \tag{1.94}$$

The electric field vector of an RCP wave propagating in the z-direction is given in Eq. (1.17); hence, the normalized Jones vector of an RCP is given by (neglecting a common phase factor)

$$|\text{RCP}\rangle = \frac{1}{\sqrt{2}} \begin{pmatrix} 1 \\ -i \end{pmatrix} \tag{1.95}$$

Similarly, the normalized Jones vector representing the LCP wave is given by

$$|\text{LCP}\rangle = \frac{1}{\sqrt{2}} \begin{pmatrix} 1 \\ +i \end{pmatrix} \tag{1.96}$$

The normalized Jones vector for a right elliptically polarized wave would be

$$|\text{REP}\rangle = \frac{1}{\sqrt{(E_1^2 + E_2^2)}} \begin{pmatrix} E_1 \\ -iE_2 \end{pmatrix} \tag{1.97}$$

Thus, we have represented a general polarized plane wave by a (2 × 1) column vector.

Optical elements such as HWP, QWP, polarizer, and so on, act on the polarization state of the light beam and thus modify the Jones vector of the light beam propagating through them. The input to these elements would be described by a

Jones vector of the input light and the output would correspond to the Jones vector of the output light wave. Thus these elements convert a (2 × 1) column vector to another (2 × 1) column vector and hence can be represented by a (2 × 2) matrix.

The Jones matrix for a linear polarizer is given by

$$J_{LP}(\alpha) = \begin{pmatrix} \cos^2 \alpha & \sin \alpha \cos \alpha \\ \cos \alpha \sin \alpha & \sin^2 \alpha \end{pmatrix} \tag{1.98}$$

where α represents the angle made by the polarizer pass axis with the x-axis. This can be easily seen as follows: if the input Jones vector is given by

$$|in\rangle = \begin{pmatrix} E_1 \\ E_2 \end{pmatrix} \tag{1.99}$$

then the Jones vector of the output wave would be given by

$$|out\rangle = \begin{pmatrix} \cos^2 \alpha & \sin \alpha \cos \alpha \\ \sin \alpha \cos \alpha & \sin^2 \alpha \end{pmatrix} \begin{pmatrix} E_1 \\ E_2 \end{pmatrix} = \begin{pmatrix} E_0 \cos \alpha \\ E_0 \sin \alpha \end{pmatrix} \tag{1.100}$$

with

$$E_0 = E_1 \cos \alpha + E_2 \sin \alpha \tag{1.101}$$

As expected Eq. (1.100) represents a linearly polarized wave. Thus no matter what the input polarization may be, the output will always be linearly polarized.

For a QWP with its fast axis along the y-direction, the Jones matrix is given by

$$J_{QWP} = \begin{pmatrix} 1 & 0 \\ 0 & i \end{pmatrix} \tag{1.102}$$

When a linearly polarized wave (making an angle 45° with the x-axis) is incident on such a QWP, the output polarization is given by

$$J_{QWP} |45°\rangle = \begin{pmatrix} 1 & 0 \\ 0 & i \end{pmatrix} \frac{1}{\sqrt{2}} \begin{pmatrix} 1 \\ 1 \end{pmatrix} = \frac{1}{\sqrt{2}} \begin{pmatrix} 1 \\ i \end{pmatrix} \tag{1.103}$$

which represents an LCP (Figure 1.10). For a QWP with its fast axis along the x-direction, the Jones matrix is given by

$$J_{QWP} = \begin{pmatrix} i & 0 \\ 0 & 1 \end{pmatrix} \tag{1.104}$$

For a HWP with its fast axis along the y-direction, the Jones matrix is given by

$$J_{HWP} = \begin{pmatrix} 1 & 0 \\ 0 & -1 \end{pmatrix} \tag{1.105}$$

When a light beam linearly polarized at 45° to the x-axis is incident on a HWP, then the output is given by

$$J_{HWP} |45°\rangle = \begin{pmatrix} 1 & 0 \\ 0 & -1 \end{pmatrix} \frac{1}{\sqrt{2}} \begin{pmatrix} 1 \\ 1 \end{pmatrix} = \frac{1}{\sqrt{2}} \begin{pmatrix} 1 \\ -1 \end{pmatrix} \tag{1.106}$$

which is again a linearly polarized beam, polarized at $-45°$ to the x-axis.

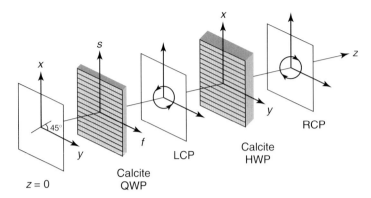

Figure 1.10 A linearly polarized beam making an angle 45° with the x-axis gets converted to an LCP after propagating through a calcite QWP; further, an LCP gets converted to a RCP after propagating through a calcite HWP. The optic axis in the QWP and HWP is along the y-direction as shown by lines parallel to the y-axis. Light polarized along the y-axis travels faster than light polarized along the x-axis; hence the symbols f (fast) and s (slow).

When an LCP is incident on such a HWP, the output polarization is given by

$$J_{\text{HWP}} |\text{LCP}\rangle = \begin{pmatrix} 1 & 0 \\ 0 & -1 \end{pmatrix} \frac{1}{\sqrt{2}} \begin{pmatrix} 1 \\ i \end{pmatrix} = \frac{1}{\sqrt{2}} \begin{pmatrix} 1 \\ -i \end{pmatrix} \quad (1.107)$$

which represents an RCP (Figure 1.10).

As the effect of each polarization component is represented by a (2 × 2) matrix, using Jones calculus becomes quite straightforward to determine the state of polarization of the light wave emerging from a series of components. Corresponding to each component, we need to multiply with the corresponding Jones matrix in the correct order. Since, in general, matrices do not commute, the output polarization state would depend on the order in which the light wave traverses them.

1.11
Lasers

Today lasers form one of the most important sources of light for various applications because of their special characteristics in terms of temporal and spatial coherence, brightness, and the possibility of extremely short pulse widths and power. Lasers spanning wavelengths from the ultraviolet to the infrared are available for various applications. Diode lasers, because of their size and efficiency, are one of the most important lasers. In this section, we discuss briefly the working of the laser and some of their properties. For more detailed discussions, readers are referred to books on lasers [9, 10].

1.11.1
Principle

Atoms and molecules are characterized by energy levels and they can interact with electromagnetic radiation through three primary processes:

1) Absorption: When radiation at an appropriate frequency falls on the atomic system atoms lying in the lower energy state can absorb the incident radiation and get excited to an upper energy level. If the energy of the two levels are E_1 and E_2 $(>E_1)$, then for absorption the frequency of the incident radiation should be $\omega_0 = (E_2 - E_1)/\hbar$.
2) Spontaneous emission: An atom lying in the upper energy level can spontaneously emit radiation at the frequency ω_0 and get de-excited to the lower level. The average time spent by the atom in the excited level is called the *spontaneous life time* of the level.
3) Stimulated emission: An atom in the excited level can also be stimulated to emit light by an incident radiation at the frequency ω_0 and the emitted light is fully coherent with the incident radiation.

When we consider an atomic system at thermal equilibrium, the number of atoms in the lower state is always higher than that in the upper state. Thus there will be more absorptions than stimulated emissions when light interacts with the collection of atoms. If the population of the upper state can be made to be higher than that of the lower state (i.e., if we have population inversion between the two levels), then there would be greater number of stimulated emissions compared to absorptions and the incident radiation can get amplified. This is the basic principle of optical amplification and is the heart of the laser (short for light amplification by stimulated emission of radiation). An external source of energy is required to bring about the state of population inversion in the collection of atoms. This could be electrical energy such as in the case of diode lasers, it could be another laser such as in the case of Ti : sapphire lasers, it could be a flashlamp such as in the case of ruby laser, or it could be an electrical discharge such as in the case of an argon ion laser. The laser uses up the energy from the pump and converts it into coherent optical energy.

With the population inversion, we obtain an optical amplifier that can amplify an incident optical radiation at the appropriate frequency. However, we need a source of light which is similar to an oscillator rather than an amplifier. The optical amplifier can be converted to a laser by providing optical feedback. This is accomplished by the use of a pair of mirrors on either side of the optical amplifier. The pair of mirrors reflects a part of the energy back into the optical amplifier and is called an *optical resonator*.

As soon as the pump takes atoms from the lower to the higher energy state, the atoms in the higher energy state start to emit spontaneously. Part of the spontaneous emission that travels along the direction perpendicular to the mirrors of the cavity is reflected back into the cavity, which then gets amplified by the population inversion generated in the cavity by the pump. The amplified spontaneous emission is reflected back into the cavity by the other mirror and this process continues back and forth. When the loss of the optical radiation gets compensated completely by the gain provided by the population inversion in the medium, the laser starts to oscillate. Thus there is a threshold to start laser action and this depends on the losses in the cavity.

Owing to various mechanisms such as the finite lifetime of the excited level, or the motion of the atoms causing Doppler shift or collisions of the atoms, a range of frequencies, rather than a single frequency, can interact with the atoms. This is referred to as *line broadening* and depends on the atomic system, the temperature and pressure in the case of a gas, the environment of atoms in the case of solids, and the energy bands in the case of semiconductors. Typical bandwidths range from a few to hundreds of gigahertz. This implies that the atomic system is capable of amplifying a range of frequencies lying within this broadened line.

The optical resonator made up of the two mirrors supports modes of oscillation within the cavity. The transverse field distributions that repeat themselves after every roundtrip within the cavity are termed as *transverse modes* and the frequencies of oscillation supported by the cavity are termed as *longitudinal modes* of the cavity. Just as in the case of modes of oscillation of a string, the various frequencies of oscillation of the laser cavity are separated by an approximate value of $\Delta \nu$ where

$$\Delta \nu = \frac{c}{2nL} \tag{1.108}$$

where c is the velocity of light in free space, n is the refractive index of the medium filling the cavity, and L is the length of the cavity.

Since the atoms forming the optical amplifier can amplify a range of frequencies, there may be many frequencies supported by the cavity that can oscillate simultaneously leading to multi-longitudinal mode oscillation. This will lead to a drastic reduction of the coherence length of the laser. For interferometry applications where the path difference between the interfering beams may be large, it is important that the laser oscillate in a single frequency only.

There are many techniques to achieve single longitudinal mode operation of the laser. One of the standard techniques is to use a tilted Fabry–Perot etalon within the cavity of the laser. Since the Fabry–Perot etalon allows only certain frequencies to pass through, by choosing an appropriate etalon and adjusting its angle within the cavity, it is possible to allow only one of the longitudinal modes to oscillate. This leads to a single longitudinal mode of oscillation with a corresponding increase in the coherence length. In the case of semiconductor lasers, an external grating or a fiber Bragg grating can be used to achieve single longitudinal mode operation. In the distributed feedback diode laser a periodic variation of the structure along the direction of lasing ensures single frequency oscillation.

Box 1.5: Transversal Modes of a Laser Beam

Although the output beam of a laser is mostly a Gaussian, there exist higher order modes inside the laser cavity. The various transverse modes of the laser can be approximately described by Hermite–Gauss functions:

$$f_{mn}(x, y) = C H_m\left(\frac{\sqrt{2}x}{w_0}\right) H_n\left(\frac{\sqrt{2}y}{w_0}\right) e^{-(x^2+y^2)/w_0^2}$$

where $H_n(\xi)$ represents a Hermite polynomial, and the lower order polynomials are

$$H_0(\xi) = 1, \qquad H_1(\xi) = 2\xi; \qquad H_2(\xi) = 4\xi^2 - 2\ldots$$

and n and m represent the mode numbers along the x- and y-directions, which are assumed to be transverse to the axis of the resonator.

The fundamental mode of the laser corresponds to $n = 0$ and $m = 0$ and is a Gaussian distribution:

$$f_{00}(x, y) = C e^{-(x^2+y^2)/w_0^2}$$

This is the mode in which most lasers operate as it has the minimum diffraction divergence and also has a uniform phase front without any reversals of phase unlike the higher order transverse modes.

Since the fundamental mode has the smallest transverse size, any aperture within the cavity can be used to set the laser to oscillate only in the fundamental transverse mode by introducing additional losses to higher order modes as compared to the fundamental mode.

1.11.2
Coherence Properties of the Laser

The temporal coherence of the laser is determined by its spectral width. For a spectral width of $\Delta \nu$, the coherence length is given by

$$l_c = \frac{c}{\Delta \nu} = \frac{\lambda^2}{\Delta \lambda} \tag{1.109}$$

As discussed earlier, when the laser is used in an interference experiment, for proper contrast, the maximum path difference between the interfering beams must be much smaller than the coherence length. Hence the coherence length is a very important parameter in applications such as holography or interference measurements. Single longitudinal mode lasers with coherence lengths of more than a few meters to a few hundred meters are commercially available.

> **Tutorial Exercise 1.12**
>
> Consider a laser operating at 800 nm with a spectral width of 10 MHz. Calculate the coherence length. What is its spectral width in wavelength?
>
> **Solution:**
>
> The corresponding coherence length would be 30 m; the spectral width in wavelength is 21 fm.

When the laser oscillates in the fundamental Gaussian mode, it has very good spatial coherence. Sometimes, the laser is coupled into a single mode optical fiber and the fiber acts as a spatial filter to filter the spatial noise in the beam. The output of the single mode fiber is approximately Gaussian.

1.12
Optical Fibers

An optical fiber is a cylindrical structure consisting of a central core of refractive index n_1 surrounded by a cladding of slightly lower refractive index n_2 (Figure 1.11). A light wave can get trapped within the core of the fiber by the phenomenon of total internal reflection. Thus such a dielectric structure can guide light from one point to another and forms the heart of today's telecommunication system. More details on fiber optics and its application to communication and sensing can be found in Ref. [11].

An optical fiber can be characterized by its NA defined as

$$\mathrm{NA} = \sqrt{(n_1^2 - n_2^2)} \tag{1.110}$$

The maximum angle of acceptance of the fiber is given by

$$\theta_{\max} = \sin^{-1}\left(\frac{\mathrm{NA}}{n_0}\right) \tag{1.111}$$

where n_0 is the refractive index of the medium surrounding the fiber.

Tutorial Exercise 1.13

Consider a fiber with an NA of 0.2 and calculate the maximum acceptance angle of the fiber when the surrounding medium is (i) air and (ii) water.

Solution:

Using Eq. (1.111) and a refractive index of 1.0 for air and 1.33 for water, the acceptance angle is

1) ~11.5° and
2) ~8.6°.

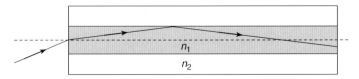

Figure 1.11 An optical fiber consists of a core of refractive index n_1 surrounded by a cladding of refractive index n_2.

Optical fibers can also be specified by a normalized frequency, also called the *V-number*, defined as

$$V = \frac{\omega}{c} a \, \text{NA} = k_0 a \, \text{NA} \tag{1.112}$$

where a is the radius of the core of the fiber.

Multimode fibers are characterized by core diameters of about 50 μm and an NA of about 0.2; they support many transverse modes of propagation. On the other hand, single mode fibers have much smaller core diameters and support only a single mode of propagation. For single mode operation, the V-number of the fiber should be less than 2.405. Thus a fiber can be single moded at a particular wavelength and if operated at lower wavelengths may have a V value of more than 2.405 and would then support more than one mode.

Since multimode fibers support many modes, the output from multimode fibers under laser illumination has a speckle pattern. Since a single mode fiber supports only a single mode, it has a well-defined transverse amplitude distribution, which can be very well approximated by a Gaussian distribution [12]:

$$\psi(r, z) = A e^{-r^2/w_0^2} e^{i(\omega t - \beta z)} \tag{1.113}$$

where w_0 is referred to as the *spot size* of the mode and β represents the propagation constant of the mode. The ratio β/k_0 is referred to as the *effective index* of the mode that lies between n_1 and n_2. An empirical expression for the spot size is given by Marcuse [7]

$$w_0 = a \left(0.65 + \frac{1.619}{V^{1.5}} + \frac{2.879}{V^6} \right) \tag{1.114}$$

If we use this fiber at a wavelength of 500 nm, then its V-number would be 2.51 (assuming that NA remains the same) and the fiber would support more than one mode. The output from the fiber would not be a Gaussian but will also contain contribution from the higher order mode.

The spot size of the mode can be increased by taking fibers with a larger core radius. However, to keep the fiber as single moded the corresponding NA has to be reduced. This implies that the refractive index difference between the core and cladding needs to be reduced. This leads to poorer guidance by the fiber and hence such a fiber is more prone to bending-induced losses.

When a laser is used to couple light into a single mode fiber, the coupling efficiency would be determined by the transverse distribution of electric field of the laser beam and the modal field distribution. The maximum coupling efficiency (assuming perfect alignment between the laser beam and the fiber) from a laser oscillating in a Gaussian mode with a spot size w_L to a single mode fiber with a spot size w_0 is given by

$$T_{\max} = \left(\frac{2 w_L w_0}{w_L^2 + w_0^2} \right)^2 \tag{1.115}$$

For unit coupling efficiency, the spot sizes of the laser beam and the fiber should be identical and the waist of the laser beam should be at the fiber end face. Any deviation from this condition would reduce the efficiency. The expression given in Eq. (1.115) is also valid for coupling efficiency between two single mode fibers with spot sizes w_L and w_0.

Standard single mode fibers do not maintain the state of polarization of light propagating through them. If linearly polarized light is launched into the fiber, then after propagating through the fiber, the state of polarization can become arbitrary because of the birefringence within the fiber induced by bending or residual stresses in the fiber. In interference experiments involving fibers, it may be necessary to maintain the state of polarization of the light. In such cases, one uses polarization maintaining fibers. Such fibers have a strong linear birefringence caused by introducing transverse stress within the fiber while fabricating them so that the fiber supports two orthogonal linearly polarized modes. Since the modes have significantly different propagation constants, only a spatially periodic perturbation with a small period (a few millimeters) can induce coupling between them. Since under normal circumstances such a perturbation does not exist, the fiber maintains the linear polarization of the propagating light beam. Such fibers are used in fiber optic sensors based on interference effects such as Mach Zehnder fiber interferometric sensor.

1.13
Summary

In this brief chapter, basic concepts in optics such as rays, image formation by optical systems, interference, and diffraction of light have been given. Laser is one of the most important sources of light and a brief outline of the working and its characteristics is also covered. Finally, a brief introduction to optical fibers has been given as optical fibers are very useful components that carry light between two different points and also provide for spatial filtering of the laser beam in experiments such as holography and interferometry.

Problems

1.1 What is the direction of propagation of a wave described by the following equation:

$$\mathbf{E} = \mathbf{y} E_0 \exp\left\{i\left(\omega t - \frac{k}{\sqrt{2}}x - \frac{3k}{\sqrt{2}}z\right)\right\}$$

Give an expression for a wave propagating in the same direction but orthogonally polarized to this wave.

1.2 Consider a piece of parabolic index medium with flat faces and of thickness d. Consider a point source placed on the axis at the entrance face of such a medium. What is the minimum value of d so that the output rays form a parallel bundle? In which direction would the parallel beam be propagating?

1.3 Consider a symmetric double convex thick lens formed by surfaces of radii of curvatures 10 cm and separated by a distance of 1 cm. If the refractive index of the medium of the lens is 1.5, obtain the system matrix for the lens and calculate the focal length of the lens. Also obtain the positions of the principal planes.

1.4 A nonreflecting film to operate at a wavelength of 600 nm is to be coated on a material of refractive index 2.2. What is the thickness of the film that would be needed? What will be the optimum refractive index so that the reflection is minimum?

1.5 Consider a Fabry–Perot interferometer made of two mirrors of reflectivities 0.9 and separated by a distance of 1 mm. Obtain the resolving power of the interferometer if it is operated at a wavelength of 500 nm.

1.6 Estimate the lateral coherence width of the sun on the surface of the earth. Assume a wavelength of 550 nm.

1.7 A lens with a transverse diameter of 10 mm and a focal length of 20 mm is used to focus a laser beam at a wavelength of 600 nm and with a transverse diameter of 5 mm. What will be the approximate area of the focused spot? If the laser beam has a power of 5 mW, what will be the maximum intensity and maximum electric field at the center of the focused spot?

1.8 A Gaussian beam emerging from a laser with a wavelength of 800 nm has a transverse diameter of 1 mm. What will be the transverse size of the beam after traversing a distance of 10 m?

1.9 Consider a uniaxial medium with $n_o = 2.26$ and $n_e = 2.20$. For an incident wavelength of 600 nm, obtain the thickness of the HWP and QWP.

1.10 A circularly polarized beam is incident on a HWP. What will be the output state of polarization? If it is incident on a QWP what will be the state of polarization of the output wave?

1.11 An interference experiment is to be conducted using a He–Ne laser. We have two lasers, one oscillating in a single longitudinal mode with a linewidth of 10 MHz and the other with two modes with linewidths of 10 MHz and separated by a frequency of 600 MHz. Estimate the minimum path difference for which the interference pattern will disappear when either of the lasers is used.

1.12 Consider a laser resonator made of a plane mirror and a concave mirror of radius of curvature 1 m. For what range of separation between the mirrors would the resonator be stable? What will be the spot size of the Gaussian mode of the laser? Where will the waist of the laser beam lie?

1.13 A He–Ne laser has a gain bandwidth of 1 GHz. For what mirror separation would the laser operate in a single longitudinal mode?

1.14 Consider a single mode fiber with $a = 2.5$ µm, NA $= 0.08$ operating at a wavelength of 600 nm. Obtain the V-number of the fiber and show that the fiber would be single moded. Obtain the corresponding spot size of the mode. What will be the angle of diffraction of the beam coming out of such a fiber? Above what wavelength will the fiber be single moded?

2
Electronic Image Sensing and Processing

Thomas Baechler

2.1
Introduction

The photoelectric effect was observed for the first time in 1839 by Alexandre Edmond Becquerel and his father. They observed that electrons were emitted from matter (metals and non-metallic solids, liquids, or gases) as a consequence of their absorption of energy from electromagnetic (EM) radiation of short wavelength, such as visible, near-infrared (NIR), or ultraviolet light. In 1886, Heinrich Hertz and his assistant Wilhelm Hallwachs conducted the first systematic examinations. Study of the photoelectric effect led to important steps in understanding the quantum nature of light and influenced the formation of the concept of wave–particle duality. A photon above a threshold frequency has the required energy to eject a single electron from an atomic orbital, creating the observed photoelectric effect. This discovery led to quantum revolution in physics and earned Einstein the Nobel Prize in Physics in 1921. Electrons emitted in this manner may be referred to as *"photoelectrons."*

The photoelectric effect is the principal physical phenomenon that is used in modern image sensors. However, there is an important phase before: the *image formation*. Its principal goal is to form the image of a distant object onto the imaging plane of an image sensor. It is important to understand how to take good pictures – still or continuous – of given scenes, which are depending on lighting conditions and different properties of their objects, such as color and surface texture, for example. This front-end part of an image acquisition chain – as generically shown in Figure 2.1 – is in charge of bringing photons coming from a scene to the right place on the image sensor where they can be detected. Section 2.2 explains the paraxial approximation, which is a first-order approximation for geometrical optics applicable to modern image sensors. As for most physical phenomena, image formation systems are not perfect and suffer from optical distortion and other limiting factors that are presented in Section 2.2.2.

Optical Methods for Solid Mechanics: A Full-Field Approach, First Edition. Edited by Pramod Rastogi and Erwin Hack.
© 2012 Wiley-VCH Verlag GmbH & Co. KGaA. Published 2012 by Wiley-VCH Verlag GmbH & Co. KGaA.

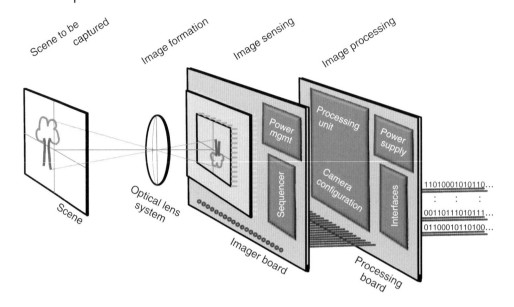

Figure 2.1 Fundamental concept of a generic electronic image data acquisition system.

The second part of this chapter is about *image sensing*. It relates the electronic charges generated by the photoelectric effect as a response to the incident photons, which is proportional to the quantity of incoming light (in a given spectral range). Section 2.3 presents the electronics part of an image data acquisition system. Its output is a digital number as a measure of the incoming light per pixel surface at the image sensor plane. If the images are taken in a continuous manner, the output is a video data stream.

The last part of this chapter is about software aspects of such a generic image data acquisition system: *image processing*. Section 2.4 gives the basic methods about how to process the captured data from the image sensing block. It is important to understand how to retrieve the relevant image information or just how to enhance images. Most of the time, data reduction is one of the major goals in the image processing block. This is the back-end part and is covered by software components, such as algorithms running on calculation units.

Figure 2.1 shows the fundamental concept of an electronic image sensing and image processing chain. To capture an object in a scenery (indicated schematically by a "tree" in the virtual object plane of Figure 2.1), adequate optical components are needed to authentically reproduce this "imaged" object into the image plane where the image sensing device is located. The image sensor translates the incoming photons into a digital number that can then be processed – mostly on a frame by frame basis – by the image data processing unit.

State-of-the art image data acquisition systems are not perfect at all. This chapter introduces and explains different limiting factors, distortion effects, and

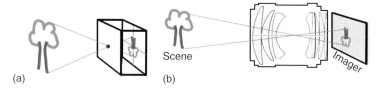

Figure 2.2 (a) Simple pinhole camera and (b) construction drawing of a complex objective.

noise sources altering the quality of the acquired image data. Many of them can be compensated for or cancelled out. Some need to be treated during image processing. Tutorial procedures and methods to overcome such limitations and degradations are given.

2.2 Image Formation

The main purpose of the front-end part of any digital image acquisition chain is to reproduce as perfectly as possible the scene to be "imaged" or captured onto the active area of the image sensor. This function is achieved through an optical system, which can be just a single hole at a certain distance of the image sensor as found in a simple pinhole camera, as represented in Figure 2.2a, or a complex objective made out of several lenses allowing modification of focus, aperture, depth of field, and field of view (FOV) as shown in Figure 2.2b.

Image formation is based on geometrical optics (see Chapter 1) using lenses and mirrors to collect light – or in more general terms, EM radiation (see Box 2.1) – from the object and to produce a geometrically similar distribution of light flux across the image plane, where the image sensor can measure it. The next paragraphs explain the basics of such imaging concepts and touch on several limiting factors. It is very important to understand the quality of such optical reproduction systems. However, this tutorial only treats the most general optical concepts. The interested reader will find more detailed explanations in [1–3].

Box 2.1: Electromagnetic Radiation

The fundamental laws describing classical EM are – along with Lorentz's force law – collected in Maxwell's equations comprising Gauss's law, Gauss's law for magnetism, Faraday's law of induction, and Ampère's law with Maxwell's correction. *Radiometry* is the scientific field studying the measurement of EM radiation, especially the optical EM radiation. It can be stated in a simplified manner that light is a sinusoidal EM wave – with wavelength λ – traveling either in material (solid, liquid, or gas) or in vacuum. The spectral range of light extends from "hard" X-rays – $\lambda > 0.01$ nm corresponding to frequencies below 3×10^{19} Hz (or 30 EHz, read exa-hertz) where γ-radiation starts – to far infrared (IR) in the THz range ($\lambda < 1$ mm, just below microwave radiation). *Photometry* is the science of light measurement in terms of perceived brightness to the human eye. The visible light is only a small fraction of the EM spectrum – with wavelength ranging from 400 to 700 nm. One of the best – and already quite old – reference books with extensive information about radiometry and photometry is [4]. Chapter 2 of [5] also gives a good introduction to this subject.

The choice of an appropriate image data acquisition system can be completely different if dedicated to machine vision where similar objects in a known and given environment working in the IR range have to be detected and identified or if needed for taking nice pictures of a colorful landscape with direct sunlight.

2.2.1
Geometrical Optics and Imaging Concepts

The main purpose of an optical system is to form an image of a distant object in the imaging plane of an image sensor. The action of the lenses and mirrors ensures that rays diverging from an object point will intersect the image plane at the corresponding image point – building the image point by point with minimum distortion. To simplify calculations in geometrical optics, it is assumed that light rays are only making small angles from or to the optical axis of the system. Angles are sufficiently small if sine and tangent of that angle can be replaced by the angle itself ($<10°$). This first-order approximation is called *paraxial approximation* and used for first-order ray-tracing and Gaussian optics. For higher precision and more complex optical systems, software tools – such as [6–8] – have to be used. Figure 2.3 shows an example of a miniaturized optical encoder with its ray-tracing plot, where light is emitted by an LED on the right side and captured by an imager on the left side. Stray-light – all undesired light rays from reflections – can also be simulated with the same tools. Such simulation plots help identify the places where stray-light needs to be shielded and whether the useful light signal gets onto the image sensing device. Stray-light can be considered as a noise source reducing the signal-to-noise ratio (SNR) of an optical system.

The number of object points is infinite and an infinite number of rays could emanate from each object point. Therefore, it is inevitable to simplify a ray-tracing diagram by drawing only a few rays from selected object points. Another simplification reduces complex lens systems (e.g., such as shown in Figure 2.2b) to a single thin lens with the same optical behavior. The thin lens approximation assumes that the thickness of a lens or a lens system along the optical axis is negligible compared to its focal length, object, or image distance. The graphical ray-tracing rules for thin lenses are [3] as follows:

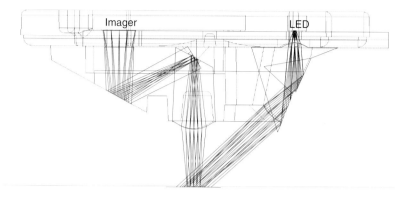

Figure 2.3 Complex ray-tracing plot of an optical encoder (light source (LED) from the right).

1) Rays entering the lens parallel to the optical axis exit through a focal point.
2) Rays entering the lens through a focal point exit parallel to the optical axis.
3) Rays that pass through the center of the lens do not change direction.

Figure 2.4 shows the basic imaging concept and introduces different notions. First, lenses have a focal point on each side, which is at distance f, called the *focal length*. Assuming the thin lens model and the same propagation medium on both sides of the lens, both focal points are at the same distance f. Focal length is positive for convex lenses (lenses that are normally thicker in the center than at the outside). In contrast, concave lenses (thinner in the center) have a negative focal length as the focal points are on the other side of such lenses.

From a geometrical point of view, the construction in Figure 2.4 explains rules (1) and (2). Rule (3), however, can only be justified by considering the region of the lens near the optical axis as a parallel plate. By the law of refraction – also known as *Snell's or Descartes' law* (see Chapter 1) – the ray passing through this region will

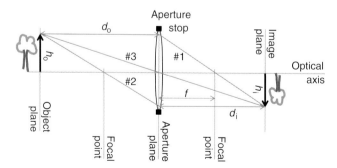

Figure 2.4 Image formation and ray-tracing rules.

be shifted parallel to its original direction. For very thin lenses, this shift is small and the ray appears to go through the center of the lens (see Tutorial Exercise 2.1).

■ **Tutorial Exercise 2.1**

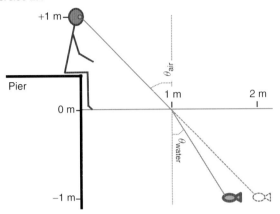

There are optical phenomena suggesting images "being apparently at a certain place." But is the fish really where it seems to be? Consider the following situation in which you are sitting on a pier with your feet just in the water. The image of a fish appears at the water surface level 1 m in front of your feet. The fish swims an estimated 1 m below the water surface. By applying the law of refraction – given as $n_{air} \cdot \sin \theta_{air} = n_{water} \cdot \sin \theta_{water}$ with $n_{air} = 1$ and $n_{water} = 4/3$ – it is seen that the fish is not 2 m from the pier, but about 37.5 cm closer.

Solution:

By assuming that you are seeing the fish at 45° angle (i.e., your eyes are 1 m above the water level at the pier border), then

$$\sin \theta_{water} = \frac{3}{4} \cdot \frac{\sqrt{2}}{2} \text{ leading to } \theta_{water} \approx 32°$$

The fish is only virtually at the 45° position, but in reality at 32°, therefore reducing the distance to the pier by $(\tan 45° - \tan 32°) \cdot 1\,m \approx 37.5\,cm$.

Applying the graphical ray-tracing rules to Figure 2.4, two algebraic equations for thin lens imaging can be developed:

$$\frac{1}{d_o} + \frac{1}{d_i} = \frac{1}{f} \tag{2.1}$$

and

$$M = \frac{h_i}{h_o} = -\frac{d_i}{d_o} \tag{2.2}$$

With Eqs. (2.1) and (2.2) the paraxial image location and size, the object location and size, and finally the focal length can be determined. If the object is located at infinity, the image is formed at the focal point ($d_i = f$). If the object is moved toward its own focal point ($d_o = f$), the image will be formed at infinity. The ratio of the image height h_i to the object height h_o – both heights are defined positive if above the optical axis (therefore the minus sign in Eq. (2.2)) – is called *linear magnification M*. If the areas of the object and the image are of interest, the area magnification M^2 has to be used, yielding

$$\frac{A_i}{A_o} = M^2 = \left(\frac{d_i}{d_o}\right)^2 \tag{2.3}$$

The imaging quality and the optical flux transfer efficiency are both affected by the size and location of apertures in the optical system under consideration.

In Figure 2.5, two special rays – the marginal ray and the chief ray – are represented. Both define two major parameters – the *f-number $f/\#$* and the *FOV(field of view)* – of an optical system and directly affect the optical flux transfer and the image quality. A large $f/\#$ and a small FOV yield a better image quality (smaller aberrations), but less optical flux reaches the image plane. On the contrary, small $f/\#$ and large FOV result in high optical flux transfer efficiency, but the overall image quality suffers. The f-number is defined as

$$f/\# = \frac{f}{d_{AS}} \tag{2.4}$$

where f is the focal length and d_{AS} is the aperture diameter delimited by the *aperture stop* defining the acceptance angle for axial rays. The axial ray through the edge of the aperture stop is called *marginal ray*.

In complex objectives, the aperture can be mechanically adapted. Therefore, a large $f/\#$ implies a small relative aperture. Such systems are termed "*slow*" referring to the fact that longer exposure time is needed to capture the same amount of light. In contrast, small $f/\#$ is termed "*fast*," producing higher image brightness with a major drawback of reduced depth of field.

Another important ray in an optical system, such as the one schematically represented in Figure 2.5, is the *chief ray*, starting at the maximum height of the object, passing through the center of the aperture stop (crossing the optical axis),

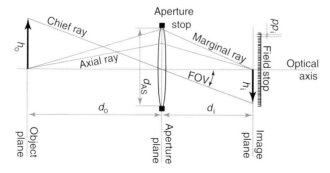

Figure 2.5 Axial, marginal, and chief rays – aperture stop, field stop, and field of view (FOV).

and going to the maximum extent of the image, which is determined by the size of the field stop. It is of highest importance to position the *field stop*, that is, the image sensor, precisely at the location of the image plane of the optical system. Any kind of deviation from this position or any lateral displacement from the optical axis will introduce image blur and, therefore, will considerably affect image quality.

The FOV is the angular coverage of an optical system (defined in Figure 2.5 as a half-angle). Owing to ray trace rule (3), the angles to define FOV in Eq. (2.5) are equivalent to

$$FOV_{\text{half-angle}} = \left|\arctan\left(\frac{h_o}{d_o}\right)\right| = \left|\arctan\left(\frac{h_i}{d_i}\right)\right| \qquad (2.5)$$

2.2.2
Optical Distortion and Other Limiting Factors

The paraxial assumption that was made at the beginning of Section 2.2.1 leads to perfect images, that is, each point in the object space maps to a point in the image space. Unfortunately, the reality is more complex and image quality suffers from several effects, such as diffraction and aberration, where points in the object space map to blur spots of finite size in the image. Intuitively speaking, better image quality can be obtained if smaller blur spots are achieved throughout the optical system. Image quality is limited by the diffraction of light passing through the lens or the lens system's finite aperture (corresponding to the aperture stop d_{AS} in Figure 2.5). Any optical system with a finite aperture can never form a point image because of the wave nature of EM radiation. Even in absence of other image defects, a characteristic minimum blur spot size exists and can be expressed as

$$d_{\text{diffraction}} \cong \frac{2.44 \cdot \lambda \cdot d_i}{d_{AS}} = 2.44 \cdot \lambda \cdot f/\# \qquad (2.6)$$

with $d_i = f$ for objects at ∞. Equation (2.6) is an approximation with the factor 2.44 derived from a calculation using the first zero of Bessel function of the resulting diffraction pattern. The diffraction-limited spot diameter is the smallest spot size a lens system can form and may be larger because of aberration effects (see Tutorial Exercise 2.2).

> **Tutorial Exercise 2.2**
>
> Calculate the diffraction limit for visible light assuming a typical ratio of focal length to lens diameter of 1.5 (e.g., lens with focal length of 15 mm with a lens aperture diameter of 10 mm).
>
> **Solution:**
>
> Using Eq. (2.6) this results in the following diffraction spot diameters: 1.5 μm for blue light at 400 nm, 2 μm for green light at 550 nm, 2.5 μm for red light at

700 nm, and 3.5 μm for NIR light at 1000 nm. With pixel sizes well below 2 μm in a state-of-the-art handy camera everybody is using, images are "artificially" blurred, especially in the red. Sunsets are romantic and some blur on such pictures can be nice, but represents low-quality from an imaging perspective. Millions (multi-mega) of small pixels will not make better pictures!

Image degradation can also occur from aberrations, such as the following:

- **Spherical aberration**: This effect occurs because spherical surfaces are not the most appropriate but the most easily manufactured lens shape. Beams parallel but at different distances to the lens axis will focus at different distances from the spherical lens, yielding an image blur (top: perfect lens; bottom: lens with spherical aberration). More expensive spheric lenses – with curvatures adapted to particular applications – minimize this blurring effect.

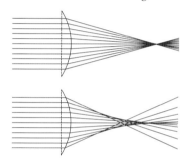

- **Chromatic aberration**: This effect is caused by the variation of the lens material's refractive index with the wavelength. Different wavelengths focus at different distances from the lens, resulting in color fringes in the image. This effect can be minimized by using an a chromat, for example, an achromatic lens doublet in which two lenses, one made out of crown glass and the other made of flint glass, are bonded together. This reduces the amount of chromatic aberration in a given wavelength range. The better the corrections, the more expensive such lens systems become.

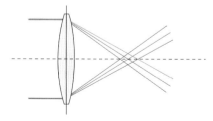

- **Coma**: This effect derives its name from the comet-like flare occurring in the aberrated image. Coma occurs for objects off the optical axis because parallel rays passing through the lens at a fixed distance from the center of the lens are focused on a ring-shaped image in the focal plane, also known as the *comatic*

circle. The sum of all these circles results in a V-shaped or comet-like appearance in the image. As with spherical aberration, coma can be minimized or even eliminated by designing the curvature of the lens surfaces to match the specific application.

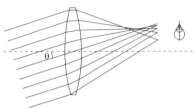

- Other kind of aberrations, such as astigmatism, field curvature, and distortion.

In summary, all these image degradations get worse with either decreasing $f/\#$ or increasing field angle, or both. But Eq. (2.8) just stated that the diffraction limit gets smaller for smaller $f/\#$, whereas aberrations increase. Object points thus map to blurred spots in the image depending on the $f/\#$ of the imaging lens and the field angle of the particular object point under consideration. Aberration estimations during the design process of an optical design – and every optical design should be adapted to the particularities of the complete system – determine a minimum allowable $f/\#$ and a maximum allowable FOV. This minimum allowable $f/\#$ will lead to a minimum pixel or detector size with respect to diffraction limits. Finally, the flux transfer has to be verified in order to get enough light onto the detector. This design process can imply several iterations before achieving a good solution. If miniaturization has to be achieved, software tools and a lot of experience are necessary to find a good solution.

2.3
Image Sensing: From Photons to Electrons

The ideal image sensor or camera has 100% quantum efficiency (QE) (every photon generates one electron) and 100% fill factor, highest frame rate, highest dynamic range, highest spatial resolution, highest analog-to-digital conversion (ADC) resolution, no power consumption, no dark current or signal, no fixed-pattern noise (FPN), and no readout or electronic noise. As everything can be perfectly integrated in one single chip, size and cost are very small. Unfortunately, this is not the reality and this section discusses the most important trade-offs and limitations of electronic image sensors and cameras.

The first part of this chapter looks into "pixelization effects" with respect to the optical image, as described in the previous sections and how many electrons from all incident photons per pixel and per frame are finally generated and measurable. The second section goes into details concerning the handling of the available electrons and the operations of state-of-the-art electronic image sensors. As there is no ideal image sensor or camera, many noise sources are present in such systems and are explained in the middle section of this chapter. Only at this stage of the entire image data acquisition chain, digital image data will be generated; fundamental shuttering and image sampling concepts are introduced in Section 2.3.3.

A photodetector is a device converting the optical image – as developed in Section 2.2 – into electrical signals that can be measured and read out. The operation of an electronic photodetector involves three steps:

1) generation of carriers or electron–hole pairs by incident light;
2) separation of the electron–hole pair in an electric field generated, for example, in a depletion zone of a photodiode; and
3) storage of either type of carriers, most commonly electrons in p-doped silicon (Si) substrate.

More detailed pieces of information can be found in Section 13.3 in [9] and Section 7.4 in [10].

The generated "photoelectrons" are either measured as a photocurrent or transferred as a charge packet onto a storage capacitor creating a measurable voltage proportional to the number of "photoelectrons." A fundamental distinction in function of the underlying technology is made. Charge-coupled devices (CCDs) – better known as *CCD imagers* – just transfer the generated "photocharges" to a simple output voltage buffer [11]. Complementary metal-oxide semiconductor (CMOS) based devices include amplification and analog readout blocks. State-of-the-art *digital CMOS imagers* additionally include the conversion of the electrical analog signal directly into digital numbers. Figure 2.6 shows the principal functionalities common to all electronic cameras:

- photon absorption
- charge-to-voltage conversion

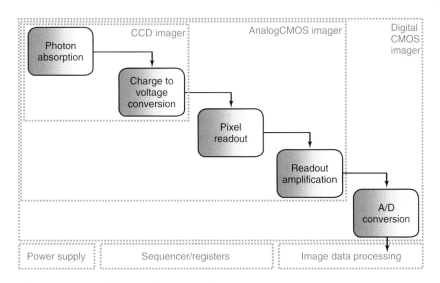

Figure 2.6 Principal functionalities in an electronic camera.

- pixel readout
- readout amplification
- analog-to-digital conversion (ADC).

Other necessary building blocks of complete electronic cameras or single-chip digital CMOS imagers are power management and different registers and memory for sequencer programs. Some systems include an image data processing unit.

Nearly all digital video or still cameras are today based on CCD or CMOS image sensors. They are based on silicon as the preferred wafer material and can be mass produced, making such devices affordable. The big advantage of CMOS is the possibility to implement all analog and digital circuitry on the same die, also called *chip*, which makes the overall camera design more compact and cheaper. CMOS image sensors are omnipresent and becoming more and more popular, especially since they achieve at least as good noise and low-light performances as CCD imagers. General references to *"solid-state image sensing"* can be found regarding CCD imagers and cameras under [5, 11, 12] and regarding CMOS imagers and cameras under [12–15].

2.3.1
Image Formation Related to Image Sensing

The normal construction process of an optical system considers the object space being imaged onto the image plane. There is a nice and useful alternative available when the image plane acts as the field stop with an image sensor positioned right there (as shown in Figure 2.5). Consider now that any image sensor has a kind of "footprint" in the object space. Especially, every individual pixel of such an image sensor can be reproduced in a "pixelized" raster in the object space. Each of these regions contributes light flux to the corresponding pixel on the image sensor. Assuming that the object is sufficiently far away ($d_o \approx \infty$ implies that $d_i \approx f$) and, consequently, image formation occurs at the focus of the lens, the relation between *pixel pitch* pp_i – that is, the constant distance between neighboring pixels (most commonly pixels are square) – and the size of the object space raster $d_{footprint}$ can be given as follows:

$$d_{footprint} = \frac{pp_i \cdot d_o}{f} \qquad (2.7)$$

For the object-at-infinity situation and for a given detector size $y_{detector}$, a lens with long focal length yields a narrow FOV, while a short focal length yields a wide FOV. Depending on the light quantities coming from the object, the incoming quantities of light onto the detector can be calculated. If there is not enough light arriving in a given exposure period on each pixel, there are basically two possibilities to overcome this limitation: (i) to increase the illumination of the scene or (ii) to choose a detector with higher sensitivity – in other words, with less noise (which is the subject of Section 2.3.3).

Tutorial Exercise 2.3

Imagine a candle on top of Lothse (fourth highest mountain on Earth, 8516 m). How many photons emitted by the candle will arrive on an image sensor on top of Mount Everest (highest mountain on Earth, 8848 m) knowing that the image sensor has an active area of 1 cm² and that the peaks of Mount Everest and Lothse are approximately 3 km apart ? – The image sensor has a pixel resolution of 100×100 pixels (therefore a pixel pitch of 10 µm) and no optics is used to focus the incoming light. How many photons will arrive during an integration time of 1 s on average per pixel?

Solution:

A candle emits 10 lux at a distance of 30 cm [4]. Regarding the correspondence between photons and unit lux, different definitions can be found; for simplicity, assume that 1 lux corresponds to 2×10^{16} photons $(m^2 s)^{-1}$. Note that these are visible photons as lux is only defined for visible light with wavelengths between 400 and 700 nm (a candle emits probably about half of its luminous intensity in the NIR range above 700 nm). The number of photons reaching the active area of the image sensor is proportional to the spherical surfaces at 30 cm – where the 10 lux of luminous flux density are emitted equally in all directions – and at 3 km. In order to determine the photon flux reaching the image sensor's active area, the projection of this 1 cm² area at 3 km distance onto the sphere at 30 cm has to be calculated as follows:

$$\frac{4\pi(3 \times 10^{-1})^2}{4\pi(3 \times 10^3)^2} \times 10 \times 2 \times 10^{16} \frac{\text{photons}}{m^2 \times s} \approx 2 \times 10^9 \frac{\text{photons}}{m^2 \times s}$$

Calculated back to 1 cm² and taking into account the integration time of 1 s, the image sensor's active area of 1 cm² will receive about 200 000 visible photons. The image sensor has 10 000 pixels yielding only 20 photons per pixel and per second. Besides the nasty environment, the image sensor has to have very high sensitivity (low temperatures prevailing on top of the highest mountain will help reduce noise in electronic imaging devices).

In any optical system, it is very important to determine how much light will reach the active area of the detector. In these flux-transfer calculations, reflections occurring at each transition from one medium (e.g., air) to another (e.g., glass in optical components) have to be taken into account. To avoid reflection losses, antireflection (AR) coatings are applied to the most critical surfaces – which could be, for example, a simple cover lid on top of the image sensor protecting the active area from dust. These AR coatings have to be adapted to the wavelength(s) used. In more extreme situations, lens materials have to be skillfully chosen in order to avoid absorptions or reflections. As an example, the top-level passivation layers and the intermediate silicon oxide layers on top of CCD or CMOS image sensors will absorb most of the UV light below 350 nm wavelength.

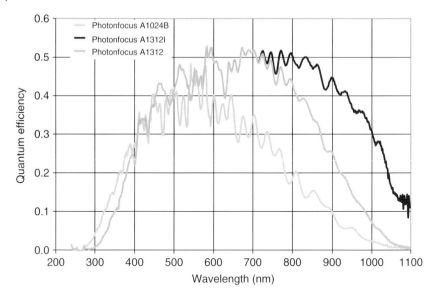

Figure 2.7 Quantum efficiency curves for state-of-the-art CMOS image sensors from Photonfocus.

Figure 2.7 shows – according to EMVA1288 standard [16] measured – QE×FF curves (courtesy: Photonfocus [17]) of state-of-the-art CMOS image sensors, where QE stands for *quantum efficiency* and FF for *fill factor*. The quantum efficiency (QE) for a given wavelength is the percentage of photons hitting the active area of a photodetector and the number of generated electron–hole pairs. It is a physical measure of the detector's electrical responsivity to incoming light over a given spectral range. Figure 2.7 shows already quite significant differences for similar image sensors, especially in the NIR range starting at 700 nm wavelength. At 800 nm, for example, imager A1024B has less than half of the sensitivity of the other two imagers and at 1000 nm imager A1312 is three times less sensitive than its NIR-optimized counterpart A1312I. QE has to be defined for each wavelength as the photon energy E_{photon} varies proportionally to the optical frequency ν or inversely proportional to the photon's wavelength λ according to

$$E_{photon}(\nu) = h\nu \text{ or } E_{photon}(\lambda) = \frac{hc}{\lambda} \tag{2.8}$$

with Planck's constant $h = 6.626 \times 10^{-34}$ J, s = 4.136×10^{-15} eVs and the speed of light $c = 299.792 \times 10^6$ m/s. Max Planck was the first scientist to presume that the EM energy is a multiple of a very small quantity, later called *quantum*. To give an idea of these magnitudes, green light at 550 nm wavelength has, for example, a frequency of about 545 T Hz, which is too high a frequency to be measured directly with state-of-the-art electronics devices, which are limited to some tens of gigahertz. Therefore, photon measurements are made for defined finite frequency spectra and the QE curves define the "conversion factor" of photons to photoelectrons for

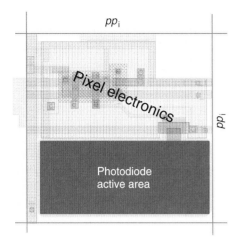

Figure 2.8 Pixel layout with active photodiode area and pixel electronics.

different detector materials and EM spectra. In order to obtain good QE, special care has to be taken to optimize the absorption of light and the collection of charges by avoiding charge recombination in the photodiode.

As can be seen from the pixel layout of Figure 2.8, fill factor FF – defined as the ratio between active photodiode area compared to the whole pixel area – is related to the pixel pitch pp_i. To design multi-mega-pixel imagers on a reasonably cheap silicon area, pixel pitch has been drastically reduced to dimensions in the micrometer range (common standards in state-of-the-art mobile phones: 1.75, 1.4, 1.2 µm and the latest at 0.9 µm). This size reduction does not leave much space for photodiode and pixel electronics (such as transfer gate, reset, select, and buffer transistors) yielding a photodiode fill factor in the 5–20% range. Microlenses on top of each pixel can focus the light onto the photodiode improving the resulting fill factor by a factor of 2–3. Another solution that is used more frequently is the backside illumination (BSI) technology, where light falls on the back of the thinned wafer. The generated photoelectrons are collected at the "frontside photodiode" and read out using standard electronics. The advantage of this more complex technology is that the photodiode can be made quite small and can still receive 100% of the incoming light (FF = 100%).

For comparison of different image sensors, it is of upmost importance to make a rough calculation of the incoming light per pixel including all optical devices and interfaces. Sometimes, image sensors with larger pixels and, therefore, reduced resolution, will produce better images. Especially, pixel sizes below 2.5 µm suffer from diffraction effects (see Section 2.2.2). Therefore, "the more pixels the better" – sometimes also called *"mega-pixel-race"* – is just a marketing gag. Professional imaging systems are optimized and adapted to the corresponding environment(s) and object(s).

> **Box 2.2: Electromagnetic (EM) Spectral Ranges**
>
> The physical implementation of the photon absorption block of Figure 2.6 can differ in function of the spectral ranges to be observed and the underlying applications, such as IR spectroscopy and astronomy. Some of the denominations for the EM spectral ranges used are as follows:
>
> - visible (VIS): 400–700 nm
> - ultraviolet (UV): 10–400 nm (Si-based imagers: 250–400 nm)
> - X-ray: <10 nm (for direct detection or with scintillators generating visible light, which can then be detected and measured with standard CCD or CMOS imagers)
> - NIR: 700–1100 nm (for Si-photodiodes) or 700–1600 nm (for SiGe)
> - different IR ranges:
>   - SWIR (short-wavelength infrared): 1400–3000 nm (1.4–3 µm)
>   - MWIR (mid-wavelength infrared): 3000–8000 µm (3–8 µm)
>   - LWIR (long-wavelength infrared): 8–15 µm
>   - FIR (far infrared): 15–1000 µm (1 mm)
> - different IR ranges according to CIE (Commission Internationale de l'Éclairage):
>   - IR-A: 700–1400 nm (sometimes also called NIR)
>   - IR-B: 1400–3000 nm
>   - IR-C: 3000–10 mm
>
> The dominant spectral region for conventional (C-band) long-distance telecommunications is at 1530–1560 nm, which can be covered by SiGe photodiodes. For the huge area of "IR" detectors, cooled or uncooled, special detector materials (CdHgTe, InSb, InGaAs, GaAs, AlGaAs, InAs, GaSb, amorphous Si, etc.), microbolometers, and pyroelectrical detectors are used.

2.3.2
Operating Modes of State-of-the-Art Electronic Image Sensors

As already mentioned, CMOS photodiodes have two quite distinct roles in the photoelectric transformation: (i) the electron–hole pair generation as a function of the wavelength of the incoming photons (according to QE-curves as shown in Figure 2.7) and (ii) the separation of the electron and the hole in the depletion region of the photodiode. A *depletion region* forms across a pn junction. The physical phenomenon behind this is that carriers diffuse from regions with higher concentration to regions with lower concentration. If pieces of p-type and n-type semiconductors are placed together to form a pn junction, electrons migrate to the p side and holes to the n side. As this migration effect leaves behind a positive donor ion (at the n side) and a negative acceptor ion (at the p side), these charged ions adjacent to the pn interface build a depletion region with no mobile carriers – resulting in an intrinsic electrical field, which is used to separate the electrons from the holes. In a standard CMOS process, the p type substrate is connected to ground and therefore holes are drained to this negative reference, whereas the electrons are pushed toward the positive potential and collected in the capacitive well of the photodiode. The maximum number of electrons that can be stored inside the photodiode is called *full well* (FW) and depends on the photodiode type, the doping profiles, and the size of the photodiode. Figure 2.9 shows a cross-section of a more complex photodiode, the buried photodiode (BPD) – also called pinned photodiode (PPD) – which will serve to illustrate the operation of such a photodetector. Charge transfer – similar to CCDs – from the photodiode into a sense node can be explained.

Figure 2.9a represents the cross-section of a PPD where the pn-interface of the photodiode is between $p_{substrate} - n_{diode}$ and $p_{pinning} - n_{diode}$. As $p_{substrate}$ and $p_{pinning}$ are electrically connected, they can be considered as one single anode – with just one minor difference in their respective doping concentrations. The purpose of the $p_{pinning}$ layer is to "bury" n_{diode} and to push the depletion region – shown with the dashed line – away from the semiconductor surface. This semiconductor surface interfaces the oxide layer and contains many lattice defects that create undesired noise effects and leakage currents. "Burying the diode" reduces these effects to

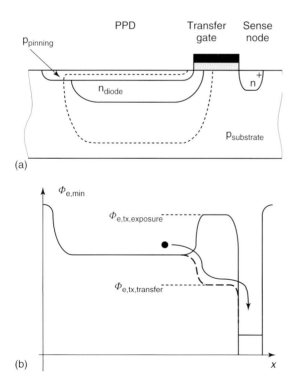

Figure 2.9 Cross-section and potential diagram of a pinned photodiode (PPD).

a minimum (2–3 orders of magnitude of leakage current reduction can easily be achieved compared to conventional photodiodes or photodetectors).

The photodiode structure shown in Figure 2.9a is just missing a contact to connect the diode to the pixel readout circuitry. As this contact has to be at the semiconductor surface, the previously described leakage effects may appear. Therefore, an additional "device," the transfer gate, has been added. The goal is to keep the charges buried in the semiconductor during exposure and to transfer the charges by activating the transfer gate as illustrated in Figure 2.9b. The transferred charges thus reach the sense node. At this point, the charge-to-voltage conversion occurs on the basis of the following physical relation:

$$V = \frac{1}{C} \cdot Q \tag{2.9}$$

where the voltage V is proportional to the charges Q. If the sense node capacitance C is small, the conversion factor will increase, but the FW – that is, the number of charges that can be stored on the sense node – will be small. This means that the photodiode will be saturated faster for the same amount of light. There is always a trade-off between conversion factor and FW, which then defines or limits (for small FW) the dynamic range of the pixel (which is normally in the range of 40–70 dB).

Among other operation modes used in image sensors, shutter operation and the corresponding readout sequence are very important. In an image sensor array with several millions of pixels, two fundamental imaging operations exist: (i) *rolling shutter* operation and (ii) *global shutter* operation. Rolling shutter is a method of image acquisition in which each frame is recorded not from a single point in time, but rather by scanning – or just "rolling" – across the frame. In other words, not all parts of the image are recorded at exactly the same time, even though the whole frame is displayed at the same time during playback (see Tutorial Exercise 2.4). In contrast, global shutter exposes the entire frame at exactly the same time. However, the readout sequence still has to be sequential, leading to important delays between exposure and readout, thereby varying the junction leakage components from one area of a captured image to another. As these events are apart in time, correlated sampling operations are not easily applicable, leaving the dominant *kTC*-noise from the global shutter and reset switch in the resulting image.

Section 2.3.3 explains the principal noise sources in image sensors and cameras. It is impossible to dispose of noise, but by employing intelligent noise-correlation operations, noise can be considerably reduced. Regarding rolling or global shutter operation, there is a major trade-off between noise performance and impact on the image quality or image distortion, especially if there are fast moving objects in the scene.

2.3.3
Noise Sources in State-of-the-Art Electronic Image Sensors

Electronic image sensors suffer from noise of various natures at different levels. Figure 2.10 contains a list of the most common noise source components and indicates where they appear in an electronic image sensor or camera [15];

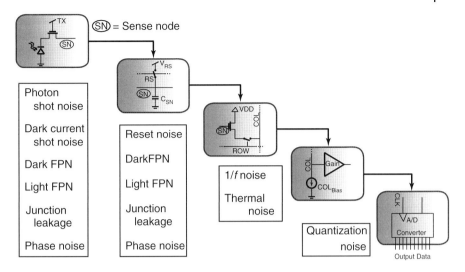

Figure 2.10 Principal noise sources in a state-of-the-art CMOS image sensor.

sections 2.3 and 2.5 of [14] contain detailed explanations of the most important noise sources and their origin.

Statistical variation in the optical signal – even before being detected by the imaging system – is called *photon shot noise* and is the natural benchmark for the imaging system's noise performance. If photon shot noise is the predominant noise source, the imaging system is said to be well designed as it works at the physical performance limit. Light sources are emitting photons in a random manner representing a statistical variation of the number of emitted photons in a given time interval – such as the integration time of an image sensor – following approximately a Poisson distribution [1]. The statistical standard deviation of the noise charge is $\sqrt{N}$, where N is the mean number of photogenerated charge carriers in a photodetector. For 1 000 000 incoming photons (on a pixel with a large FW), a root mean square (rms) shot noise of 1000 electrons is expected. Most presumably, all other noise components will become irrelevant. Such image sensing systems are said to be "shot noise limited." Thus, for a mean signal of one single photoelectron, one rms shot noise electron is expected. This "single-electron detection" requirement, however, is very ambitious for an electronic imaging system.

The most important noise sources to be controlled in CMOS image sensors are (i) the dark current or leakage current shot noise of the photodiode – and of the sense node junction in the case of a PPD, (ii) electronic noise of the readout and signal processing circuitry with 1/f-noise and thermal noise, and possibly (iii) quantization noise from the ADC stage.

As stated previously, photon shot noise is a physical phenomenon that cannot be avoided. However, state-of-the-art CMOS imagers or CCD/CMOS cameras should be carefully designed such that all other noise sources are becoming less important.

Only at the lowest illumination levels, the imaging system will be limited by its proper internal noise sources. This limit is called *sensitivity* and is defined as the minimum illumination signal required for producing a specified output signal with a specified SNR.

Box 2.3: Technology-Dependent Noise Sources

Some of the noise sources listed in Figure 2.10 – junction leakage and dark current shot noise – are semiconductor processing dependent. Technologists in CCD or CMOS fabs (short name for semiconductor fabrication sites) can improve and optimize such parameters by varying process step parameters, such as doping concentration, implant energy, implant angles, thermal annealing functions, and many more. The same is applicable to *1/f-noise* in CMOS transistors because of trapping and release of charge carriers in semiconductor–oxide interface defects, called *traps or slow states*, resulting in a fluctuation of the number of mobile electrons in the channel. This "low-frequency" noise component is also called *flicker noise or pink noise*.

Other noise sources can be reduced or eliminated by design or by compensation. *Thermal noise* in resistors or channels of CMOS transistors describes a random voltage between the corresponding terminals due to random motion of carriers as a result of thermal energy in the semiconductor material. Thermal noise is also called *Johnson–Nyquist noise* and has a white spectrum. In a first approximation, the thermal noise generated in the readout path of a CMOS image sensor can be reduced through bandwidth limitation techniques. This also limits the frame rate as settling in such "slower" electronic blocks takes much more time. In summary, thermal noise can be traded against power consumption or speed and therefore *frame rate*. Signal settling times in the analog readout path define the imager's row time t_{row} and thus determine the imager's *frame time* as #rows $\cdot t_{row}$, which at the end fixes the number of frames per second (fps) as $\frac{1}{\text{frame time} + \text{frame overhead}}$. For example, video rate imagers provide 60 fps, which defines the frame time to be about 16 ms, and the row time for a mega-pixel imager will therefore be about 16 µs.

Finally, the term *fixed pattern noise* usually refers to a particular noise pattern on electronic image sensors often noticeable during longer exposure times where particular pixels are susceptible to giving brighter intensities above the general background noise. One parameter is called the *dark signal nonuniformity* (DSNU), which is the deviation from the average across the imaging array without any external illumination. The second parameter related to FPN is called *photo response nonuniformity* (PRNU) describing variations in gain, that is, ratio between optical power on a pixel versus the electrical signal output, which may in practice become a visible defect and degrade the image quality. Although FPN does not change appreciably across a series of captures, it may vary with integration time, imager temperature, imager gain, and incident illumination. FPN can be compensated through proper calibration and dark frame subtraction.

> **Box 2.4: Noise Cancellation Methods**
>
> As already mentioned in relation with global shutter operation, one of the dominant noise components is thermal noise on sense node capacitors – also known as *kTC*-noise from the reset switch transistor. This thermal noise component is sometimes called *sense node reset noise* or just *reset noise*. The temporal amplitude is frozen on the sense node when the reset switch transistor is entirely turned off at the end of the sense node reset phase. However, in rolling shutter operation, reset noise can be cancelled by applying *correlated double sampling* (CDS) or digital correlated double sampling (DCDS). The principle is quite simple to explain, but practical implementation is not straightforward. The basic idea for CDS and DCDS is to first sample the reset value onto a sampling capacitor (or in the case of DCDS to directly readout the reset value) and only subsequently to transfer the signal on top of the reset value. In a second sampling phase – hence the name "double sampling" – the reset value plus signal is sampled onto another sampling capacitor (or, in the case of DCDS, read out again). This technique yields the difference of these two sampled signals (or in the case of DCDS to calculate the difference from the two read out values) as $\text{sample}_2 - \text{sample}_1 = (v_{\text{signal}} + v_{\text{reset}}) - v_{\text{reset}} = v_{\text{signal}}$. For ultrasensitive imagers it is indispensable to have CDS operation with reduced sampling time difference between the two samples. The dominant noisy device in a CDS configuration remains the source-follower transistor with its important thermal and 1/f-noise. The only drawback is that it takes twice as much time to convert both values and, therefore, high-speed imaging with frame rates over 100 fps using CDS or DCDS is difficult to achieve besides the fact that rolling shutter can induce artifacts that are undesirable in some high-speed imaging applications (see Tutorial Exercise 2.4).

2.3.4
From Electrons to Digital Image Data

State-of-the-art CMOS image sensor readout paths – as schematically represented in Figure 2.6 – contain a gain stage, which should be as close to the pixel as possible. All noise contributions from the amplifier stage downwards are divided by the gain of the amplifier or amplifier stages. It can be stated that all noise sources from the pixel fully contribute to the overall noise figure. However, contributions from the readout circuitry including ADC stage will be divided by the amplifier gain and can thus be neglected in the case of high amplifier gain. Most advanced techniques even integrate such an amplifier in each pixel [14], which eliminates the most dominant noise sources in the source-follower transistor. As a rule of thumb, the following approximate noise figures can be given in equivalent noise electrons (input referred noise voltage can be calculated through the conversion factor):

- Global shutter operation (mainly kTC-noise): $\sim$30–100 e$^-$
- Rolling shutter with CDS operation (mainly source-follower noise): $\sim$3–10e$^-$
- In pixel amplification with CDS operation: $\sim$1 e$^-$.

> **Box 2.5: Imager Formats**
>
> Electronic image sensing has a long history that started in 1969 with the invention of the "bucket brigade device" at Philips Research Labs and in 1970 with the first publication of a CCD at Bell Labs [11]. In the following years many papers about CCD physics and technologies appeared, but only a few commercial applications made it to the market. Once CCDs entered the consumer market – especially in camcorders and still cameras in the 1980s – they ruled over the solid-state imaging world (imaging technologies mainly based on silicon). Only in the 1990s, the first successful attempts with active-pixel sensors (APSs) based on CMOS technologies were made, leading to a second revolution in the consumer market with complete camera integration including simple optics on the same die for mobile phones and webcams, among others. There are still historic remains to indicate image sensor format coming from the pre-CCD times. One of the most popular formats in conventional photography is the full-frame 35 mm format with geometrical dimensions of 36 mm × 24 mm. Another common denominator in APS imager formats is the sensor's diagonal. However, some inch-based format names are not standardized and exact dimensions may vary, too. A good summary of image sensor format can be found in [18]. In general, the "rule of 16" applies to approximate imager diagonals expressed in millimeters to inch-based dimensions (e.g., the 1/5-in. sensor times 16 yields a 3.2 mm diagonal). The following (incomplete) table helps to understand this relationship:
>
> | Format name (in. (mm)) | Imager diagonal (mm) | Dimensions imager active area (mm × mm) |
> | --- | --- | --- |
> | 1 (25.4) | 16.0 | 12.8 × 9.6 |
> | 2/3 (16.9) | 11.0 | 8.8 × 6.6 |
> | 1/2 (12.7) | 8.0 | 6.4 × 4.8 |
> | 1/3 (8.5) | 6.0 | 4.8 × 3.6 |
> | 1/4 (6.35) | 4.0 | 3.2 × 2.4 |
> | 1/5 (5.08) | 3.2 | 2.6 × 2.2 |
> | 1/6 (4.23) | 2.7 | 2.2 × 1.6 |
> | 1/7 (3.63) | 2.3 | 1.9 × 1.5 |

In general, data sampling is the process of converting a continuous-space/time signal into a discrete-space/time data signal. Sampling theory and specifically continuous-space/time signal sampling is an important part of mathematics and is addressed in depth in [19, 20] (and there are many others good books available). However, this paragraph gives a feel of the process of signal sampling – temporal and spatial – and the necessity to sample the signal at the right time and sufficiently densely.

Tutorial Exercise 2.4

Special care has to be taken in temporal sampling of fast moving objects. Could you explain why the golf club is "unnaturally warped forward" in this rolling shutter image?

Solution:

The rolling shutter is running row-wise from top to bottom of the image (see horizontal lines in the right image). At time zero, the top row is integrated first and then each row will be taken an instant later (precisely a row time later). As the golf club moves during rolling shutter, the top part of the golf club will be taken "earlier" in the same frame than the bottom part of the golf club. Assuming that the frame time is 1/60 s and that the head of the golf club is moving about 50 cm during this frame, the speed of the head of the golf club is about 30 m s^{-1} or a little bit more than 100 km h^{-1}. Another famous example of such image distortion effect is the rolling shutter image of a spinning propeller of an airplane.

Video camera systems have already done the temporal sampling by taking frames equally spaced in time. For example, standard video frame rate of 60 fps means that the temporal distance from one image to the next image is 16.6 ms. As already mentioned in the previous chapter, global and rolling shutter modes do not have the same temporal sampling behavior. The time difference of two consecutive samples of the same pixel is identical for all pixels in both modes (16.6 ms in the case of 60 fps), but the sampling time point is not the same. In the global shutter mode, all pixels start the light integration at the same time and, therefore, the temporal sampling is the same for all pixels in a given frame. In the rolling shutter mode, however, this is only true for the same frame row, whereas each following row is temporally shifted by its row time t_{row}. Taking the example of a mega-pixel image sensor having 1000 rows per frame at 60 fps, the temporal sampling point will be shifted from one row to the next by a row time of 16.6 µs.

2.4
Image Processing

The image sensor delivers a stream of digital numbers – the image raw data – which has to be processed to get single images or video streams. Such arranged data formats can easily be visualized and shared – most of the time in a compressed format – with other users. It is likely to extract relevant details for decision-making or feature extraction from such images or video streams. To do so, it is necessary to transform, enhance, and process a huge quantity of raw data. This section on image processing will present some of the most common methods and algorithms.

The raw image data is a record of the light intensity signals incident on the image sensor at a given time. The ADC defines the number of bits of the light intensity

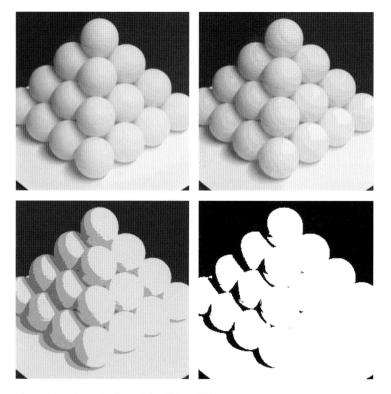

Figure 2.11 Quantization of the 256 × 256 image "ping-pong balls": 8, 4, 2, and 1 bit(s) per pixel.

value per pixel and per image. Standard ADC resolutions are in the range of 8–16 bits for external conversion components and mainly 6–12 bits for on-chip ADCs (e.g., 12 bits yield 4096 gray levels). Color images can be considered as "multivalue" images and require individual or collective quantization; for example, a three-color image can be represented by either 3 individual (most of the time interpolated) images or with 24 bits of color precision per pixel.

A quantized image can be considered as a stacked set of "bit planes," one per bit of the corresponding gray-level resolution depths. The less "bit planes" are available, the less image details can be resolved with the only advantage of reducing the amount of image data. At 8 bits, the image of the ping-pong balls of Figure 2.11 is visually acceptable, whereas stripes – similar to Easter egg painting – appear in the 4-bit version. The 8- and 4-bit images take 65.5 and 32.8 kB of data space respectively. Significant information is lost in the 2-bit or 1-bit image, making it difficult to recognize the exact configuration of the ping-pong balls, especially to the right side of the image. Even without compression, the quantity of data is only 16.4 kB (2-bit version) and 8.2 kB (1-bit only version) respectively. This is always the important trade-off between the level of detail in an image needed for clear distinction of the relevant pieces of information and the amount of information to be stored or to be processed in a given period, that is, the required complexity, speed, and power consumption of the calculation unit.

Box 2.6: Quantization Error

In ADC stages, the difference between the actual analog value and the quantized digital value – expressed in fractions of LSBs (least significant bits) – is called *quantization error* and is defined as

$$\frac{1}{\sqrt{12}} \times \text{LSB} = \frac{\text{FS (full-scale)}}{\sqrt{12} \cdot 2^{\#bit}}$$

This error is due to rounding or truncation and is considered an additional random signal called *quantization noise*. As the ADC stage is at the end of the readout path and normally after at least one gain stage, its noise components is often insignificant.

2.4.1
Image Histograms

Before getting into more detailed data processing operations, some basic notations have to be defined. Most of the presented point, algebraic, and geometric operations are defined on images of any dimensionality including video data; only 2D images will be considered for simplicity. The extension to three or higher dimensions is not difficult and most operations are independent of dimensionality. Of these, only monochromatic images are considered here, since extensions to color or other spectral ranges are rather trivial: The same operations are applied identically to each spectral range (e.g., R, G, B, UV, NIR, etc.).

A monochromatic image $f(img)$ is assumed to be quantized to k levels $[0, \ldots, K-1]$ in which $K = 2^{(ADC\ resolution)}$. Each pixel can take an integer value in this range, also referred to as gray levels for simplicity. The single-valued image $f(img)$ is defined on a two-dimensional discrete-space coordinate system $img = (m, n)$. The image is of finite dimension $[0, M-1] \times [0, N-1]$ and can be contained in a matrix of dimensions $M \times N$ (rows $\times$ columns).

The most fundamental tool used in designing point operations on digital images is the *image histogram* H_f, which is the plot of the frequency of occurrence of each gray level in f. The histogram is given by $H_f(k)$, meaning f contains exactly $H_f(k)$ occurrences of gray level k for each $k = [0, \ldots, K-1]$. An algorithm to compute the image histogram implies a simple counting of gray levels, which can easily be done during image acquisition or scanning. Owing to a reduction of dimensionality relative to the original image f, image information is lost. The image f cannot be reproduced from the histogram H_f. In fact, the histogram H_f contains no spatial information of f. Nevertheless, this information is very helpful in deriving useful image processing operations. Every image processing development environment and software library contains basic histogram computation, manipulation, and display functions.

The *average optical density* (*AOD*) is a simple way to estimate the center of an image's gray-level distribution. This basic measure of an image's overall brightness can be directly computed from the image or from its image histogram:

$$\text{AOD}(f) = \frac{1}{NM} \sum_{n=0}^{N-1} \sum_{m=0}^{M-1} f(n, m) = \frac{1}{NM} \sum_{k=0}^{K-1} k H_f(k) \tag{2.10}$$

A target value for the AOD may be defined in order to designate a point operation to modify the overall gray-level distribution of an image f (such operations will be explained in the following sections). In terms of histograms, the following notions can be commonly distinguished: "underexposed" and "overexposed" images or images with "good or poor visual contrast."

By usual convention, image gray levels with lower (higher) numbers indicate darker (brighter) pixels. Therefore, the histogram of Figure 2.12d corresponds to a predominantly bright image. This may occur if the image f was initially overexposed and/or the process of digitization was performed improperly (Figure 2.12c shows a histogram of an underexposed image). More generally, an image may make poor usage of the available gray-scale range. For example, an image with a compact histogram (Figure 2.12b) will most often have a poor visual contrast or a kind of "washed-out" appearance. However, an image taken under low-light condition can show a correct appearance, but the histogram can be skewed and will be of the "underexposed" type. It is up to the viewer to decide whether an image appears useful. Section 2.4.2 presents specific point operations effectively expanding the gray-scale distribution of an image, thereby improving the image quality.

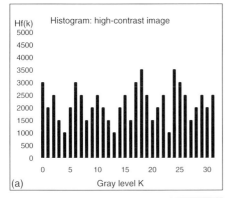
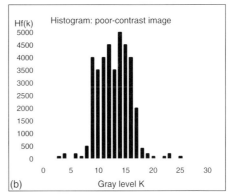
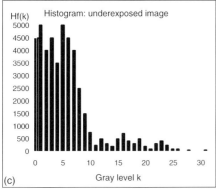
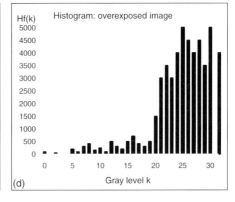

Figure 2.12 (a) Image histograms of well-distributed (higher contrast), (b) "poor visual contrast," (c) "underexposed," and (d) "overexposed" image samples.

2.4.2
Linear Point Operations

A point operation on an image $f(img)$ is a function p of a single variable applied the same way to each pixel in the image. The result is a modified or transformed image $t(img)$. At each coordinate of the image, the following relationship is applied:

$$t(img) = p[f(img)] \qquad (2.11)$$

As $t(img)$ is a function of a single pixel gray value with the (linear) function p, the effects obtained by point operations is limited. No spatial information is used and no change is made in the spatial relationships between pixels in the transformed image $t(img)$. Spatial positions or shapes of objects in the original image are not affected. The simplest point operations are linear point operations where p is chosen to be a simple linear function of gray level:

$$t(img) = A \cdot f(img) + B \qquad (2.12)$$

Linear point operations are means to provide a *gray-level additive offset B* and a *gray-level multiplicative scaling A* of the image f. If possible, saturation conditions $|t(img)| < 0$ and $|t(img)| > K-1$ are to be avoided, since gray levels are not properly defined and can lead to wrong processing or displaying of the resulting image $t(img)$. The easiest way to prevent distortion by values falling outside of gray-level range is to clip those values to the respective end point values (0 or $K-1$ respectively). Such a result is no longer a linear point operation in strict mathematical terms and care has to be taken since image information is lost in the clipping operation.

Additive image offset has the form $t(img) = f(img) + B$. If $B > 0$ ($B < 0$), then $t(img)$ will be the brightened (dimmed) version of the original image $f(img)$. The histograms of the two images have a simple relationship:

$$H_t(k) = H_f(k - B) \tag{2.13}$$

An offset B corresponds to a shift of the histogram by an amount B to the right (B positive) or to the left (B negative). Additive image offset is used to calibrate images to a given average brightness level; for example, comparison of images taken under different daylight conditions (from an outdoor security camera system) requires offset compensation in order to facilitate or improve image post-processing. A simple approach is to use the AOD of the images $f_1, f_2, \ldots, f_n$. To center all histograms around $K/2$, which is a reasonable AOD in most cases, the equalized images $t_1, t_2, \ldots, t_n$ are given by

$$t_i(img) = f_i(img) - AOD(f_i) + K/2 \tag{2.14}$$

The practical version of the *multiplicative image scaling* is given as

$$t(img) = INT[A^* f(img) + 0.5] \tag{2.15}$$

with A being positive since $t(img)$ has to be positive and $INT(x)$ delivers the nearest integer being less than or equal to x (the function $INT(x)$ truncates all digits behind the comma).

Multiplicative image scaling either stretches ($A > 1$) or compresses ($A < 1$) the image histogram, which can be written as

$$H_t(INT[A \cdot k + 0.5]) = H_f(k) \tag{2.16}$$

By making better use of the gray-scale range, histogram extension often yields more distinctive images, provided that saturation effects can be avoided or neglected. Histogram contraction normally leads to the opposite result.

Another simple linear point operation using both scaling and offset is the *image negative* by applying $A = -1$ (which flips the histogram) and $B = K - 1$ (required to get positive values to be in the allowable gray-scale range) to formula (2.13). The transformed histogram becomes

$$H_t(k) = H_f(K - 1 - k) \tag{2.17}$$

This operation is very useful in viewing the positive image of a scanned negative. Another practical application can be found in detecting holes in an object under through-light conditions. In the negative image, holes appear as dark objects on a bright background, which can be more easily recognized.

The last and one of the most useful linear point operations is the *full-scale histogram stretch* (FSHS), also known by the term "contrast stretch" as this operation tends to improve the contrast of the image. The function to expand the image histogram to fill the entire available gray-scale range is sometimes called *"automatic gain control* (AGC)" on commercially available still and video cameras. Assuming that image f has a compressed histogram with minimum gray-level value $H_{\min}$ and maximum gray-level value $H_{\max}$, the FSHS maps the gray levels $H_{\min}$ and $H_{\max}$ in the original image f to gray levels 0 and $K-1$ in the transformed image according to

$$\text{FSHS}(f) = \left(\frac{K-1}{H_{\max} - H_{\min}}\right) \cdot (f(\text{img}) - H_{\min}) \quad (2.18)$$

The FSHS operation can provide impressive improvements in the visual quality of an image suffering from a poor and narrow gray-scale distribution. Such resulting histograms can show characteristic gaps of an expanded discrete histogram. If the original image f already has a broad gray-level range, the FSHS may produce little or no effect. Such images might benefit from gray-level density redistribution, which are nonlinear transformations (see Box 2.7 for some additional explanations).

Box 2.7: Nonlinear Point Operations

Let's assume that the transformation function p, as described in Eq. (2.11), is a nonlinear function, which widely opens the range of possible applications, such as absolute value, square, and square root functions. Another such simple and commonly used nonlinear function is the *logarithmic point operation*, which can be defined as $t(\text{img}) = \text{FSHS}\{\log[1 + f(\text{img})]\}$.

The logarithmic point operation is an excellent choice for improving important features obscured in the dark. The transformed histogram is significantly spread at these low-light levels. Adding unity in the above equation eliminates the possibility of calculating the logarithm of zero. All gray levels of the original image are compressed to the range $[0, \ldots, \log(K)]$ with brighter gray levels compressed much stronger than dimmer gray levels. The subsequent FSHS operation linearly expands the log-compressed gray levels to the full gray-scale range. Dim objects in the original image f are allocated a much larger part of the gray-scale range in the transformed image t, thus improving their visibility.

In scientific and astronomical imaging, some stars and bright galaxies tend to dominate the visual perception of the image, while most of the interesting pieces of information have low-light levels. The procedure described above helps to bring up the visually interesting details.

Another important nonlinear point operation is *histogram equalization* or *histogram flattening*. This operation consists in uniformly distributing the

> image's gray values over the complete gray-scale range. The goal is not only an FSHS, but also a flat histogram, which contains the largest possible amount of information from such a transformed image.

2.4.3
Multi-image Operations for Noise Reduction

As already explained in Section 2.3, there are many noise sources in the image acquisition chain with the principal components coming from the photon shot noise and the image sensor – whichever is more significant under the given lighting conditions and sensor configuration. In a simplified approximation, all noise contributions can be merged into one single additive noise component Q. This can be written as

$$f(img) = f_{noiseless}(img) + Q(img) \qquad (2.19)$$

with $f_{noiseless}(img)$ being the perfect image and $Q(img)$ being an $N \times M$ matrix with elements of random noise values. "Adding up consecutive noise events" is not just a simple addition of each random noise variable, but taking the square root of the sum of each squared random noise variable. In a first-order approximation, noise can be reduced by a factor $\frac{1}{\sqrt{n}}$. Therefore, the average of 10 consecutive images reduces the noise by more than a factor of 3 and 100 images yield already a noise reduction factor of 10. The drawback is the n-times smaller frame rate.

With the simplified assumption of a noise average of zero and without getting into too many details of noise theory and the associated mathematical background, it can be stated that the average of n independently occurring noise events Q_i tends toward zero with n growing larger ($n \to \infty$). This can be rewritten more generally for an $N \times M$ matrix of noise events as

$$\left(\frac{1}{n}\right) \sum_{i=1}^{n} Q_i(img) \approx 0 (\text{zero matrix}) \qquad (2.20)$$

By taking n consecutive images $f_1, f_2, \ldots, f_n$ of the same scene, only the temporal noise contribution varies from image to image. This leads to

$$\left(\frac{1}{n}\right) \sum_{i=1}^{n} f_i = \left(\frac{1}{n}\right) \sum_{i=1}^{n} f_{noiseless} + \left(\frac{1}{n}\right) \sum_{i=1}^{n} Q_i \approx f_{noiseless} \qquad (2.21)$$

For a large enough number of frames averaged together, the resulting image should be nearly noise-free and corresponds to the original scene. The noise reduction effect can be quite significant – under the assumption that the noise model is accurate enough and that there is no motion or variation in the scene itself. This could partially be corrected by calculating motion vectors, by spatially translating (shifting) specific parts of the images and, finally, by applying the above averaging algorithm to these "spatially adapted" sections.

2.4.4
Morphological Image Operations

Geometric image operations modify the spatial position and spatial relationships of pixels, but are generally not changing their gray-level values. These operations are normally complex and computer intensive. Only a few of the most common "image transformation and distortion" algorithms are presented in this chapter.

Image translation is the most basic geometric transformation function. Such simple shift – defined by two integer constants c_x and c_y – maps each coordinate $img_f = (m_f, n_f)$ of the original image f to new coordinates $img_t = (m_f + c_x, n_f + c_y)$ of the transformed image t. This simple algorithm is used in display systems to move images on the screen or in image convolution operations where images are continuously shifted against each other. Such transformations do not need coordinate interpolation steps to readjust the resulting coordinates onto a given discrete grid. This is, however, the case for *image rotation* and *image zoom* transformations, where the new coordinates are not integers any more. To rotate an image counterclockwise around $(0,0)$ by angle θ, new coordinates can be calculated as follows:

$$img_t = (m_f \cdot \cos\theta - n_f \cdot \sin\theta, m_f \cdot \sin\theta + n_f \cdot \cos\theta) \qquad (2.22)$$

As the rotation center is not defined at the center of the image, the obtained coordinate pairs can fall outside the original range of coordinates and need some additional image translation to "re-center" the rotated image. To magnify or demagnify an image – or alternatively, to zoom out or zoom in an image – the following mapping function can be applied:

$$img_t = (A_x \cdot m_f, A_y \cdot n_f) \qquad (2.23)$$

where $A_x > 1$ to achieve horizontal (x-direction) and $A_y > 1$ to achieve vertical (y-direction) magnification. For $A_x \neq A_y$, the zoom effect will not be the same in horizontal and vertical directions. A special case of image zoom is *digital binning*, where the gray-level values of 2×2 neighboring pixels are "averaged" into a single pixel. This corresponds to an image zoom with $A_x = A_y = 0.5$ applied to Eq. (2.23). Similar to Section 2.4.4, noise reduction through multipixel averaging can be achieved (in the case of 2×2 binning, a factor 2 noise reduction per new binned pixel) with the only drawback of image resolution reduction to 25% of the original image.

It is important to understand that image zoom does not add new information to the image, although a magnified image seems to be easier to see or to interpret. Image zoom operation is only an interpolation of known information and rescales the image size. It can introduce some undesired effects, such as distortion artifacts – also known as *aliasing effects* – resulting from reconstruction of signal samples that are different from the original continuous signal. Many different methods and algorithms to enhance image quality (e.g., smoothness, sharpness, etc.) are documented and are part of standard image processing libraries.

In the morphological operations presented above, one big issue consists in coordinate mapping between the original image $f(img_f)$ and the transformed image $t(img_t)$. Such coordinate mapping function $\{map(img_f \to img_{t'})\}$ normally does not deliver integer coordinates, which implies the definition of an interpolation operation of (intermediate) noninteger coordinates $img_{t'}$ to integer values, such that $t(img_t)$ can be expressed in a standard row–column format. The simplest of such interpolation operations is the *nearest neighbor interpolation*, which truncates the calculated coordinate values of $img_{t'}$ to the closest integer by applying the function $img_t = INT(img_{t'})$. This rather simplistic method is easy to implement, but has a major drawback. Several coordinates can be mapped to the same coordinate, yielding structures that are physically meaningless. This is particularly noticeable for image areas with "edges" (i.e., sudden intensity changes).

Much smoother interpolation can be achieved by applying a *bilinear interpolation* algorithm. If the coordinates of the intermediate image $t'(img_{t'})$ are "off-grid," the idea is to interpolate the gray-level value of each pixel sitting "on-grid" on the img_t coordinate system by using the "least squares" fitting method of four – normally 2×2 – closest neighboring pixels on $img_{t'}$: $t'(m_{00'}, n_{00'})$, $t'(m_{10'}, n_{10'})$, $t'(m_{01'}, n_{01'})$, and $t'(m_{11'}, n_{11'})$. The least square method minimizes the sum of squared residuals, a residual being the difference between an observed gray-level value and the fitted value. The transformed image t can be calculated as follows:

$$t(m_t, n_t) = A_0 + A_1 \cdot m_t + A_2 \cdot n_t + A_3 \cdot m_t \cdot n_t \tag{2.24}$$

which is a bilinear function at the coordinate (m_t, n_t). The bilinear weighting factors A_0, A_1, A_2, and A_3 are calculated by solving the following matrix:

$$\begin{bmatrix} A_0 \\ A_1 \\ A_2 \\ A_3 \end{bmatrix} = \begin{bmatrix} 1 & m_{00'} & n_{00'} & m_{00'} \cdot n_{00'} \\ 1 & m_{10'} & n_{10'} & m_{10'} \cdot n_{10'} \\ 1 & m_{01'} & n_{01'} & m_{01'} \cdot n_{01'} \\ 1 & m_{11'} & n_{11'} & m_{11'} \cdot n_{11'} \end{bmatrix}^{-1} \begin{bmatrix} t'(m_{00'}, n_{00'}) \\ t'(m_{10'}, n_{10'}) \\ t'(m_{01'}, n_{01'}) \\ t'(m_{11'}, n_{11'}) \end{bmatrix} \tag{2.25}$$

These weighting factors express the two-dimensional distance of (m_t, n_t) to each of the four respective known points $(m_{t'}, n_{t'})$ – the closer the distance, the higher the weighting factor. There are other interpolation methods, such as cubic, bicubic, trilinear (for 3D), spline, and stair step interpolation, which can help move transformed image points onto a constant scale and maintain image quality.

2.4.5
Introduction to Feature and Motion Detection

The simplest image information abstraction process is called *image thresholding*. This extreme method of gray-level quantization is achieved through simple comparison of each pixel's gray-level value with a threshold value T using the following rules:

$$t(img) = \begin{cases} \text{'1'} & \text{if } f(img) \leq T \\ \text{'0'} & \text{if } f(img) < T \end{cases} \tag{2.26}$$

The "1"-level corresponds to the highest gray level or quantization level $K - 1$ (see Section 2.4.2). Different threshold values can yield different transformation images. Therefore, it is important to have first a look at the original image histogram to determine the most appropriate threshold value *T* in order to achieve the most meaningful binary image for simplified processing, interpretation, or display. Finally, the data of the resulting image is drastically reduced (see Section 2.4.1 and the 1-bit quantized image in Figure 2.10).

Feature extraction simplifies the amount of required resources to analyze complex image data. These operations reduce the number of inherent variables, but still describe image data with sufficient accuracy. It is of utmost importance to ingeniously choose the application-dependent features and the decision criteria to be applied.

Another simple data reduction approach for feature extraction is the *contrast image*. The transformed image is built by taking the difference of gray-level values of neighboring pixels in the image. There are different ways of calculating the transformed image that may modify the number of pixels in the new image. One of the easiest way to determine the contrast gray-level value of a pixel is to take four (or even all eight) neighboring pixels and to calculate the average of the differences of the two (or four) pairs of geometrically – diagonally or horizontally and vertically or all of them – opposing pixels. Image thresholding can be used for better detection or isolation of various features or shapes in a digitized image or even video stream – where data reduction is very helpful in order to handle huge quantities of real-time image data. Finally, image *motion detection* is often of interest, especially in video sequences. Detecting and analyzing changes in the same scene at different times – in contrast to the above described method of differences between pixels in the same image and, therefore, at the same time – is very useful in applications such as target tracking, moving object recognition, or moving object trajectory determination for video compression algorithms. The basic technique is the image difference expressed as $t_i(img) = |f_i(img) - f_{i-1}(img)|$. Such difference images could be processed by change detection algorithms that separate "change" from "nonchange" by application of a threshold value and by counting the number of "high-change" pixels.

There are many other binary image processing operations closely related to gray-level morphological operations (chapters 4 and 13 of [19]). These operators for windowing, dilation, erosion, open–close and close–open filters, and some logical operations (NOT, AND, OR, XOR, and MAJ) modify the shapes of the objects in an image. These functions are useful to expand, shrink, or smoothen object shapes, to eliminate too small features, or to detect boundaries. An image processing library normally contains all these functions. There are many other known algorithms adapted to many different applications and situations.

2.5
Conclusions

The scope of this chapter *"Electronic Image Sensing and Processing"* was to give a short introduction into such various technical fields as *image forming, image sensing,* and *image processing*. Modern cameras contain all the three components to form, to capture, and to process images. Many state-of-the art image data acquisition systems are available and ready-to-use, but are not perfect.

This chapter explained simple concepts and introduced different limiting factors, distortion effects, and noise sources in order to understand where the limitations of such image data acquisition systems are. Limited depth of field, limited optical resolution, reduced QE, several material absorption and reflection effects, distortion effects – such as diffraction, aberration, and transistor mismatch effects – and noise mechanisms – such as photon shot noise, dark current shot noise, reset or kTC-noise, flicker or $1/f$-noise, thermal, and quantization noise – alter the quality of the acquired image data.

At all stages, more or less sophisticated and time- and power-consuming noise reduction methods are applicable and have been presented. The choice of the "correct" and best suited electronic image data acquisition system is not trivial and many aspects have to be taken into consideration depending on the underlying applications.

Problems

2.1 An incident light ray in air ($n_{air} = 1$) enters a glass plate ($n_{glass} = 1.5$) at an angle of $30°$ (from the vertical). Calculate the angle of the light ray inside the glass. Remember from Tutorial Exercise 2.1 that the angle of the light ray is bent toward the vertical line as the light ray is entering a medium with higher refractive index.

2.2 Take the same configuration as for Problem 2.1, but assume that the light ray starts in glass at the same angle of $30°$ and exits into the air. Calculate the angle of the light ray in the air. Do the same calculation with a starting angle of $45°$. Prove that the light ray will be completely reflected in the glass for angles $>41.8°$. This phenomenon is commonly used in glass fibers, which are found in all high-speed communication channels.

2.3 Calculate the angle θ of the light ray leaving the prism ($n_{prism} = 1.55$) as shown below:

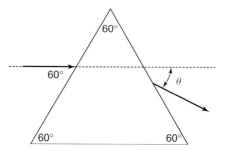

2.4 A flattened glass sphere with radius $R = 25$ mm ($n_{sphere} = 1.5$) is on top of a drawing of a 10-mm-long arrow as shown in the figure below. The shortest distance from the flattened surface to the center of the sphere is $r = 15$ mm.

An observer looks vertically into the sphere. By using an approximation of the lens maker's equation $\frac{1}{f} = \frac{(n-1)^2 \cdot 2R}{n \cdot R^2}$, find out where the arrow appears. Is it mirror-inverted? And what is the magnification value?

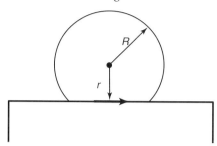

2.5 A digital camera is equipped with an objective with a fixed focal length of 50 mm. Inside the camera, a CMOS image sensor with 1000 × 1000 pixels (1 Mpix) with an overall active area of 10 mm × 10 mm is implemented. If you take a picture of the north face of Mount Eiger – while sitting at 2 km from KleineScheidegg – how many pixels in this image are covered by a climber trying to get to the top of Eiger? How can this quite low number of pixels be improved to get more details of the climber?

2.6 A manufacturer of O-rings needs an optical measurement system to carry out a visual inspection of the final products. Typical O-rings have a diameter of 10–30 mm, a thickness of 1–5 mm, and an optical reflection coefficient of 5% for black up to 50% for yellow O-rings. The faults to be detected are changes in local reflectance of less than 10% of the mean value, and in an area of 0.2 mm of diameter or larger. A suitable conveyor belt passes up to 1000 O-rings per second in front of the inspection system. The O-rings are well separated and well centered. A white-light illumination system based on halogen lamps provides the fault contrast as required above, yielding 10 000 lux in the O-ring plane. Suggest an optical setup for measuring the required quantities, including image sensor characteristics, imaging optics, and the required exposure times. Sketch suitable system timing(s), taking into account the available light, the necessary spatial resolution, and the motion blur (the conveyor belt is permanently running). Are there alternative measuring principles?

2.7 Imagine an inspection system with the goal to determine the internal and external diameter of metal rings (e.g., shaft collars). Comment on such measurement system setups and focus on the measurement precision (down to 10 µm) and the required data processing flow (data rate in function of the measurement algorithm and the data resolution). Hint: think of pure black-and-white pictures to start with.

2.8 Perfect electronic image data acquisition systems do not exist. Image data can be missing, which can appear (i) on a pixel-by-pixel basis (only single pixels are missing), (ii) in pixel clusters (two or more neighboring pixels are missing), or (iii) in complete image rows or columns. Think of different algorithms to restore missing pixel information for each of these three situations and give the level of uncertainty regarding the restored data.

2.9 Electronic data and signal acquisition chains are noisy. Consider the following building blocks as found in CMOS image sensors:

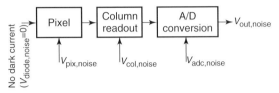

The output noise (in bandwidth Δf) is calculated as a mean-square value by adding the individual mean-square contributions in the chain (by neglecting the noise contributions from the photodiode). As the noise sources are uncorrelated, the following formula can be applied:

$$\overline{v^2_{out,noise}} = \overline{v^2_{pix,noise}} + \overline{v^2_{col,noise}} + \overline{v^2_{adc,noise}}$$

By assuming that all three noise sources have the same value, for example, $1mV_{rms}$, the overall output noise without any internal amplification will become $\sqrt{3} \times 1mV_{rms}$. By adding an amplifier into the column readout block with an amplification factor of 9, re-calculate the overall output noise. Do the same exercise with an amplification stage inside the pixel – again with an amplification factor of 9. Finally, both pixel and column readout have each an amplifier implemented. What conclusions can be drawn?

2.10 Transferring charges is an essential operation in CCD and CMOS image sensors. Consider the following situation with a simple circuit made of two equivalent capacitors C and a switch connected between them. At the beginning one capacitor contains the charge Q yielding a voltage V_i, whereas the second capacitor is empty (no charge and therefore no voltage). In terms of energy, this represents $E_i = \frac{1}{2} \times C \times V_i^2$ for the charged capacitor.

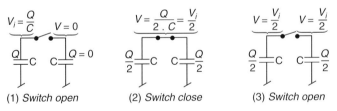

By closing the switch, the charges are distributed, halving both charges and voltage. By opening the switch again, the energy can be calculated for each capacitor: $E_{cap} = \frac{1}{2} \times C \times \left(\frac{V_i}{2}\right)^2$, yielding a total energy after the charge transfer $E_f = 2 \times E_{cap} = \frac{C \times V_i^2}{4} = \frac{E_i}{2}$, which is half of the initial energy E_i. How and where did the other half get lost? – Hint: the switch is not perfect and represents a finite resistor when closed.

3
Phase Decoding and Reconstruction

Jan Burke[1]

3.1
Introduction

Light, with its very short wavelength, is a very sensitive length gauge; but with the speed of light being what it is, the price to pay is a very high frequency. A useful rule of thumb for the center of the visual range is that a wavelength of 548 nm (green) has a frequency of 548 THz. This is not directly measurable with any electronic detector known today, so how can we track and determine the phase of a light wave that gives us access to a nanometer scale? For optical frequencies, the only viable detector is again light.

If a coherent reference wave of equal or very similar frequency[2] is mixed with the test wave, we can create a beat signal to "freeze" the rapid oscillation in time (in the case of equal frequency) or slow it down sufficiently to measure it electronically. Good coherence is the key for this to work, which is why interferometry has really only blossomed as a research field after sufficiently strong sources of light with unprecedented coherence length (i.e., lasers) became available.

For phase detection, the signal we use is an interference modulation of the resulting light intensity that reveals the phase difference $\varphi(x, y)$ between the two waves, and therefore allows the calculation of a phase map of the wavefront under test. The method of "phase shifting" that we will learn about in this chapter is really the art and science of how to record efficiently a small number of irradiance samples as a proxy for a continuously modulated signal, and how to get the most reliable phase decoding from the dataset. As few as three phase-shifted interferograms are sufficient to reconstruct the phase of the tested light wave.

Since the intensity modulation is cosinusoidal, unfortunately the relationship between phase and path difference is cyclic, not bijective, and the phase can be reconstructed only in a $[0, 2\pi)$ unambiguous range; this is also called *modulo* 2π. The phase "wraps" around and starts at 0 again after one full 2π cycle. The initial

---

1) The author is now with the Bremer Institut für Angewandte Strahltechnik, Klagenfurter Str. 2, 28359 Bremen, Germany (Burke@BIAS.de).

2) Shifted by a few Gigahertz at the most, more typically tens of Megahertz.

Optical Methods for Solid Mechanics: A Full-Field Approach, First Edition. Edited by Pramod Rastogi and Erwin Hack.
© 2012 Wiley-VCH Verlag GmbH & Co. KGaA. Published 2012 by Wiley-VCH Verlag GmbH & Co. KGaA.

result after the phase reconstruction is a "wrapped" phase map. Usually, ascending phases are displayed on screen as ascending gray levels, creating a "sawtooth" image that leaps from white back to black whenever the phase crosses 2π. The wrapping is clearly a loss of information. However, if we know that the wrapped phase map belongs to a continuous and smooth surface, we can restore the fringe order by adding suitable multiples of 2π to different fringes in the sawtooth image, so that continuous surfaces give a continuous phase map again. This process is known as *phase unwrapping*.

Once we have a correct map of relative phases (absolute phases are usually irrelevant, unless coordinate mapping corrections are made in a measurement volume), the final step is to calculate the optical path difference, which requires knowledge of the shape of the reference wavefront and the geometry of object and reference waves, that is, the sensitivity vector. In smooth-surface interferometry, if we can work in reflection at normal incidence, this conversion is

$$h(x, y) = \Phi(x, y) \frac{\lambda}{4\pi} \tag{3.1}$$

where $\Phi(x, y)$ is the unwrapped phase map, λ is the wavelength, and $h(x, y)$ is the height map. A tilt of the reflecting surface by ϑ will entail a loss in sensitivity by $\cos \vartheta$. This can vary spatially, such as in simple null tests of aspherics, but at least the incident angle equals the emergent angle and we only have one sensitivity factor per ray to deal with. If scattering is involved, as in speckle techniques and fringe projection, any combination of angles is possible and the conversion of unwrapped phase to surface coordinates can become much more complicated. The three stages of phase reconstruction are depicted in Figure 3.1.

In this chapter, we will learn about methods to get from Figure 3.1a to 3.1c, that is, to convert a set of intensity distributions into a phase map, and to "unwrap" the cyclic phase into the quantity we want to measure. Also, the phase-measuring process is so well understood today that we will even be able to learn how to tailor our own custom phase-shifting methods and optimize them for a given measurement problem.

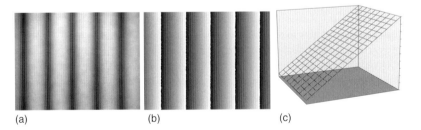

(a) (b) (c)

Figure 3.1 (a) Interferogram with fringes following a cosine intensity profile. (b) Wrapped phase map with saw-tooth fringe profile, where black represents $\varphi(x, y) = 0$, and white represents $\varphi(x, y) = 2\pi$ minus the smallest discrete increment. (c) Unwrapped phase map (resembles height profile if conversion factor is constant).

3.2
Basic Concepts

This section familiarizes you with a few basic equations and techniques, and takes particular care to point out the underlying and well-known theory of information processing that has enabled the rapid development of the field and to place the phase-shifting technique in that context.

The two wavefronts we are mixing are described by

$$W_T(x, y) = A_T(x, y) \exp\left(i\varphi_T(x, y)\right)$$
$$W_R(x, y, p) = A_R(x, y) \exp\left(i\varphi_R(x, y, p)\right), \quad (3.2)$$

where W_T is the tested wavefront, A_T is its amplitude, and φ_T is its phase. All of these quantities are usually space-dependent. For convenience, we ignore the z-direction here and write the wavefront as it appears on a two-dimensional (2-D) light detector array. (We will often drop the (x, y) dependence in the following discussion for readability, but it is always implied.) The superimposed reference wavefront W_R has an extra parameter p. It can change the phase of W_R as a step function, or a linear function, and can do so in time or space, or both.

Now, thanks to the frequencies of the two waves being very similar, the resulting interferogram

$$I(x, y, p) = \left|W_T(x, y) + W_R(x, y, p)\right|^2 \quad (3.3)$$

oscillates slowly enough to be measurable. $I(x, y, p)$ denotes the spatial irradiance distribution (the fringe pattern) on the sensor. This is frequently named "intensity," but strictly speaking, in our context it would be more appropriate to use the term "*irradiance*" (and the unit is W m^{-2}). It should be understood that "intensity," where used, denotes this quantity in our context.

The information about the optical path difference (and thereby, the difference between W_T and W_R, bearing information about surface shapes and/or deformations) is now encoded in this quasistatic or slowly progressing signal, which we can evaluate with Eq. (3.2) as

$$I(x, y, p) = I_b(x, y) + M_I(x, y) \times \cos\left(\varphi_T(x, y) - \varphi_R(x, y, p)\right), \quad (3.4)$$

where $I_b(x, y)$ is the bias irradiance ($A_T^2 + A_R^2$, i.e., the individual amplitudes squared) and $M_I(x, y)$ is the irradiance modulation ($2A_T A_R$, i.e., the product of the amplitudes), so that we have an irradiance that oscillates between its maximum and minimum in a cosinusoidal manner, depending on the instantaneous local phase difference.

Tutorial Exercise 3.1

Derive Eq. (3.4) and show that $A_T^2 + A_R^2 \geq 2A_T A_R$.

Solution:

The minimum for Eq. (3.3) is $(A_T - A_R)^2 = A_T^2 - 2A_T A_R + A_R^2$, but this is a square and cannot be negative, which proves the inequality. Further, when $A_T = A_R$, $I_{max} = 4 A_T$ and $I_{min} = 0$.

It is customary in interferometry to recast Eq. (3.4) into a slightly more convenient form:

$$I(x, y, p) = I_b(x, y) + M_1(x, y) \times \cos(\varphi(x, y) + \alpha(p)) \tag{3.5}$$

where we have lumped the generic phase difference between the two wavefronts together as $\varphi(x, y)$ which is the quantity that we are usually interested in, but isolate the phase-shift term $\alpha(p)$, as this term is the one we will be influencing to obtain enough information to reconstruct $\varphi(x, y)$. It is worthwhile noting here that the cosine, being an even function, would not suffice to give us the sign information about the phase difference without the extra phase term $\alpha(p)$. For example, a concave or convex phase difference function can generate exactly the same fringe pattern. In practice, opticists obtain extra information by pressing on the workpiece or its support in the interferometer with a finger and thereby introducing an α of known sign – a very early form of "digital phase-shifting interferometry"! – and, by observing which way the fringes move, the ambiguity is overcome.

Since there are three unknowns (I_b, M_1, and φ) in Eq. (3.5), we need at least three linearly independent equations to solve for φ. This is what $\alpha(p)$ is used for: by deliberately changing the phase difference between the waves, we can modulate the irradiance in a cosinusoidal manner. If we now record at least three interferograms, and track the relative phases of the modulation at each pixel in the image, we will know where on the cosine curve each of them is sitting, and can therefore determine φ. Consider Figure 3.2, an example of temporal phase shifting (TPS) on two different pixels, to see how this principle works.

We mark in white the irradiance modulation of pixel 1, or (x_1, y_1), as we vary α from 0 to 4π. Obviously, the contrast is <1, which means the amplitudes of the individual waves are not equal. This is the usual case in interferometry; a contrast >0.9 is seldom encountered in reality. Further, we have $I_b(x_1, y_1) = 0.6\, I_{max}$, where I_{max} is some arbitrary value, which is most usefully set to the irradiance reading that saturates the camera. Further, we can read off the plot that $M_1(x_1, y_1) = 0.3\, I_{max}$. Consider a second pixel (x_2, y_2, marked in black), which has a phase different from the first pixel and is darker on average, at $I_b(x_1, y_1) = 0.5\, I_{max}$, and whose modulation is also lower: $M_1(x_2, y_2) = 0.2\, I_{max}$. This, too, is a case frequently encountered in practice: as we approach the edge of the image, the amplitudes of the test and reference wave can diminish, following the natural power profile of a laser beam. Now imagine that we took our irradiance readings for the two pixels at settings of α_1, α_2, and α_3. At α_1 and α_2, we get the same irradiance readings for both the pixels, so if these readings are all we have, we cannot determine anything

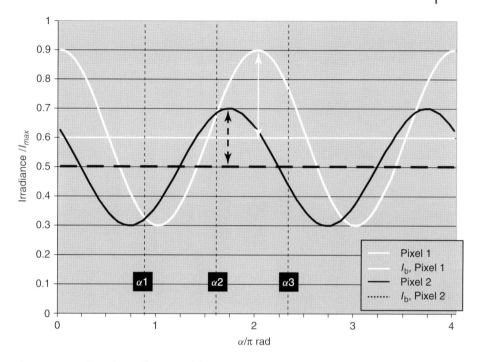

Figure 3.2 Tracking the irradiance modulation at two different pixels (see text).

yet! The situation is resolved by taking a third reading at α_3, and we now get vastly different readings for our two example pixels, which allows us to determine all the quantities in Eq. (3.5) for each pixel individually. And this is really what phase-shifting interferometry is all about.

The underlying idea of the phase-shifting technique is to detect the phase of a periodic signal by mixing it with another periodic signal (i.e., encoding it on a temporal or spatial carrier frequency sideband) and retrieving the signal with the knowledge of the modulation that was deliberately introduced. This is not to be confused with the first step of phase detection, which was to encode phase as irradiance by utilizing a coherent reference. Now it is the phase of the reference that is modulated to manipulate the irradiance signal. Figure 3.3 gives an overview of this in the temporal domain. The individual interference images, in this case four, have been recorded at different times, for example, as consecutive camera frames, and are stored in a computer memory.

Note how the spots marked "pixel A" and "pixel B" will each record a cosinusoidal irradiance modulation, but with slightly different phase offsets, as the reference phase is shifted. The visual impression during phase shifting will be a movement of the fringe pattern from left to right, which is why the technique was called *fringe scanning* by early authors. From each pixel's recorded modulation, a modulo 2π phase value can then be calculated and displayed as a gray value on a wrapped phase map (see the center of Figure 3.3).

88 | 3 Phase Decoding and Reconstruction

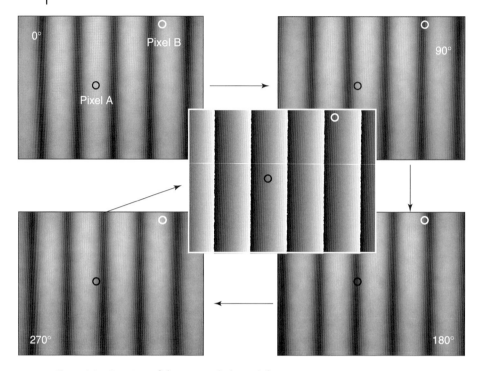

Figure 3.3 Overview of the temporal phase-shifting process. Clockwise from top left: reference phase increments 0°; 90°; 180°; 270°. Center: wrapped phase map.

Simple though this looks, the technique is built on volumes of signal-processing knowledge. This is evident in the early names that have been given to this technique: synchronous detection, heterodyne interferometry, phase-lock interferometry, phase-biased interferometry, fringe-scanning interferometry, and the author's personal favorite, quadrature multiplicative moiré (explained below).

If the phase is instead modulated in the spatial domain, the irradiance modulation will become encoded in one image on the sensor; in other words, we will adjust the interferometer such that there is always a fringe pattern on the sensor, similar to Figure 3.3 but much denser, where the phase advances linearly across the sensor. In Figure 3.3, the linear fringe pattern is the result of our measurement; for spatial phase shifting (SPS), a known "bias" or "carrier" frequency is introduced to begin with. Let us consider a pictorial example in Figure 3.4 to clarify the concept.

By introducing a tilt between plane waves, or an offset perpendicular to the optical axis for spherical waves, a linear or quasilinear phase progression is created in the interferogram. It is important to realize that any signal present amounts to a phase modulation of the wavefront and, therefore, to a spatial frequency modulation of the carrier fringe pattern. In practice, we make the carrier frequency as large as we can to accommodate a wide range of signal frequencies and to keep the relative

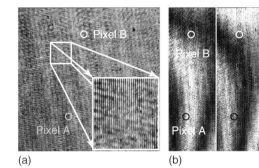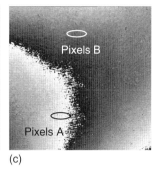

Figure 3.4 Example of a spatially phase-biased interferogram, and its evaluation. (a) Spatially phase-biased interferogram, with magnified portion showing the dense carrier fringe pattern. (b) Image columns reordered (every third column, with offsets 0, 1, and 2 columns), showing that three phase-shifted images are interlaced in the interferogram. (c) Phase (sawtooth) image calculated from the phase-shifted subimages, and stretched to original image width. (From Ref. [1].)

frequency variations small. If the bias fringe period is adjusted to $P_0 = 4$ pixels in horizontal direction, then $\alpha(x) = 90°$/column; if we choose $P_0 = 3$ pixels, as in Figure 3.4, then $\alpha(x) = 120°$/column, that is, we sample each carrier fringe with only three pixels, and a three-step phase-shifting scheme becomes applicable.

The great advantages of putting all the information needed in one interferogram are simplicity and speed: the phase bias is preadjusted geometrically, no moving parts are involved, and a phase measurement can be carried out in one snapshot. If the camera's exposure time is very short, or if a pulsed laser is used, very fast phase measurements can be made with this technique, and vibration isolation becomes unnecessary. However, as always, we do not get something for nothing. If we minimize the time to record, we have to give up spatial resolution in exchange. This is why the circular spots in the original interferogram expand to ellipses in the sawtooth image: each of the phase-shifted images is only one-third of the total image width, and thus to calculate $\varphi(x, y)$, we need $I(x, y)$, $I(x - 1, y)$, and $I(x + 1, y)$, and have to assume that $\varphi(x - 1, y) \cong \varphi(x, y) \cong \varphi(x + 1, y)$.

> **Box 3.1: Multiple Reflections**
>
> In Figure 3.4, there is an intensity ripple on the individual interferograms, which comes from parasitic interference between the camera sensor and the protective glass cover. Since it is of high spatial frequency, it creates a slight ripple in the calculated wrapped phase map as well.
>
> This is a very common problem when using video cameras in interferometry and the light source has a coherence length of more than a few millimeters. It is very worthwhile asking the manufacturer whether a camera can be delivered

with no cover glass on the sensor or the cover glass can be removed by oneself (both of which will void all warranties and require the user to keep all dust and aerosols away from the sensor), or to verify whether the camera will work in the intended application as is. Reducing the spatial coherence of the illumination, for example, by a rotating diffuser, is usually a good way to mitigate the problem.

Box 3.2: Phase Decoding by Quadrature Multiplication

If an analog video camera is used to record interferograms as in Figure 3.4, the video lines are read out as a standard TV line signal, so that the spatial oscillation becomes a time-varying electronic signal, which can be processed in real time to yield the phase, as shown in Figure 3.5.

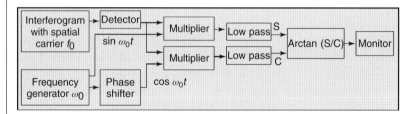

Figure 3.5 Analog circuitry for real-time phase calculation from one interferogram [2]. ω_0 must be set to match the product of the phase shift per column and the pixel readout rate.

Again, the principle is to create a beat frequency between two signals— here, the video line signal and the local oscillator frequency ω_0. The multiplication of signals will create sum and difference frequencies; the phase difference between the signals will be encoded in the very low difference frequency, and is therefore extractable with a low-pass filter. This is nothing but an electronic moiré technique! If we create a second artificial signal, with a phase shift of 90° (therefore labeled $\cos\omega_0 t$ in Figure 3.5), the beat signal will be 90° out of phase with the first one. Once we have a sine and a cosine, that is, two signals in *quadrature*, we can retrieve the phase unambiguously. The concept of quadrature is very important in our discussion of phase-shifting techniques; we will see that the overarching objective is to synthesize two functions that have equal amplitude and are 90° out of phase. Thus, Figure 3.5 gives a beautiful example of why this phase-decoding technique has been named *quadrature multiplicative moiré* [3].

Quadrature is a universal concept to define a phase angle, and is ubiquitous in quadrature position and rotation encoders. The most fascinating technical

> application the author knows of is the NSU "Ultramax" overhead camshaft drive used in German motorcycles and cars in the 1950s. It uses a pair of eccentrics offset by 90° on a driving shaft, a pair of connecting rods, and another similar pair of eccentrics on the camshaft, to transfer rotation without a chain, cogs, or belt (*http://www.nsu-max.com/Schubstangen.html*).

3.2.1
Deriving a Generic Phase-Shifting Formula

Let us now see how we can create two signals in quadrature from our cosinusoidal irradiance modulation. In order to do this, a phase shift of the reference wave, $\alpha(p) = 90°$ per sample, being the phase lag that turns a "sine" into a "cosine" fringe pattern, is a natural choice. We suppress spatial and/or temporal dependencies to keep the discussion general. Our set of irradiance readings then becomes

$$I_0 = I_b + M_I \times \cos(\varphi)$$
$$I_1 = I_b + M_I \times \cos(\varphi + 90°)$$
$$I_2 = I_b + M_I \times \cos(\varphi + 180°)$$
$$I_3 = I_b + M_I \times \cos(\varphi + 270°) \tag{3.6}$$

which we can rewrite as

$$I_0 = I_b + M_I \times \cos(\varphi)$$
$$I_1 = I_b - M_I \times \sin(\varphi)$$
$$I_2 = I_b - M_I \times \cos(\varphi)$$
$$I_3 = I_b + M_I \times \sin(\varphi) \tag{3.7}$$

Consequently, we have the desired cosine and sine functions to work with. Eliminating I_b, and therefore making our signals zero-mean, can now be done by

$$I_3 - I_1 = 2M_I \times \sin(\varphi)$$
$$I_0 - I_2 = 2M_I \times \cos(\varphi) \tag{3.8}$$

and we have what we want: two signals of equal amplitude and with a 90° phase lag, which we can use as proxies for the sine and cosine of the wavefront phase. Note here that this only works if we keep our signals within the detector's dynamic range. We must avoid saturated or too dark (and therefore noise-dominated) readings, or else the sine and cosine will not be restored properly. We can then solve for φ by arranging the terms as

$$\frac{I_3 - I_1}{I_0 - I_2} = \frac{2M_I \times \sin(\varphi)}{2M_I \times \cos(\varphi)} = \frac{\sin(\varphi)}{\cos(\varphi)} = \tan\varphi \tag{3.9}$$

where it is necessary for M_I to be significantly larger than zero to get a good signal-to-noise ratio (SNR), and it follows immediately that

$$\varphi \bmod \pi = \arctan\frac{I_3 - I_1}{I_0 - I_2} \tag{3.10}$$

3 Phase Decoding and Reconstruction

for the usual $(-\pi/2, \pi/2)$ definition range of the arctangent. However, since we have access to the numerator and denominator separately, we can use the knowledge of the signs to extend the unambiguous range:

$$\varphi \bmod 2\pi = \arctan \frac{I_3 - I_1}{I_0 - I_2} \qquad (3.11)$$

and with this we have derived our first phase-shifting formula – a four-step 90° formula. With just four irradiance measurements, we can calculate the phase of a light wave. Indeed, even this is one more measurement than needed (Figures 3.2 and 3.4) – the extra measurement can be used to determine another quantity, for example, $\alpha(p)$ if it is unknown, or to optimize noise rejection or other parameters. We will discuss this in more detail below.

Tutorial Exercise 3.2

Write down a table accounting for the signs of $\sin \varphi$ and $\cos \varphi$, to extend the range of the arctangent to the full 2π.

Solution:

	$\varphi = 0$	$0 < \varphi < \pi/2$	$\varphi = \pi/2$	$\pi/2 < \varphi < \pi$	$\varphi = \pi$	$\pi < \varphi < 3\pi/2$	$\varphi = 3\pi/2$	$3\pi/2 < \varphi < 2\pi$	$\varphi = 2\pi$
$\sin\varphi$	0	+		+	0		−		0
$\cos\varphi$	+		0		−		0		+

For all phase-shifting schemes of this type, suppression of I_b requires that

$$\varphi \bmod 2\pi = \arctan \frac{\sum_n b_n I_n}{\sum_n a_n I_n} \quad \text{with} \quad \sum_n a_n = \sum_n b_n = 0 \qquad (3.12)$$

That is, all the sample weights must add up to zero – only then will the numerator and denominator represent a zero-mean signal and be able to emulate a sine or a cosine. This is a necessary condition for the phase decoding to work properly.

From dealing with a pair of signals, one representing a sine, the other a cosine, it is only a small (and extremely useful) step to combine them in a single, complex number

$$z = \sum_n a_n I_n + i \sum_n b_n I_n \qquad (3.13)$$

with φ then being the argument of z. This unlocks the power of complex signal and filter analysis, and has in the last two decades enabled tremendous advances in phase shifting. We will see examples of this in Section 3.4. If the amplitudes of our $\sin(\varphi)$ and $\cos(\varphi)$ proxy functions are matched, we can write $\text{Re}(z) = |z| \cos(\varphi)$ and $\text{Im}(z) = |z| \sin(\varphi)$, with the former being even and the latter being odd about $\varphi = 0$, and the phase-shifting formula becomes

$$\varphi \bmod 2\pi = \arg \frac{|z| \, i \sin \varphi}{|z| \cos \varphi} = \arg \frac{i(I_3 - I_1)}{I_0 - I_2} \qquad (3.14)$$

The other piece of information in a complex number is its magnitude. In the case considered here, this is

$$|z| = \sqrt{(I_3 - I_1)^2 + (I_0 - I_2)^2} = 2M_I \qquad (3.15)$$

which gives us M_I almost as a by-product of the phase calculation. This figure can be important in classical interferometry to reject badly modulated pixels, and is indispensable in speckle interferometry to guide signal-enhancement algorithms such as filtering and unwrapping [4] (more about this in Section 3.6). Graphing the sine against the cosine will give a Lissajous figure that can be used to detect the phase-shift errors and nonlinearities. Also, if a threshold $M_{I,\min}$ is defined, below which a measurement is considered unreliable, all such measurements will lie inside a circle centered on (0, 0). The "phasor" interpretation is sketched in Figure 3.6.

In all digital phase-shifting schemes, the irradiance readings are digitized before processing. This happens sometimes after transmission to a frame grabber, but now typically within the camera itself, because the length and layout of cables (T connectors, incorrect or missing termination, proximity to power cables, etc.) can create significant noise and signal distortion when analog transmission is used.

The digitization will create a relatively sparse array of discrete values for numerator and denominator out of infinitely many values for the arctangent function. The granularity depends on the number of intensity readings that are combined in the sine and cosine terms, but, more importantly, on the modulation M_I. We will study the effect of too coarse digitization in Section 3.5.5.

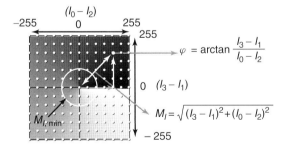

Figure 3.6 Two-dimensional display of sine term against cosine term. Assuming 8-bit digitization (gray levels from 0 to 255), both terms can range from −255 to +255 in discrete steps, and can therefore be assigned a precalculated phase value. The underlaid gray-level distribution shows how the phase readings are converted to a saw-tooth image.

The complex interpretation also makes it very easy to integrate the subtraction of a phase reference into the phase-shift formula. In dynamic processes, large numbers of fringes are likely to accumulate over time, and the reference image to be subtracted must be updated from time to time. Since a phase-angle subtraction is accomplished by a complex division, we can express a phase difference between an initial object state i and final state f by

$$\Delta\varphi = \arg(z_f) - \arg(z_i) = \arg\frac{z_f}{z_i} \tag{3.16}$$

which after some simple arithmetic according to the rules for complex division, yields [5]

$$\arg\frac{z_f}{z_i} = \arctan\frac{(I_3 - I_1)_f (I_0 - I_2)_i - (I_0 - I_2)_f (I_3 - I_1)_i}{(I_3 - I_1)_f (I_3 - I_1)_i + (I_0 - I_2)_f (I_0 - I_2)_i} \tag{3.17}$$

if we apply the four-step formula derived in Eq. (3.11). An advantage of this technique over simply applying the phase calculations separately, $\Delta\varphi = \varphi_f - \varphi_i$, is that the complex calculation preserves the information about signal strength and therefore reliability.

It should be obvious from the general nature of the preceding discussion that calculating M_I and $\Delta\varphi$ in this manner is possible for all other phase-shifting formulae that are composed in this "sine/cosine" scheme; and fortunately this includes almost all phase-shifting schemes in use today.

Tutorial Exercise 3.3

Derive I_b from Eq. (3.7).

Solution:

$I_b = \frac{I_0 + I_1 + I_2 + I_3}{4}$ or $I_b = \frac{\sqrt{(I_0 + I_2)^2 + (I_1 + I_3)^2}}{2\sqrt{2}}$ – of course I_b can also be determined by measuring $|W_T|^2$ and $|W_R|^2$ individually (with the respective other beam obscured) and then adding the results; but using all the available irradiance measurements the I_i is more reliable and conveniently comes out as a by-product of the phase-shifting sequence.

3.2.2
Phase Shifting with Three Steps

If we assume phase shifts of $\{-\alpha, 0, \alpha\}$, and label our samples I_{-1}, I_0, and I_1, to reflect the phase shifts used, we can write down our irradiance readings as

$$I_{-1} = I_b + M_I \times \cos(\varphi - \alpha)$$
$$I_0 = I_b + M_I \times \cos(\varphi)$$
$$I_1 = I_b + M_I \times \cos(\varphi + \alpha) \tag{3.18}$$

and by using trigonometric identities, we arrive at

$$\varphi \bmod 2\pi = \arctan\left(\frac{1-\cos\alpha}{\sin\alpha} \frac{+I_{-1}-I_1}{-I_{-1}+2I_0-I_1}\right)$$
$$= \arctan\left(\tan\left(\frac{\alpha}{2}\right) \frac{+I_{-1}-I_1}{-I_{-1}+2I_0-I_1}\right) \tag{3.19}$$

> **Tutorial Exercise 3.4**
>
> Show how Eq. (3.19) follows from Eq. (3.18).
>
> **Solution:**
>
> $$I_{-1} = I_b + M_1\left(\cos\varphi\cos\alpha + \sin\varphi\sin\alpha\right)$$
> $$I_0 = I_b + M_1 \times \cos(\varphi)$$
> $$I_1 = I_b + M_1\left(\cos\varphi\cos\alpha - \sin\varphi\sin\alpha\right)$$
> $$I_{-1} - I_1 = 2M_1 \sin\varphi \sin\alpha$$
> $$I_{-1} + I_1 = 2I_b + 2M_1\left(\cos\varphi\cos\alpha\right)$$
> $$2I_0 - (I_{-1} + I_1) = 2M_1 \cos\varphi\left(1 - \cos\alpha\right)$$
> $$\frac{I_{-1} - I_1}{2I_0 - (I_{-1} + I_1)} = \frac{2M_1 \sin\varphi \sin\alpha}{2M_1 \cos\varphi\left(1 - \cos\alpha\right)} = \frac{\sin\varphi \sin\alpha}{\cos\varphi\left(1 - \cos\alpha\right)}$$

Note that this is a general description and is valid for all phase-shift angles except $0°$ and $180°$, where the formula becomes undefined. This is physically sound because, if we do not shift the phase, or just invert the fringe pattern with each step, we cannot measure φ. A good approach is to try and keep the numerator and denominator terms reasonably balanced, and generally α will be between $45°$ and $120°$. The most stable behavior usually results from $\alpha = 90°$, since that is the "natural" phase-shift interval as argued above, and is also farthest away from the two extremes.

With the help of Eq. (3.19), we can derive

$$\varphi \bmod 2\pi = \arctan\left(\frac{+I_{-1}-I_1}{-I_{-1}+2I_0-I_1}\right) \tag{3.20}$$

for $\alpha = 90°$, and

$$\varphi \bmod 2\pi = \arctan\left(\frac{\sqrt{3}\left(+I_{-1}-I_1\right)}{-I_{-1}+2I_0-I_1}\right) \tag{3.21}$$

for $\alpha = 120°$. Note here how a simple rescaling of the numerator has shifted our detection frequency from four samples per scanned fringe ($\alpha = 90°$) to three samples per scanned fringe ($\alpha = 120°$): we have already encountered a first example of a tunable phase-shifting formula! As a general rule, tuning and error suppression are accomplished by changing the relative weights of the irradiance samples, and in Section 3.4 we will learn advanced but easy methods for how to do this.

3.2.3
General N-Step Method

While deriving the generic four-step formula above, he have already seen how irradiance samples can be weighted with the sine or cosine of the phase-shift angle, which is $n \times \alpha$ for the nth sample. This can be done with phase steps other than $90°$ as well:

$$\varphi \bmod 2\pi = \arctan \frac{-\sum_{n=0}^{N-1} I_n \times \sin \alpha_n}{\sum_{n=0}^{N-1} I_n \times \cos \alpha_n}, \quad \text{with} \quad \alpha_n = n \times \frac{2\pi}{N} \quad (3.22)$$

and $N \geq 3$. For $N = 4$, Eq. (3.10) follows immediately, and for $N = 3$, we get Eq. (3.21) again in a slightly different guise.

> **Tutorial Exercise 3.5**
>
> Calculate the formula for $N = 3$ according to Eq. (3.22) and prove that it is the same as Eq. (3.21).
>
> **Solution:**
>
> $$\varphi \bmod 2\pi = \arctan \frac{\frac{\sqrt{3}}{2}(-I_1 + I_2)}{+I_0 - \frac{1}{2}I_1 - \frac{1}{2}I_2}$$
>
> $$= \arctan \frac{\sqrt{3}(-I_1 + I_2)}{+2I_0 - I_1 - I_2} = \arctan \frac{\sqrt{3}(-I_1 + I_{-1})}{+2I_0 - I_1 - I_{-1}}$$
>
> The last step is because $I_2 = I_{-1}$ for $\alpha = 240° = -120°$.

In Eq. (3.22), α can also be given an offset, that is, the start phase can be varied anywhere between 0 and 2π, so that the sample weights will vary according to the $-\sin \alpha_n$ and $\cos \alpha_n$ prescriptions. Taking a given formula and just permutating or re-balancing the coefficients will leave the performance of a phase-decoding formula unchanged. We should distinguish here between the *type* of a phase-shifting formula, and its *representation*. As an example, Eq. (3.21) can be offset-shifted to read

$$(\varphi - 15°) \bmod 2\pi = \arctan \frac{-0.259 I_0 - 0.707 I_1 + 0.966 I_2}{+0.966 I_0 - 0.707 I_1 - 0.259 I_2} \quad (3.23)$$

or Eq. (3.11) can be modified into

$$(\varphi - 45°) \bmod 2\pi = \arctan \frac{\frac{1}{\sqrt{2}}(-I_0 - I_1 + I_2 + I_3)}{\frac{1}{\sqrt{2}}(+I_0 - I_1 - I_2 + I_3)}$$

$$= \arctan \frac{-I_0 - I_1 + I_2 + I_3}{+I_0 - I_1 - I_2 + I_3} \quad (3.24)$$

These operations do not affect the calculated phases, other than adding an offset and thereby shifting the positions of the black-white edges in the sawtooth image.

The discovery that some phase-shifting formulae could be represented with integer coefficients led to worthwhile increases in processing speed; in the twenty-first century however, the situation is almost reversed, in that most processors can handle floating-point numbers well, and integer values are often automatically converted to a floating-point format anyway.

With the scheme of sampling and weighting as in Eq. (3.22), the numerator/denominator represent the discrete implementation of a Fourier sine/cosine transform [6]. Therefore this type of formula has been given the label "DFT (discrete Fourier transform) formula" [7] to reflect the fact that it represents a Fourier transform of extremely short length.

Of course, we could record, say, 1024 intensity readings at $\alpha_n = n \times 2\pi/1024$, apply an actual Fourier transform, and then just look at the phase of the signal at $\nu = 1$ cycle/(1024 samples); again, we can see how all that we are trying to do is to retrieve the phase of a certain frequency with as little effort as possible. The critical reader may now wonder what happens at multiples (harmonics) of the carrier frequency (between 2 and 512 cycles per 1024 samples) in the case of our hypothetical fast Fourier transform (FFT), or 2 or 3 cycles/(4 samples) in case of $N = 4$. We will come to this, as is it very important in practice; for now, suffice to say that a very short sampling sequence of delta pulses, whichever way they are weighted, does not make a very selective filter in frequency space, and this is indeed the main reason that the phase-decoding technique is susceptible to various errors. This behavior is equivalent to the uncertainty principle in Fourier analysis.

It is not generally necessary to confine α to $[0, 2\pi]$: some methods use two full cycles of irradiance modulation ($0 \leq \alpha \leq 4\pi$), and fractional values are possible at no penalty (see Section 3.2.5). Large amounts of measurement data can be collected for lower uncertainty; but as the number of samples grows, the measurements will take longer, which will have an impact because of parameter drift.

3.2.4
Symmetrical N + 1 Step Formulae

One very simple way to introduce some error tolerance to the phase calculation is to define α as in Eq. (3.22) but let α run for one full period from 0 to 360°, which adds an extra irradiance sample. This last sample cannot enter the calculation with equal weight. However, if we halve the weights of the last sample, we introduce a simple form of averaging, and the four-sample formula Eq. (3.10) becomes the 4 + 1 formula [8, 9].

$$\varphi \bmod 2\pi = \arctan \frac{-I_1 + I_3}{+\frac{I_0}{2} - I_2 + \frac{I_4}{2}} = \arctan \frac{2(-I_1 + I_3)}{+I_0 - 2I_2 + I_4} \qquad (3.25)$$

which has been shown to suppress errors from incorrectly calibrated phase shift quite well, that is, it also works reliably for $\alpha \neq 90°$ within reasonable limits of about ±5°. This is one of the more popular methods in phase shifting, probably because practitioners have decided that it finds the "sweet spot" between simplicity and suppression of errors that are relevant in practice.

We can also extend the general three-step method in the same way:

$$\varphi \bmod 2\pi = \arctan \frac{\sqrt{3}\,(+I_1 - I_2)}{+I_0 - I_1 - I_2 + I_3} \qquad (3.26)$$

In this case, we find slightly more error tolerance for too small sampling frequencies, but no improvement for higher sampling frequencies. Hence, the $N + 1$ technique is not generally a technique to create the best possible error suppression, but rather a generic technique to make a formula "symmetrical," meaning that the sample weights are odd about the center for the sine term, and even about the center for the cosine term:

$$a_n = +a_{N-n}$$

$$b_n = -b_{N-n} \qquad (3.27)$$

Note that n runs from 0 to N here and thus encompasses $N + 1$ samples. Also, Eq. (3.27) is valid for odd and even N, which leads us immediately to $a_{N/2} = 0$ when N is even, $N + 1$ therefore odd, and there is a sample point in the center, where we can take α to be zero. We have already seen an example of this in Section 3.2.2. When N is odd and $N + 1$ is even, there is no actual sample point in the center, but we still take α to be zero in the middle of the sequence, that is, between the central two samples. In that case, the bracketing phase-shift values are $\pm\alpha/2, \pm 3\alpha/2$, and so on.

The distribution of sample weights as in Eq. (3.27) has become known as the *Hermitian form* of a phase-shifting formula [10], because one can also write

$$c_n = a_n + ib_n \qquad (3.28)$$

That is, we can interpret our series of coefficients as a series of complex numbers, whose imaginary part is odd and forms the sine, and the real part is even and forms the cosine. In that interpretation,

$$c_n = c_{N-n}^* \qquad (3.29)$$

where * is the complex conjugate, and as described above, $\alpha = 0$ is taken to be in the center of the sampling sequence. It is well known [6] that the Fourier transforms of odd functions are purely imaginary, and the Fourier transforms of even functions are real – and this is really useful for investigating the responses of the sine and cosine terms in the frequency domain, because it means that the responses will always be in quadrature! Of course, things can still go wrong, that is, the phase lag can be $+90°$ or $-90°$, or the amplitudes of the terms may not match; but once we have found a Hermitian representation for a phase-shifting formula, we can ignore the phase transfer functions in frequency space.

3.2.5
Extended Averaging

The reason that the $N+1$ method worked very well for $\alpha = 90°$/sample is that phase-shift errors give rise to a cyclic error in the phase that oscillates at twice the fringe frequency (see also Section 3.5.1). This is very easy to see intuitively from Figure 3.6: if the Lissajous figure becomes an ellipse (because the terms have different amplitudes or are not in quadrature), the error will go through two identical cycles per full turn of φ. Thus, when we average data that are 90° out of phase, the error will be 180° out of phase, and will be canceled to a very good approximation. Residual errors will then come from the imperfection of the offset (for when the phase shift is different from 90°, so is the offset between successive sample sequences!), and the fact that the error is slightly asymmetrical.

Let us now see how this works in detail. Starting from Eq. (3.11), we write down two consecutive phase measurements with an offset of 90° as

$$\varphi_0 \bmod 2\pi = \arctan \frac{I_3 - I_1}{I_0 - I_2} := \frac{N_0}{D_0} = \frac{\sin \varphi}{\cos \varphi}$$

$$\varphi_1 \bmod 2\pi = \arctan \frac{I_4 - I_2}{I_1 - I_3} = \frac{\sin(\varphi + 90°)}{\cos(\varphi + 90°)} \qquad (3.30)$$

and we then adapt the second formula to yield φ instead of $\varphi + 90°$:

$$\varphi_1' \bmod 2\pi = \arctan \frac{I_3 - I_1}{I_4 - I_2} = \frac{\sin \varphi'}{\cos \varphi'} := \frac{N_1}{D_1} \qquad (3.31)$$

where we have used

$$\sin \varphi = -\cos(\varphi + 90°)$$
$$\cos \varphi = \sin(\varphi + 90°) \qquad (3.32)$$

Again, when constructing the phase average, it is better to average the N_n and D_n terms before executing the arctangent operation, as opposed to averaging φ_0 and φ_1 after separate arctangent operations, because all terms are phasors, and adding them before carrying out the arctangent operation preserves the relative weighting and thus improves the signal-to-noise ratio (SNR). Therefore, the N and D terms are averaged according to [8, 9, 11]

$$\overline{\varphi} \bmod 2\pi = \arctan \frac{N_0 + N_1}{D_0 + D_1} \qquad (3.33)$$

which results in

$$\overline{\varphi} \bmod 2\pi = \arctan \frac{2 \sin \varphi}{2 \cos \varphi} = \arctan \frac{2(I_3 - I_1)}{I_0 - 2I_2 + I_4} \qquad (3.34)$$

and is Eq. (3.25) again.

If we apply the technique similarly to

$$\varphi \bmod 2\pi = \arctan\left(\frac{I_1 - I_2}{I_1 - I_0}\right) \tag{3.35}$$

a very simple three-step formula, which is an offset version of Eq. (3.20) (see also Eq. (3.63) below), we arrive at [11]

$$\bar{\varphi} \bmod 2\pi = \arctan\frac{2(I_1 - I_2)}{-I_0 + I_1 + I_2 - I_3} \tag{3.36}$$

which has intrinsic error compensation with only four samples.

With these two types of formulae, the averaging can of course be repeated [12], where for each phase step added, more error compensation is gained. Averaging Eq. (3.36) again gives

$$\bar{\varphi} \bmod 2\pi = \arctan\frac{+3I_1 - 3I_2 - I_3 + I_4}{-I_0 + I_1 + 3I_2 - 3I_3} \tag{3.37}$$

(note also how averaging will create symmetrical and asymmetrical formulae in alternation).

Tutorial Exercise 3.6

Write down the averaged formula derived from Eq. (3.37), following the rules of Eqs. (3.30–3.33).

Solution:

$$\bar{\varphi} \bmod 2\pi = \arctan\frac{+4I_1 - 4I_2 - 4I_3 + 4I_4}{-I_0 + I_1 + 6I_2 - 6I_3 - I_4 + I_5}$$

Note how this is symmetrical, but the a_n and b_n are of the form $\frac{-\cos\varphi}{-\sin\varphi}$ because the central reference point for the phase is now advanced by 90° with respect to Eq. (3.36).

The important feature of extended averaging is that the rectangular envelope from which we are starting (with all sample coefficients equally weighted according to sine/cosine) morphs into a triangular function first, and then into a bell-shaped function [13–15]. If we take, for example, the $|c_n|$ in Eq. (3.37), we get {1, 3.16, 4.24, 3.16, 1}. Using the envelope for the complex phasor magnitudes provides more insight than just looking at the a_n or b_n, and demonstrates in this example that the envelope is still symmetrical, even though the individual sample functions are not.

3.2.6 Summary

In the preceding discussions, we have seen how the phase of a wave(front) can be retrieved. By creating a beat signal with a reference wave(front), taking a few readings of the resulting irradiance modulation at predefined phase shifts, and processing these into two quadrature signals, we can decode the phase of the wavefront under test.

There are several great advantages to this technique over trying to infer the wavefront information from only one fringe pattern.

- We do not have to introduce a fringe pattern on the camera at all: we can remove the tilt between the object and reference wavefronts completely (that is called a *fluffed-out fringe*) and just track the intensity modulation in each pixel.
- Each pixel is a separate interferometer with its own values for I_b and M_I, and, as long as these are reasonably above the noise level, the phase reading will be quite accurate. This is a significant improvement over fringe-profile evaluation, where the peak irradiance is a function not only of the phase but also of parameters like the irradiance profile of the beams, or even local changes in the state of polarization.
- The information about the sign of the phase function is an automatic result of the phase decoding, and therefore closed fringes are allowable in TPS. In SPS, the condition is still that the spatial bias frequency be high enough to prevent closed fringes, that is, it must not change sign.
- Taking several readings introduces some suppression of random noise by virtue of averaging, and also allows some freedom in designing the best evaluation method. If more (or different) error suppression is desired, more samples can be added to the measurement and/or the relative sample weights can be altered.

3.3
Methods of Phase Shifting

As we have seen, the phase shift can take place in either the temporal domain, or the spatial domain, or even both. This gives the user some freedom to design an implementation suitable for the purpose; however, all possibilities involve trade-offs. If information is to be recorded, it can be done at full spatial resolution – but extra time will be needed to record the information – or as one snapshot – but some spatial resolution must then be given up.

To extend the unambiguous measurement range, more information is required, and synthetic wavelength techniques can be used that implement phase-shifting at two or more different wavelengths or fringe periods to create a "vernier." This increases the complexity of the measurement and the evaluation, but is sometimes necessary to find the correct solution, for example, in step-height or shape measurement, where the fringe pattern is discontinuous and/or absolute fringe orders must be found.

3.3.1
Temporal Phase Shifting (TPS)

If we shift the phase of the reference wave in time, we can rewrite Eq. (3.4) as

$$I(x, y, t) = I_b(x, y) + M_I(x, y) \times \cos\left(\varphi(x, y) + \alpha(t)\right) \tag{3.38}$$

and the necessary phase shift is most often introduced by varying the optical path length in the reference arm. When a reflecting surface in the reference beam path is attached to a piezoelectric transducer (a lead–zirconium–titanium material, abbreviated PZT for $(PbZr_xTi_{1-x})O_3$), the optical path length can be varied by the interferometer control unit, which outputs digital or analog control signals to a high-voltage amplifier, in synchronization with the data recording. The voltage change causes a length change in the PZT unit (typically a few nanometers per volt) and thereby adjusts the desired phase shift. There are two caveats when using PZTs to shift the phase:

1) The material exhibits creep and hysteresis, so that the speed of the phase shift and the immediate voltage history make a difference to the actual phase shifts obtained. When feeding a linear sequence of steps or a voltage ramp into a PZT, one will almost certainly get a small non-linearity in the phase shift.
2) The expansion of a PZT stack is not always uniform. Ideally, one would want to obtain only a piston-type motion; but frequently the PZT also tilts a little as it expands, so that the phase shift will depend on the image coordinate. This behavior is not covered in standard phase-shifting schemes, and one can use either a self-calibrating formula that makes allowance for $\alpha(x, y)$ as another unknown variable [16], or one can design a phase-shifting scheme to suppress this type of error. We will treat this problem in detail in Section 3.4. Figure 3.7 presents a few common phase-shifting interferometers that lend themselves to mechanical phase shifting.

The Twyman–Green interferometer (Figure 3.7a) functions very much like a Michelson interferometer; however, one of the arms is used to test an optical surface. The beam does not have to be expanded, but the possibility exists to use different telescopes and/or transmission components, creating flat or spherical wavefronts to suit different measurement problems. Since the path lengths can be quite different in this type of setup, it is important to have sufficient coherence length and a very stable wavelength; otherwise, the phase will drift as a result of small wavelength changes. Often, a wavelength-stabilized HeNe laser is used. The precision attainable with this type of interferometer is limited by the aberrations accumulating on the way through the object arm and back. A reference measurement can be provided by measuring a calibrated test surface, then changing to the actual test piece, and subtracting the reference result from the current measurement.

This type of problem is almost completely eliminated by using a Fizeau interferometer (drawn in Figure 3.7b). The layout is such that the test and reference surfaces are close together and the actual interferometer, where a difference in optical paths gives rise to a fringe pattern, is quite short. Both waves then travel

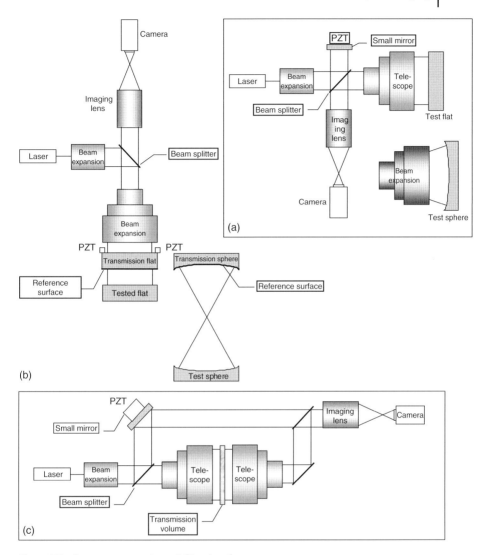

Figure 3.7 Some common phase-shifting interferometers.
(a) Twyman–Green interferometer. (b) Fizeau interferometer.
(c) Mach–Zehnder interferometer. See text for descriptions.

on a common path to the camera. This is the most frequently used precision interferometer type, and theoretically errors are introduced only in the gap between the two surfaces which are as close together as possible. The Fizeau interferometer can also conveniently be used with large apertures, or, instead of a transmission flat, a transmission sphere can be fitted, which allows testing of spherical surfaces as well. Since the transmission optics is relatively large, the phase shifting is often realized by three PZT stacks at 120° angles around the aperture. If these have slightly different responses and are not calibrated separately (more on calibration in Section 3.5.1), again a tilt will occur during phase shifting. Moreover, the masses to be moved mechanically grow larger very quickly with the aperture diameter, so that it has been suggested to utilize a very different approach to phase shifting and to move the reference surface in a smooth sinusoidal vibration, instead of stepping or linearly ramping the phase and resetting after each phase-shifting sequence [17].

The Mach–Zehnder interferometer, pictured in Figure 3.7c, is more suitable for transmission studies. If TPS is used, the transmission objects will be optical components, or other phase distributions that change slowly (in contrast to the example in Figure 3.9). Since the moving mirror reflects the reference wave at an angle here, the phase shift must be increased to compensate for the non-normal incidence. The transmitted wave can be expanded if desired.

Box 3.3: More Mechanical Phase-Shift Devices

When fiber optics is used to deliver the laser light, it is possible to wind a length of fiber onto a PZT cylinder and glue it in place. Applying a voltage to the PZT will then stretch the fiber and therefore alter its optical path length. However, fibers of extended length are difficult to handle in phase-shifting applications, because temperature and pressure changes will also cause phase shifts and, unless high-birefringence fiber is used, stretching a fiber will also alter the polarization state of the transmitted light. In most cases, extra phase stabilization must be added to achieve sufficiently accurate phase shifts.

Another way to shift the phase of a wave is to shift a diffraction grating in one of the beams, where the undiffracted wave undergoes no phase shift and the diffracted orders experience phase shifts of

$$\alpha = 2\pi m \frac{d}{P} \tag{3.39}$$

where m is the diffraction order, d is the translation of the grating perpendicular to its transmission or optical thickness modulation, and P is the grating period.

In acoustooptic modulators, the grating is constantly moved in one direction as a constant wave of refractive index modulation propagates through a crystal. This leads to a phase shift that keeps growing with time, and is therefore best understood as a frequency shift. Typical frequencies for this application are in the tens of Megahertz, and are often used in high-speed quadrature detection schemes such as laser vibrometry (where the required 90° phase shift between the two signal channels is created by polarization optics). In these so-called heterodyne techniques, the phase decoding is more often done with analog electronics than digital circuitry, as we have already seen in Figure 3.5.

Rather than moving a component mechanically, the reference path length can also be varied by electrooptical means. Liquid-crystal phase modulators can be used as transmissive components to shift the phase; however, care needs to be taken with the linearity and possible changes in polarization and, therefore, coherence and interferometric modulation.

A very convenient way of adjusting phase differences has become available in the form of tunable single-mode laser diodes. Whenever a difference exists in optical path lengths between the test and reference arms of the interferometer, a change in wavelength will cause the two arms to change their length in waves at a different rate, and therefore introduce a phase shift. Early applications of wavelength tuning utilized the diode current to shift the wavelength, but an undesirable side effect of this was that the laser power, and therefore I_b and M_I, also changed with the current – which in turn led to the development of phase-shifting methods that could compensate for some change in the background intensity [18]. The situation was significantly improved with tunable external resonators involving a rotating diffraction grating (the so-called Littman–Metcalf layout), which allowed tuning of a single mode with no mode hops and only small power fluctuations.

The formula for phase shifting by wavelength tuning is

$$\frac{L}{\lambda_f} - \frac{L}{\lambda_i} = N \quad \text{or} \quad \frac{2\pi L}{\lambda_f} - \frac{2\pi L}{\lambda_i} = \Delta\alpha \qquad (3.40)$$

where L is the path length difference (note that traversing an air gap twice, as in a Fizeau interferometer, will add twice the length of that air gap), and N is the number of fringes of phase shift. Sometimes $2\pi/\lambda$ is replaced by k, the wavenumber, to obtain a linear scale; but for the small wavelength shifts that we are concerned with, this is quite inconsequential, and the approximation

$$\Delta\lambda = \frac{\lambda_0^2}{L} N \qquad (3.41)$$

is often made, in which λ_0 stands for the central wavelength. As a consequence, the frequency of the signal modulation depends on the separation of the surfaces involved.

Tutorial Exercise 3.7

Calculate the wavelength shift that is needed to cause a 4π phase shift in a 25 mm air gap, where the object wave traverses the air gap twice. Assume the central wavelength to be 690 nm. What is the corresponding frequency shift?

Solution:

Using Eq. (3.41), we obtain

$$\Delta\lambda = \frac{690^2 \times 10^{-18} \text{ m}^2}{0.05 \text{ m}} 2 = 0.019 \text{ nm}$$

$$\Delta\nu = c\left(\frac{1}{\lambda_2} - \frac{1}{\lambda_1}\right) = c\left(\frac{\lambda_1 - \lambda_2}{\lambda_2 \lambda_1}\right) = 3 \times 10^8 \frac{\text{m}}{\text{s}} \left(\frac{0.019 \times 10^{-9} \text{ m}}{690^2 \times 10^{-18} \text{ m}^2}\right) \cong 12 \text{ GHz}$$

$$(3.42)$$

Even a tiny wavelength shift corresponds to a large frequency shift!

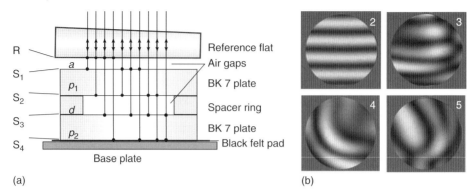

Figure 3.8 Application of wavelength shifting. (a) Layered object with glass plates and air gaps, creating 10 two-beam interference patterns in single reflections alone. (b) Interferograms from increasing numbers of quasiparallel surfaces, as given in the labels. Intensities have been renormalized for this display – adding more interference terms will increase the maximum brightness significantly. From three-beam interference upward, the fringe pattern will not "shift," but "boil" as the wavelength is scanned, because of the many contributions with different frequencies.

This opens up a way to separate signals from more than two parallel surfaces [19, 20], where tunable phase shifting methods with tens of samples are used to extract the frequency of interest from the mixture of frequencies in the resulting multibeam interferogram.

It is possible to see from Eq. (3.22) how this might be done: if α is decreased to smaller values of, say, $10°$/sample, then $\sin \alpha_n$ and $\cos \alpha_n$ are represented by 36 samples per cycle, and quite oversampled; but we can then "tune" the formula by using weights $\sin (m \times \alpha_n)$ and $\cos (m \times \alpha_n)$, where m is the harmonic order to be detected, and we would reach the Nyqvist limit only at $m = 18$.

To illustrate the usefulness of this approach, Figure 3.8 presents an example for a layered object that cannot be measured with mechanical phase shifting and requires a path-selective method such as wavelength shifting (another option would be to decrease the coherence length enough to suppress multiple-beam interference, for example, white-light scanning interferometry).

Using phase-shifting formulae with a sharp frequency response *and* some error tolerance is very important in this application, because the gaps between the surfaces cannot always be adjusted to make the frequency ratios into exact integers. Also, if dispersive materials are involved, the group refractive index of the material will cause a slight detuning of the phase shift. However, even this technique cannot separate equal or very close frequencies. In Figure 3.8, if $p_1 = p_2$, the individual thickness maps of the BK7 plates cannot be measured directly.

In confocal spherical cavities as shown in Figure 3.7b, wavelength tuning is the only way to achieve uniform phase shifts. Consider that, if a spherical reference surface is translated mechanically, we have

$$\alpha(\vartheta) = \alpha_{\max} \cos \vartheta \tag{3.43}$$

where ϑ is the off-axis ray angle. At higher numerical apertures (where the *numerical aperture* is defined as $n_0 \sin \vartheta$, n_0 being the refractive index of the medium), this becomes a severe systematic error. However, if wavelength shifting is used, the sphericity of the interferometric cavity works in our favor in that all ray path lengths are equal. Thus, a wavelength shift would introduce equal phase shifts over the entire surface, regardless of the numerical aperture, and the systematic error disappears.

However, the distance between the surfaces is preset by the radius of curvature, and for longer radii this often leads to problems with turbulences and draughts in the air gap.

Box 3.4: Phase Ramping

Besides phase stepping, there is also the possibility of phase ramping, that is, the fringes are not advanced, then stopped for recording, and advanced again, but we can just move the fringes at a constant rate so that each recorded image will be offset by α. Clearly, this "washes out" the moving fringes to some extent. We can calculate the loss in signal as

$$I_n = \frac{1}{\alpha} \times \int_{\alpha_n - \frac{\alpha}{2}}^{\alpha_n + \frac{\alpha}{2}} I_b + M_I \times \cos(\varphi + \alpha_n + \alpha') d\alpha'$$

$$= I_b + M_I \times \frac{2\sin(\frac{\alpha}{2})}{\alpha} \times \cos(\varphi + \alpha_n) \quad (3.44)$$

The modulation loss factor is 0.9 when $\alpha = 90°$, and 0.83 for $\alpha = 120°$, so that the overall effect of the ramping approach is a slight decrease in the modulation of the data. It is of the form $\sin(x)/x$ and is, unsurprisingly, the Fourier transform of a rectangular function, which is called the *modulation transfer function* in this context. We also see from Eq. (3.44) that, as long as the integration interval is symmetrical, the measured φ remains the same. A positive side effect of phase ramping is that it attenuates unwanted signal harmonics much more than the actual signal.

It is sometimes advantageous, for technical reasons, to ramp the phase instead of stepping; if M_I must be maximized and sufficient laser power is available, the integration window can be narrowed again by decreasing the detector's exposure time.

3.3.2
Spatial Phase Shifting (SPS)

While TPS requires digital storage of measurement data, SPS in its early versions has been implemented electronically. Figures 3.4 and 3.5 give examples of how this worked: on electronic readout, a spatial carrier frequency becomes a temporal carrier frequency. Decoding schemes such as quadrature multiplicative moiré, or even a method for 120° phase shift, can then be used [21]. Thus, Eq. (3.38) changes to

$$I(x, y) = I_b(x, y) + M_1(x, y) \times \cos\left(\varphi(x, y) + \alpha(x, y)\right) \qquad (3.45)$$

where we have allowed the general case of a phase shift in the x- and y-directions, which would result in an oblique carrier fringe pattern. The phase can then be calculated either by

- de-interlacing the images, as shown in Figure 3.4 for a purely horizontal phase shift, and treating each of them like a TPS interferogram (but rescaling for lost resolution in the evaluation); or
- applying the "sine" and "cosine" correlation spatial filter masks that are implicitly defined by the numerator and denominator of the phase-decoding formula

$$S_k = \sum_{n=0}^{N-1} I\left(x_{k+n}, y_{k+n}\right) b\left(x_n, y_n\right) = I\left(x_k, y_k\right) \otimes S(n) \propto \sin\varphi\left(x_k, y_k\right)$$

$$C_k = \sum_{n=0}^{N-1} I\left(x_{k+n}, y_{k+n}\right) a\left(x_n, y_n\right) = I\left(x_k, y_k\right) \otimes C(n) \propto \cos\varphi\left(x_k, y_k\right)$$

$$\text{and } \varphi\left(x_k, y_k\right) \bmod 2\pi = \arctan\frac{I\left(x_k, y_k\right) \otimes S(n)}{I\left(x_k, y_k\right) \otimes C(n)} \qquad (3.46)$$

where $\otimes$ denotes a correlation or convolution and a_n and b_n are replaced by $a(x_n, y_n)$ and $b(x_n, y_n)$ to account for the possibility of horizontal and vertical phase shifts, but we do not treat oblique phase shifts and possible offsets of φ explicitly, in order to keep the expressions readable. Obviously, the $S(n)$ and $C(n)$ are the sampling functions, or sequences of weighted δ pulses, that we have already seen as numerator and denominator of various phase-decoding formulae. They define spatial filter windows here that are applied to every pixel in the image and include information from neighboring pixels – which is of course the price to pay for being able to record a phase-shifting dataset in one image.

In analogy to Eq. (3.28), one can regard the two quadrature signals as one complex digital correlation:

$$Z\left(x_k, y_k\right) = I\left(x_k, y_k\right) \otimes \left(C(n) + iS(n)\right)$$

$$\text{with } \varphi\left(x_k, y_k\right) \bmod 2\pi = \arg\left(Z(x_k, y_k)\right) \qquad (3.47)$$

which makes it unnecessary to create two correlation images.

In the example of Figure 3.4, where the phase shift is only along the x-direction, with $\alpha = 120°$/column, the filter masks are $S(n) = \{\sqrt{3}, 0, -\sqrt{3}\}$ and $C(n) = \{-1, 2, -1\}$, according to Eq. (3.21). Most image processing programs allow the definition of custom filters (which, in this case, would be band-pass filters). It is easy to enter these coefficients and confirm that the filtering steps create two new fringe patterns 90° out of phase, and the phase is then calculated as usual by the arctangent operation.

Exactly how much spatial resolution does this correlation operation cost us? The answer depends very much on what the phase gradients of the tested wavefront are. The higher the phase gradient of the tested wavefront, the harder is it for a correlation window a few pixels wide to track changes adequately, and phase-shift errors will cause artifacts. Most wavefronts change slowly on a pixel scale, so that the spatial carrier frequency is several times larger than the highest phase gradient in the measured wavefront. If this is not the case, the options are to align the spatial carrier with the lowest phase gradient found (because we can maintain the full spatial resolution in the direction normal to the phase shift); to magnify the measured wavefront and measure only a portion of it, so that the sampling becomes adequate; or to quantify the errors and accept them. We can calculate the spectral responses of the phase-shift formula (and we will learn how to do this in Section 3.4) and can in this way come to an estimate of the errors resulting from steep object phase gradients. Adding more samples to the phase-shift sequence is not helpful here: it may suppress some errors, but the correlation window gets even wider, and one could arrive at very smooth phase maps lacking most of the relevant detail.

The most important practical reasons for utilizing SPS are convenience and speed. Figure 3.9 shows an example of a high-speed recording in which a flickering flame was situated just under one of the beams in a Michelson interferometer and SPS was implemented by a slight tilt between the mirrors. The video frames were recorded with a standard PAL video camera which reads out the odd and even image lines in alternation at 25 Hz. The odd and even "fields," as they are known, are separated by 20 ms in time, together constituting a complete video frame. With an exposure time of 100 µs, very rapid changes in the wavefront can be measured with low noise.

Besides taking snapshots of fast phenomena, the short acquisition time can also be used for fast averaging of nominally constant phase distributions, for example, in high-precision interferometers [22], where high-speed processing is used to accumulate thousands of measurements of a wavefront for an average with low uncertainty. If the waves arriving at the detector are spherical (or have a spherical

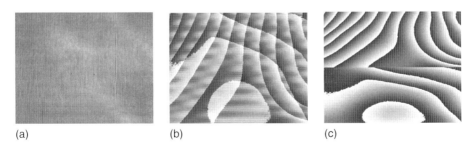

Figure 3.9 (a) Phase-shifted interferogram of strong turbulence. (b) Reconstructed sawtooth image. (c) Deinterlaced fields of video frame.

phase envelope, such as in speckle interferometry with imaging systems, where the iris can be taken to be the source point of the test wave), a spatial phase bias is introduced via a lateral offset of the source point of the reference wave from the optical axis [23, 24]. This requires a beam combiner to couple in the reference wave, or a modification of the iris in the imaging lens.

Sometimes the restriction to static phase-shift angles of 90°, 120°, or some other value between pixels is not practical. However, since the sensor now samples many cycles of the carrier wave, we can interpret this as an extended signal, which is long enough (or rather, wide enough, since we are working in the spatial domain) to apply a Fourier transform to its evaluation. Going back to Eq. (3.2), we can rewrite a spatial-carrier interferogram as

$$W_T(x, y) = A_T(x, y) \exp\left(i\varphi_T(x, y)\right)$$
$$W_R(x, y) = A_R \exp\left(i2\pi\left(\frac{x}{P_x} + \frac{y}{P_y}\right)\right) \quad (3.48)$$

where P_x and P_y are the nominal periods of the carrier fringes in x- and y-direction, respectively. Note how we have assumed a uniform spatial modulation for what was formerly φ_R. Writing $A_T(x, y)\exp(i\varphi_T(x, y)) = \mathbf{A}_T(x, y)$ for convenience, where $\mathbf{A}_T(x, y)$ is a complex phasor, the resulting irradiance is then

$$I(x, y) = I_T(x, y) + I_R(x, y) + \mathbf{A}_T^*(x, y)A_R \exp\left(2\pi i\left(\frac{x}{P_x} + \frac{y}{P_y}\right)\right)$$
$$+ \mathbf{A}_T(x, y)A_R \exp\left(-2\pi i\left(\frac{x}{P_x} + \frac{y}{P_y}\right)\right) \quad (3.49)$$

where we would usually combine the last two terms to a cosine. In this case, we keep them separate and write the Fourier transform as

$$F(\nu_x, \nu_y) = \tilde{I}_T(\nu_x, \nu_y) + I_R\delta(\nu_x, \nu_y) + \tilde{\mathbf{A}}_T(\nu_x, \nu_y) \otimes A_R\delta(\nu_x + \nu_{0x}, \nu_y + \nu_{0y})$$
$$+ \tilde{\mathbf{A}}_T^*(\nu_x, \nu_y) \otimes A_R\delta(\nu_x - \nu_{0x}, \nu_y - \nu_{0y}) \quad (3.50)$$

where ν_{0x} and ν_{0y} are the spatial carrier frequencies, and $\sim$ denotes the Fourier transform. The uniform carrier frequency becomes a pair of δ pulses on Fourier transformation, appearing at its positive and negative coordinates in the frequency plane; and the convolution of $\tilde{\mathbf{A}}_T(\nu_x, \nu_y)$ with a δ pulse will reproduce $\tilde{\mathbf{A}}_T(\nu_x, \nu_y)$, centered at these frequency coordinates [6]. Thus, if we isolate one of the complex sidebands, we break the symmetry that is still implicit in Eq. (3.50) (the sum of the mixed terms is still a real cosine), and we can then determine the phase from an inverse Fourier transform of the filtered signal $F_f(\nu_x, \nu_y)$. Yet, it is much more convenient to remove the carrier frequency first: in analogy to the low-pass stages in Figure 3.5, the filtered signal band is shifted back to zero in the frequency plane, so that the actual wavefront phase can be retrieved from an inverse Fourier transform (IF) of the filtered and shifted signal band $F_{fs}(\nu_x, \nu_y)$:

$$F_f(\nu_x, \nu_y) = \tilde{\mathbf{A}}_T(\nu_x, \nu_y) \otimes A_R\delta(\nu_x + \nu_{0x}, \nu_y + \nu_{0y}) \quad F_{fs}(\nu_x, \nu_y) = \tilde{\mathbf{A}}_T(\nu_x, \nu_y) \otimes A_R\delta(\nu_x, \nu_y)$$
$$IF_f(\nu_x, \nu_y) = \mathbf{A}_T(x, y)A_R \exp\left(-2\pi i\left(\frac{x}{P_x} + \frac{y}{P_y}\right)\right) \quad IF_{fs}(\nu_x, \nu_y) = \mathbf{A}_T(x, y)A_R$$
$$(3.51)$$

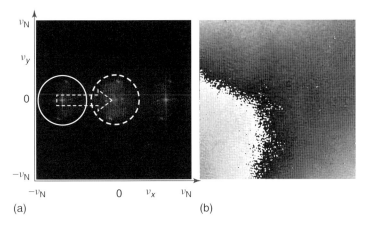

Figure 3.10 (a) Fourier spectrum of Figure 3.4a. Filtering is indicated by the solid white circle; shifting of the filtered sideband is sketched by dashed outlines. (b) Wrapped phase map calculated from Fourier-transform evaluation of Figure 3.4a.

Figure 3.10 illustrates the process with the familiar phase-shifted image from Figure 3.4.

The Fourier transform was calculated on 512 × 512 pixels, and the carrier frequency $\nu_{0x} = 1$ fringe/(3 pixels) therefore appears at 171 fringes/(512 pixels) in the frequency plane. The borders of the frequency plane are at the Nyqvist frequencies ν_N of 256 fringes/(512 pixels) in both coordinates. The sidebands created by the carrier frequency are clearly visible, and one of them is then taken from the complex Fourier transform image, centered at zero frequency (all other signals are deleted), and inverse-transformed. The phase map, including the spurious ripple, looks very similar to Figure 3.4.c

For a phase shift only in x-direction, we would not normally need a 2-D Fourier transform. But for adjusting and aligning the carrier frequency properly and identifying spurious fringe patterns, this should be done anyway. Also, using a 2-D frequency filter window will give a matched spatial resolution in both coordinates. Its size should be chosen so that both the Nyqvist frequency and an overlap with the zero-frequency band are avoided. This is especially important in speckle interferometry, where the speckled sidebands, and the speckle pattern itself, have considerable bandwidth regardless of what is actually being measured.

Note that the carrier-frequency approaches are the spatial version of phase ramping, and the calculation given in Section 3.3.2 for the loss in modulation applies in the same way, except that now the integration window is over a spatial distance and not over a time interval. To calculate the loss in fringe contrast, the so-called fill factor, that is, the fraction of a pixel area that will actually collect the light, must be known. Some sensors use microlenses to utilize all the light reaching the sensor plane, and concentrate it on the light-sensitive area of the pixels, so that the phase-shift integration interval is approximately equal to the pixel pitch.

As we have seen in the discussion of Fourier evaluations, a unique feature of SPS is that each of the coordinates can have a carrier frequency. In that case, the integration window is no longer a rectangle, but it will remain symmetrical. A very useful choice of phase shift(s) is $\alpha = 90°$/column and $\alpha = 90°$/row, leading to a 45° slant in the fringe pattern. This utilizes the spatial-frequency plane very efficiently [24], and considerable research has been carried out on the best spatial filter kernels to use for stable and low-error phase decoding [25].

Tutorial Exercise 3.8

Following the example in Eq. (3.44), Box 3.4, calculate the modulation loss in SPS when the fringes are slanted 45° on the pixel grid and the phase shift is 90° per column and 90° per row.

Solution:

The problem can be decomposed into the two coordinates:

$$I_n = I_b + M_I \times \frac{2\sin(\frac{\alpha_x}{2})}{\alpha_x} \times \frac{2\sin(\frac{\alpha_y}{2})}{\alpha_y} \times \cos(\varphi + \alpha_n)$$

Alternatively, the window can be scanned along the resultant direction of the phase shift only; in either case, the window function is a symmetrical triangle, and the modulation loss factor is $4\sin^2(\alpha_x/2)/\alpha_x^2$ (because of $\alpha_x = \alpha_y$) which is the modulation transfer function of a triangle. For angles between 0 and 45°, the window function will be trapezoidal, but still symmetric.

3.3.3
Spatiotemporal Phase Shifting (STPS)

The phase-shifting technique was first applied to interferometry, because interference naturally creates the periodic signal required. However, the technique is equally applicable to all other periodic signals. A good example is the measurement and digitization of shapes and distortions with projected white-light patterns, as shown in Figure 3.11 (Chapter 7).

A height difference Δz between an object and a reference plane is converted to a lateral offset Δx of the fringe pattern recorded by the camera. The relationship between these quantities is approximately

$$\Delta x = M \sin\theta \, \Delta z \tag{3.52}$$

where θ is the angle between illumination and viewing, and M is the optical magnification of the imaging system. Once the system is calibrated for geometry and distortion, it is then possible to determine absolute object coordinates [26].

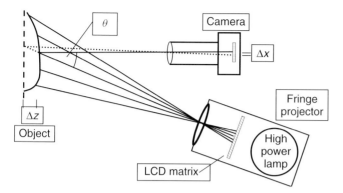

Figure 3.11 White-light fringe projection system. An LCD projector creates a fringe pattern on a 3-D object, whose surface is imaged by a camera.

Early implementations of this technique used "coded light," with black and white patterns in a coding sequence, going from very coarse to fine, which allowed the restoration of the 3-D distribution of points, limited by the granularity of the finest projected fringe pattern.

With liquid-crystal display (LCD) projectors becoming more widely available, is was realized quickly that the black and white fringe patterns could be replaced by cosine patterns, which made the technique amenable to the use of phase shifting by just stepping the projected fringe pattern column by column across the LCD matrix.

The fringe offset Δx is related to the detected phase offset by

$$\Delta \varphi \bmod 2\pi = 2\pi \frac{\Delta x}{P} \bmod P \tag{3.53}$$

where P is the nominal fringe period in x (note that in a system with divergent illumination P will change slightly across the field of view). Again, the phase calculation is initially modulo 2π and unwrapping will have to be applied to restore a continuous surface. As P gets smaller, the phase change that a certain Δx will introduce gets larger, and thus, more easily detectable. So one can say that adjusting P is a way to tune our wavelength or sensitivity in this technique. The phase-shift increment will be exact whenever the phase shift can be implemented by advancing the fringe pattern through an integer number of pixels, and there are no stability issues except in the relative positions of the projector and the camera. Therefore, simple phase-shift formulae are typically sufficient here, and Eq. (3.11) is often used.

More important is the proper calibration of the LCD array, which typically has strong nonlinearities, that is, the phase calculation will be impaired by the presence of strong harmonics. Even after a proper calibration, this problem is much more important in fringe projection than in laser interferometry, as Figure 3.12 demonstrates.

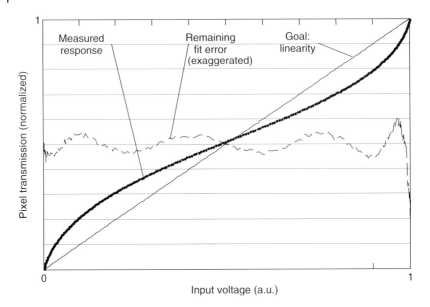

Figure 3.12 Brightness response and calibration curve for a typical LCD matrix used in fringe projection (Bothe, Personal communication, used with permission).

Therefore, the priority in fringe projection is to use a phase-shifting formula that suppresses harmonics, rather than effects of phase-shift errors. It is also desirable to defocus the fringe projector a little: the averaging over neighboring pixels on the LCD matrix makes the intensity profile smoother; moreover, moiré effects between the LCD matrix and the detector pixel grid must be avoided.

White-light fringe projection thus allows measurement of continuous (but scattering) surfaces in a manner quite analogous to interferometry. However, various additional complications can occur: there may be steps in the surface; the object consists of several unconnected regions; or shadowing (being inevitable in either the projection or the imaging) does not allow obtaining a valid reading.

As is done in two- or multiwavelength interferometry on discontinuous objects (typically in speckle interferometry), a synthetic wavelength must be created to assign absolute height readings and unwrap the entire height map. This is quite easy with an LCD matrix by varying P as described above, and we will discuss procedures to do this in Section 3.6.3.

The accuracy and reliability of the fringe-projection method has to be paid for with longer measurement times, because it is really a spatiotemporal approach that makes use of both domains. Information is collected by TPS; then, the spatial period/wavelength of the fringe pattern is altered and the measurement sequence is repeated. For this reason, quite some effort has gone into finding the most efficient way to collect the required data, not more (or much more), and certainly not less! Of course, the necessary amount of information will vary with the given measurement problem and the noise level. Sometimes, it would suffice to make one phase-shifting measurement *or* use a spatial evaluation of one fringe pattern (preferably somewhat denser, even though closed fringes cannot occur here due to shadowing, and the number of fringes in the measurement can never exceed the number of fringes projected). In practical problems, the number of fringe patterns with different spatial periods ranges from 3 to 5, a phase-shifting sequence is measured or collected with each (or just with the last one), and measurement times are usually several seconds. The complexity of the surface to be measured dictates the complexity of the technique to be used.

3.4
Designing and Analyzing Phase-Shift Methods with the Complex-Polynomial Method

Since we sample the irradiance modulation only at very few selected points along the cycle as explained in Section 3.2, it is clear that the phase decoding will not be very frequency-selective. The usual uncertainty relation applies: if we measure or sample a signal for only a few cycles (in our case, sometimes less than one cycle!), we cannot be too sure about its frequency or phase.

The combination of intensity samples, in operators like the "sine" and "cosine" terms, constitutes a digital filter, which of course has a defined amplitude and phase response. This analysis was initially carried out in the Fourier domain, and it was observed that the sampling functions are discrete, the filter spectra are therefore continuous [6, 27], and matching the gradients of the "sine" and "cosine" response functions is a way to obtain error tolerance in the phase calculation [27, 28]. The Fourier method is useful to analyze phase-shift formulae, but does not offer a straightforward way to design new ones.

This has been remedied in the past decade with the full application of digital filter theory to the phase-shifting method. The formalism introduced below is so useful, and so easy and fun to apply, that the author considers the familiarity with it as indispensable for anything other than routine work. It literally will allow you to design a custom phase-shifting formula with the properties you want in 10 min, and less if you have computer assistance.

Since we are concerned with the properties of filter operators, we can apply to this problem the tools afforded by the Laplace transform [29], which is a mighty instrument to analyze anything governed by differential equations, most notably electronic circuitry, and does so by probing a filter with exponentials and periodic functions.[3] Since our filter is discrete, we are dealing with difference equations, and can therefore use the Z transform, which provides similar analysis capabilities for digital signal processing [29].

---

3) The Fourier transform is the "vibration" subset of the Laplace transform. As such, the Laplace transform is the analytic continuation of the Fourier transform onto the complex plane.

3 Phase Decoding and Reconstruction

Box 3.5: The Z Transform

The bilateral (also two-sided) Z transform of a discrete time signal or filter $I[n]$ is

$$Z(I[n]) = \sum_{n=-\infty}^{\infty} I[n]\zeta^{-n} \qquad (3.54)$$

where n is a discrete index (such as the number of our irradiance sample) and ζ is a "test signal"

$$\zeta = R\exp(-i\omega) \qquad (3.55)$$

R being its magnitude and ω its angular frequency. This maps the filter response onto the complex plane and allows analysis of the filter's response to any possible input signal (including linear combinations). Exponentially increasing input is modeled by $R > 1$ outside the unit circle; decaying input has $R < 1$; and constant input is associated with $R = 1$. Oscillation is added to this envelope by letting $\omega \neq 0$. Hence, sinusoidal input signals (such as the irradiance modulation of a moving fringe pattern) will be characterized by $R = 1$ and $\omega \neq 0$. In the Z transform plane, then, the conventional frequency axis is mapped onto the unit circle – this reflects the fact that positive frequencies above the Nyqvist limit appear as negative frequencies, and approach zero again as the frequency is increased further. Figure 3.13 explains the mapping pictorially.

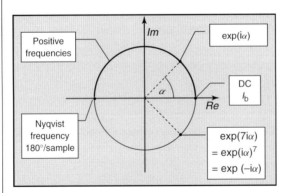

Figure 3.13 Sketch of the frequency axis after a discrete Z transform. The positive frequencies range from 0 (DC) to the Nyqvist limit. If we assume a phase shift of $\alpha = +45°$/sample, the fourth harmonic will be 180°/sample. Frequencies will be aliased back toward zero after this, and the seventh harmonic will be the same as $-45°$/sample. All possible α and harmonics can be graphed in this diagram.

The Z transform of a digital filter can be expressed as a polynomial containing the filter zeroes (where the filter does not detect any signal), divided by a polynomial containing the poles (where the filter output becomes infinite).

3.4 Designing and Analyzing Phase-Shift Methods with the Complex-Polynomial Method

> Poles (usually placed near zeroes) are a means of adjusting filter slopes, but for our purpose of phase detection and error suppression, only zeroes are needed and used.
>
> Usually the $I[n]$ are input signals and therefore real; in that case, the Z transform has mirror symmetry with respect to the real axis. In our application, the $I[n]$ denote the c_n as defined in Eq. (3.28) and the filter responses are *not* the same for positive and negative frequencies. This is a necessary condition for phase detection.

Let us now see how the formalism of the Z transform can be gainfully applied to phase decoding. The periodic fringe signal with its harmonics (if present) can be written as a Fourier series:

$$I(\varphi) = \sum_{m=-\infty}^{\infty} A_m \exp(im\varphi) \tag{3.56}$$

where A_m is the coefficient of the mth harmonic, which is generally complex and thereby includes a phase offset. For pure cosines (zero phase offset implied), $A_m = A_{-m}$, and A_m is real. If the reference phase is now advanced to α, or a multiple of it, we have

$$I(\varphi + n\alpha) = \sum_{m=-\infty}^{\infty} A_m \exp(im\varphi) \exp(imn\alpha)$$

$$= \sum_{m=-\infty}^{\infty} A_m \exp(im\varphi) \exp(im\alpha)^n \tag{3.57}$$

That is, harmonics, if present, undergo a multiplied phase shift according to their harmonic order. Note how the phase shifts appear as nth powers here. Using the phase-stepping procedure given in Eq. (3.12) in the complex notation of Eq. (3.28) gives

$$z = \sum_{n=0}^{N-1} c_n \sum_{m=-\infty}^{\infty} A_m \exp(im\varphi) \exp(im\alpha)^n \tag{3.58}$$

This z is exactly the complex phasor that we have already encountered above in Eq. (3.13). Now we can separate the signal from the filter and the phase-stepping by rearranging Eq. (3.58) to

$$z = \sum_{m=-\infty}^{\infty} A_m \exp(im\varphi) \sum_{n=0}^{N-1} c_n \exp(im\alpha)^n$$

$$= I(\varphi) P\left(\exp(im\alpha)\right) \tag{3.59}$$

with P being a complex polynomial of the form

$$P(x) = \sum_{n=0}^{N-1} c_n x^n \tag{3.60}$$

This is called the *characteristic polynomial* [7] and defines the complex filter that we apply to our irradiance readings. It is unique for each type of phase-shifting formula, and is the Z transform of the discrete complex filter defined by the c_n. With $x = \exp(im\alpha)$, our polynomial will operate on the unit circle in the complex plane, and advancing the phase by n steps is the same as raising $\exp(im\alpha)$ to the nth power. Recall here that the phases of harmonics are being advanced by multiples of α.

The phase-decoding filter will reject any signal content of $I(\varphi)$ for which the polynomial Eq. (3.60) has a zero, and we can find the zeroes easily because every complex polynomial can be broken down, or factorized, into a product of monomials, which are the roots of the polynomial.

As a first example, suppression of I_b as discussed above requires $(x - 1)$ to be a root in the characteristic polynomial of every phase-shifting formula. That puts a zero on $(1 + 0i)$ and blocks the DC part of the signal. Further, the negative signal frequency $\exp(-i\alpha)$ must be suppressed so that only $\exp(+i\alpha)$ is detected. This is the minimal set of conditions to obtain a working phase-shift formula, and if we use these two conditions, we get

$$P(x) = (x - 1)(x - \exp(-i\alpha))$$
$$= (x - 1)(x + i) \quad \text{for} \quad \alpha = 90°/\text{sample} \tag{3.61}$$

Multiplying these roots leads to a polynomial of second order, and sorting the terms into imaginary part (for the b_n) and real part (for the a_n) gives us the coefficients

$$(x - 1)(x + i) = x^2 - x + ix - i$$
$$\Rightarrow b_n = \{-1, +1, 0\}$$
$$a_n = \{0, -1, +1\} \tag{3.62}$$

which is indeed the recipe for the very simple phase-shifting formula

$$\varphi \bmod 2\pi = \arctan\left(\frac{-I_0 + I_1}{-I_1 + I_2}\right) \tag{3.63}$$

We can see immediately that the terms will suppress I_b, as required, and that they have a phase lag of 90°, if $\alpha = 90°$. But how does this formula compare with Eq. (3.20), the "other" minimalistic three-sample 90° formula? With three irradiance samples, and α defined, there are no degrees of freedom left – so why do the equations look different? The answer lies in the phase offset of the formulae (cf. Eqs. (3.23) and (3.24)). Using a phase factor $r = \exp(i - 3\pi/4)$, we alter the phase offset and obtain

$$(x - 1)(x + i) \exp\left(i\frac{-3\pi}{4}\right) = \frac{1}{\sqrt{2}}\left(-x^2 + 2x - 1 + i(-x^2 + 1)\right)$$
$$\Rightarrow b_n = \{+1, 0, -1\}$$
$$a_n = \{-1, +2, -1\} \tag{3.64}$$

after dropping the scaling factor, and we do have Eq. (3.20) now.

3.4 Designing and Analyzing Phase-Shift Methods with the Complex-Polynomial Method

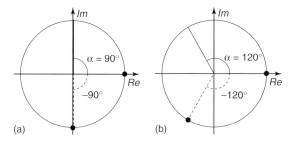

Figure 3.14 Root diagrams of complex polynomials. Roots are marked by black dots on the unit circle. (a) Diagram for Eqs. (3.20) and (3.63). (b) Diagram for Eqs. (3.21) and (3.23).

Of course, just multiplying the polynomial with a real or complex constant does not change its zeroes, and we get the same zeroes for both formulae, as shown in Figure 3.14a.

From the diagram in Figure 3.14a, it is evident that, to detect $+\alpha$, we need to suppress $-\alpha$, and this is why the second root of the polynomial must be at $-i$. In Figure 3.14b, we plot another minimalistic polynomial, for $\alpha = 120°$/sample. The "tuning" of the phase detection is accomplished by shifting the root along the frequency axis, that is, by changing its argument in the polynomial.

Tutorial Exercise 3.9

Write down the polynomial for the diagram in Figure 3.20b and show that it leads to Eq. (3.21) with a phase shift of $-120°$, and to Eq. (3.23) with a phase shift of $-255°$.

Solution:

There are two zeroes in the graph at positions $(1 + 0i)$ and at $-120°$. So the polynomial is

$$(x-1)\left(x + \left(\frac{1}{2} + \frac{\sqrt{3}}{2}i\right)\right) = \frac{1}{2}\left(2x^2 - x - 1 + i(\sqrt{3}x - \sqrt{3})\right)$$

$$\Rightarrow b_n = \{-\sqrt{3}, +\sqrt{3}, 0\}$$
$$a_n = \{-1, -1, +2\}$$

from which follows Eq. (3.21) by permuting the samples, or with

$$(x-1)\left(x + \left(\frac{1}{2} + \frac{\sqrt{3}}{2}i\right)\right)\exp\left(i\frac{-2\pi}{3}\right) = \frac{1}{2}\left(-x^2 + 2x - 1 + i(-\sqrt{3}x^2 + \sqrt{3})\right)$$

$$\Rightarrow b_n = \{+\sqrt{3}, 0, -\sqrt{3}\}$$
$$a_n = \{-1, +2, -1\}$$

For the $-255°$ phase offset

$$(x-1)\left(x+\left(\frac{1}{2}+\frac{\sqrt{3}}{2}i\right)\right)\exp\left(i\frac{-17\pi}{24}\right) = -0.259x^2 - 0.707x + 0.966$$
$$+ i(0.966x^2 - 0.707x - 0.259)$$

It is also very instructive to draw the c_n as phasors in a complex plane to understand how the offsets come about.

A recent addition to phase-shifting analysis is to plot the magnitude of the characteristic polynomial on the frequency axis as a Cartesian diagram [30, 31]. Inserting $z_c = \exp(i\alpha)$ into Eq. (3.60), we can plot the filter responses for any α, as shown in Figure 3.15. Again, the plots will look the same for every possible representation/phase offset of a given formula. In digital-filter parlance, the Hermitian representation is known as a *linear phase* design [32].

From these plots, the behavior of the phase-shifting formula can readily be seen: in Figure 3.15a, we see immediately the zero responses at 0 and -90 (i.e., $+270$) degrees per step. The highest sensitivity occurs at $135°$/step; this is not optimal, but the formula will work as designed in the absence of spurious signals. In Figure 3.15b, the $120°$ formulae can be recognized by the rejection of $-\alpha = -120°$/sample; and in this case, the highest sensitivity is at the design frequency.

We can now multiply additional roots into the characteristic polynomial, thereby increasing its order, where each extra root requires one more phase step. (In a more conventional interpretation, this is equivalent to obtaining more equations and being able to solve for more unknowns.) Suppose we want to create a $90°$ phase-detection formula that is insensitive to the second harmonic and offers some tolerance of phase-shift error. For suppressing harmonics, we must remember that we have a positive and negative frequency for a given Fourier component, and so we must suppress both of these with extra zeroes. When $\alpha = 90°$/sample, we find that the second harmonic is at $\pm 180°$/sample, which is the same point on the frequency circle, so that we need only one extra zero at -1 to suppress the second

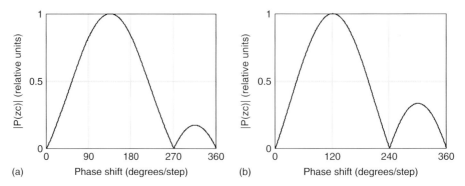

Figure 3.15 Magnitude plots of the complex polynomial for (a) Eqs. (3.20) and (3.63) and (b) Eqs. (3.21) and (3.23).

harmonic:

$$(x-1)(x+i)(x+1) = x^3 - x + i(x^2 - 1)$$
$$\Rightarrow b_n = \{-1, 0, +1, 0\}$$
$$a_n = \{0, -1, 0, +1\} \tag{3.65}$$

which gives the phase-shifting formula

$$\varphi \bmod 2\pi = \arctan \frac{I_2 - I_0}{I_3 - I_1} \tag{3.66}$$

This is an offset-shifted version of Eq. (3.11) and can be converted into Eq. (3.11) or (3.24) by suitable phase factors.

To introduce phase-shift error tolerance, the polynomial should remain close to zero near $\exp(-i\alpha)$, which ensures correct phase detection even when the phase shift is slightly different from the nominal α. (Note how this tolerance does not exist in either diagram in Figure 3.15: the response around the respective zero loci rises quickly.) In complete analogy to how other polynomials work, we can achieve a lower sensitivity around $-\alpha$ if we make the zero there into a second-order zero, thus ensuring that the derivative of the polynomial in its argument, $\exp(im\alpha)$, is also zero at α. So all we have to do is square the monomial defining the root at $-i$ in this case:

$$(x-1)(x+i)^2(x+1) = x^4 - 2x^2 + 1 + i(2x^3 - 2x)$$
$$\Rightarrow b_n = \{0, -2, 0, +2, 0\}$$
$$a_n = \{+1, 0, -2, 0, +1\} \tag{3.67}$$

which leads us exactly, and very quickly, to Eq. (3.25). Root diagrams for Eqs. (3.66) and (3.25) in the customary simplified form are displayed in Figure 3.16, where the double root is symbolized by a dot and a surrounding circle. For a triple root, we would add another circle, and so on.

This simple plotting method is the key to efficient and rapid digital filter design: we can now write down the properties that our phase-shifting formula should have, where for each new requirement we must multiply another monomial (or possibly

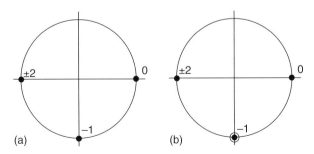

Figure 3.16 More root diagrams of complex polynomials. (a) Diagram for Eq. (3.66), where only one root is needed to suppress the second harmonic. (b) Diagram for Eq. (3.25), where the root at $-i$ is doubled.

two) with the appropriate zero(es) into our formula – and this can reduce again if we choose α so that the highest possible number of monomials come to coincide. The rules are as follows:

1) We always need to suppress the bias irradiance, therefore we always start with $(x - 1)$.
2) To define the positive detection frequency, we must factor in a zero for the negative fundamental.
3) For each harmonic we want to suppress, we must factor in two zeroes (positive and negative frequency) – if these coincide, one extra zero is sufficient.
4) For insensitivity to an nth order phase shift error at a given frequency, we must raise the corresponding root of the polynomial to the power $n + 1$. For example, to compensate for quadratic phase-shift nonlinearity at a given frequency, the polynomial must have a triple root there.
5) To suppress phase-shift errors in the harmonics, we must multiply zeroes into the positive and negative frequencies – again, if these coincide, one extra zero is sufficient. Judicious choice of α can help here as well in minimizing the number of irradiance samples required.

Figure 3.17 gives the frequency-axis plots corresponding to the diagrams of Figure 3.16.

Note also that here, in contrast to Figure 3.15a, the highest sensitivity occurs at the design frequency because the response is pinned to zero at 0 and 180°/step. The use of extra zeroes for nulling the response wherever we want has been demonstrated in Ref. [33].

Depending on the combination of zeroes, sometimes the method generates a symmetrical formula, and sometimes a formula with a phase offset. It is very useful to symmetrize any phase-shifting formula before use, because this provides a very

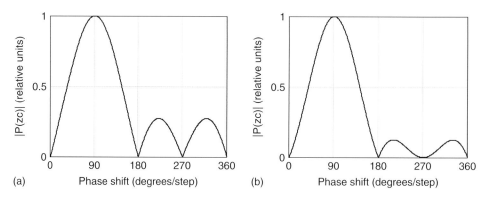

Figure 3.17 Magnitude plots of the complex polynomials for (a) Eqs. (3.11), (3.24), and (3.72), where only one extra root is needed to suppress the second harmonic at 180°/sample and (b) Eq. (3.25), where the root at $-i$ is doubled and zero is touched with zero slope, so that the range of small sensitivity around 270°/step is distinctly wider.

simple check on whether the formula has been coded correctly, and also it is then easy to see whether the sign of the result will be correct.[4] A method to symmetrize any phase-shifting sequence has been given in Ref. [34], which is again very easy to understand from the c_n as complex phasors. The value for the phase factor r results from

$$r = -\arg\left(c_{\frac{N-1}{2}}\right) \quad \text{for} \quad N \text{ odd}$$
$$r = -\arg\left(c_{\frac{N}{2}-1} + c_{\frac{N}{2}+1}\right) \quad \text{for} \quad N \text{ even} \tag{3.68}$$

where we count the c_n from zero. As an example, for Eq. (3.63) we have $N = 3$ and $c_1 = (-1 + i)$, which has the argument $(3\pi/4)$. Multiplication of the polynomial with r then "rotates" all the coefficients and makes the middle coefficient purely real, as required for a symmetrical representation of the phase-detection formula. To symmetrize Eq. (3.66), we take $c_1 + c_2 = (-1 + i)$, so that the same rotation is applied, and we arrive at Eq. (3.24), except that the signs of numerator and denominator are inverted, which means that the phase outputs of the formulae are off by 180°. This comes from the definition of $\varphi = 0$, which was taken to be at the start of the sequence in Section 3.2.3, and is taken to be in the center of the sequence for the symmetrical representation. When the c_n are Hermitian (Eq. (3.29)) and the middle a_n is/are positive, we have a generalized representation of a phase-shifting formula, for which the attribute "canonical" has been suggested, as there is only one of these for each type of phase-shifting formula.

With the toolkit described here, you will be able to tailor phase-shifting formulae exactly adapted for your specific measurement task. In many cases, the error-compensating five-sample formula (3.25) is suitable; another popular formula has 90° phase shift, eight samples, and a quintuple zero at the negative fundamental, and you can now derive and symmetrize it yourself. It is very good at suppressing phase-shift errors in the absence of harmonics.

Tutorial Exercise 3.10

Write down a 90°, eight-sample phase shift formula with a quintuple zero at $-\nu_0$, second-harmonic insensitivity, and convert it to the canonical form.

Solution:

The first zero at $(1 + 0i)$ must be maintained. The phase angle of 90° corresponds to the signal frequency being at $(0 + 1i)$ in the Z transform. Hence a factor $(x + i)$ must appear to the power of 5 for a quintuple zero at the negative signal frequency. Finally, insensitivity to second harmonics means that a zero should appear for $2 \times 90° = 180°$, which translates into a zero at $(-1 + 0i)$. Putting it

---

4) Different commercial phase-shifting interferometers adopt different conventions here, and the author has seen phases coming out inverted after a software upgrade! In modern instruments, the decoding formulae can be defined by the user, so all that needs to be done if the phase comes out inverted, is to invert the b_n in the formula. Then, when all other formulae used are also entered in the symmetrical form, they can all be treated the same way, which will help to prevent errors and will make maintenance and trouble-shooting easier.

all together yields

$$(x-1)(x+i)^5(x+1) = x^7 - 11x^5 + 15x^3 - 5x + i(5x^6 - 15x^4 + 11x^2 - 1)$$

$$\Rightarrow c_{\frac{N}{2}-1} + c_{\frac{N}{2}+1} = 15 - 15i$$

$$\Rightarrow r = \frac{\pi}{4}$$

$$(x-1)(x+i)^5(x+1)\exp\left(i\frac{\pi}{4}\right)$$
$$= \frac{1}{\sqrt{2}}\left(\begin{array}{c}x^7 - 5x^6 - 11x^5 + 15x^4 + 15x^3 - 11x^2 - 5x + 1 \\ +i(x^7 + 5x^6 - 11x^5 - 15x^4 + 15x^3 + 11x^2 - 5x - 1)\end{array}\right)$$

$$\Rightarrow \arctan\varphi \bmod 2\pi = \frac{-I_0 - 5I_1 + 11I_2 + 15I_3 - 15I_4 - 11I_5 + 5I_6 + I_7}{+I_0 - 5I_1 - 11I_2 + 15I_3 + 15I_4 - 11I_5 - 5I_6 + I_7}$$

Plotting $|P(z)|$ will reveal why this formula is so resistant to phase-shift errors.

3.5
Sources and Removal of Errors

We will now discuss a few errors that can impact on the reliability of phase-shifting measurements and also come to an assessment of how important they are in practice. As a general rule for interferometry, air movement (particularly turbulence) must be kept to a minimum. This is done by making the beam path as short as possible and shielding it from airflow. Further, the instrument must be calibrated if accuracies equal to or better than the reference surface are desired.

3.5.1
Phase-Shift Miscalibration and Calibration

We have seen above that a large part of the behavior of phase-shifting formulae can be understood in the frequency domain, and we can safely say that phase-shift miscalibration is the most important error source. A general and exact form of the error is [35]

$$\Delta\varphi = \arctan\frac{s\cos\varphi - c\sin\varphi}{1 + c\cos\varphi + s\sin\varphi} \tag{3.69}$$

where $s = \Delta N/M_1$ and $c = \Delta D/M_1$ are the normalized errors in the numerator and denominator terms of the phase-decoding expression used. Essentially, this gives the 2φ-cyclic error that we have discussed before, and we should remember here that, to a good approximation, the following rules apply:

- If the phase-shift formula has no error tolerance, the error increases linearly with the phase-shift deviation and the peak-to-valley (PV) phase error is in the same order of magnitude as the phase-shift deviation.

- If the phase-shift formula has error tolerance, the error increases quadratically with the phase shift deviation, and is always smaller than in the uncompensated case – significantly so for practically relevant phase-shift errors of 0–20° per sample.
- The error suppression can be improved further by factoring extra zeroes into the characteristic polynomial, to third order, and so on, and the error plot becomes flatter and flatter around the nominal frequency.

If four or more irradiance samples are taken, the system of equations contains enough information to solve for the actual value of the phase shift and therefore to obtain a figure for calibration. This is implicit in the self-calibrating Carré formula [7], but can also be done very easily as a by-product of the five-frame formula (3.25) and lead to the simple expression

$$\alpha = \arccos \frac{I_4 - I_0}{2(I_3 - I_1)} \tag{3.70}$$

which is frequently used in phase-shifting interferometers for calibration. The expression leads to large errors for $I_3 \cong I_1$ and becomes altogether undefined for $I_3 = I_1$, so that in practice it is easiest to introduce a few fringes of tilt and to ignore the small regions where the calculation becomes unreliable. Figure 3.18 gives an example of some error fringes and of a phase-shift histogram for a transmission sphere of high numerical aperture. The histogram is of course much narrower for flat surfaces.

An interesting observation from Figure 3.18 is that the annulus of correct phase shift is not centered in the aperture, which indicates that besides the piston movement of the transmission sphere, a slight tilt takes place as well (see discussion in Section 3.3.1). This is almost inevitable when using PZT stacks, and will be the same whether spherical or flat surfaces are used, so that the phase shift will never be perfectly uniform across the image. Therefore, even when the mean phase shift is exactly 90°/sample, the phase-shift formula should have some tolerance of slight errors to compensate for this tilt motion.

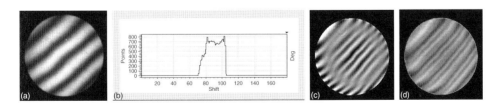

Figure 3.18 Phase-shift calibration and measurement results with NA = 0.68 transmission sphere. (a) Fringe pattern used for calibration. (b) Distribution of phase-shift angles after adjusting for an average of $\alpha = 90°$/sample. (c) $\Delta\varphi(x,y)$ from using Eq. (3.11) in pseudofringes (gray scale spans ±5 nm or about ±6° of fringe phase). Note that $\Delta\varphi(x,y)$ occurs at twice the frequency of the original fringes. In the central region, $\alpha > 90°$/sample; the gray ring with no error marks a small zone of correct phase shift. Outside that zone, $\alpha < 90°$/sample, with a sign change of $\Delta\varphi(x,y)$. (d) $\Delta\varphi(x,y)$ from using Eq. (3.37) on the same irradiance data; same gray scale, smaller errors.

Also, there may be more error sources than just linear phase-shift miscalibration. If we use Eq. (3.37), with third-order phase-shift error compensation, the error fringe pattern changes to that shown in Figure 3.18d. Clearly, the annular zone has completely disappeared and the variation in phase-shift angle does not play a role anymore. The remaining fringes indicate either the presence of a nonlinear signal (i.e., not a pure cosine but containing higher harmonics), or nonlinearity in the phase steps, or both.

3.5.2
Signal Nonlinearity and Harmonics

As should be clear from our discussion of the signal as a Fourier series, a nonlinear signal can be interpreted as a mixture of harmonics superimposed on the fundamental cosine signal. In interferometry, detector nonlinearity is not a practical issue – provided the so-called gamma curve of a charge-coupled-device (CCD) camera is set to 1, but the default is often 0.45. This is a legacy setting and has its origins in the nonlinearity of cathode ray tube (CRT) monitors which a gamma setting of 0.45 was supposed to compensate. It scales the output according to $U_{out} \propto I_{in}^{0.45}$, which may display the fringes more realistically on screen but violates the assumptions made in deriving the equations.

In recent years, complementary metal–oxide–semiconductor (CMOS) cameras have become more widespread, whose pixels are essentially photodiodes. The strong point of CMOS cameras running in instantaneous photocurrent mode used to be their logarithmic response and the resulting wide dynamic range; however, they have become available in integrating, "linearized" versions as well. It has been shown that the linearity is not always as good as that of CCDs [36], and thus it is worthwhile to characterize the camera used before relying on it as part of a linear signal chain. Other elements known to be strongly nonlinear are LCD projectors, as shown in Section 3.3.3.

All such cases can be treated by analyzing the distorted signals for their harmonic content and then using a phase-decoding scheme that is well suited to suppress the strongest harmonics. If a signal as in Figure 3.12 were to be used without calibration, the first five Fourier coefficients (normalized to the fundamental) would be {1, 0.56, 0.39, 0.30, 0.25}, and thus considerable effort would have to go into suppressing harmonics.

In the case of spurious fringes due to multiple reflections, harmonics from different reflection sequences appear on the signal. For example, when we are trying to measure a high-reflectivity surface such as a mirror in a Fizeau interferometer, we will not get just the single reflection off the mirror surface, but also a re-reflection on the reference surface, followed by another reflection off the mirror, and so on – so that in such an application it will be beneficial to pay attention to the suppression of odd (third, fifth, etc.) harmonics in particular.

It is worthwhile trying to get to know the optical system well enough to understand the spectrum of the signal, as the characteristic polynomial method then allows very selective targeting of certain error sources, and large improvements of the measurement uncertainty are thereby possible at little expense.

3.5.3
Vibrations

Vibrations pose a problem for both TPS and SPS, but much less so in the latter case. If sufficient laser power can be made available, vibration problems in the spatial techniques can be solved by making the exposure time short enough, either in the illumination (using a pulsed laser) or in the recording (using a short shutter time). In TPS, the situation is more complicated. Vibrations cause phase-shift errors in the frame sequence, which depend on the vibration amplitude and frequency. A general rule of thumb is that, if the vibration frequency is an integer multiple of the temporal sampling frequency, the problem disappears; if it is a half-integer, the errors are largest; but the details depend on the phase-shift scheme chosen. Regardless of the phase-shift formula used, low-frequency vibrations (which are unfortunately the hardest to suppress) tend to cause larger errors [16].

These rules suggest two possible approaches if vibrations cannot be eliminated by mechanical means: (i) tune the temporal recording frequency to the dominant vibration frequency, which can be done with timed triggering of the camera and matching phase-shift rate; (ii) measure the vibration spectrum beforehand, for example, with an accelerometer, and design a phase-shift formula that suppresses effectively the vibrations found. Note that, for this purpose, detection zeroes need not be at integer multiples of the phase shift: they can be put anywhere on the complex unit circle in complex-conjugate pairs. This approach may use up a lot of zeroes in the characteristic polynomial and it is therefore best tailored to the particular application. For high-frequency vibrations, it may be beneficial to increase the exposure time to average out their effect. This costs some signal modulation and is therefore applicable only to vibrations of low amplitude.

There are also phase-measurement approaches that can determine the fringe phase from a set of randomly phase-shifted fringes [37, 38], so that one can even rely on vibration alone to provide the phase shift; but these methods require a new and different computation for each measurement taken, and therefore we do not treat them further here.

3.5.4
Random Noise

Each irradiance measurement is subject to random noise, which is composed of photon shot noise and electronic readout noise. Photon shot noise is usually underestimated as a source of errors. Regardless of how low the dark current of the sensor and how good the readout electronics are, once there is signal on the sensor, that signal itself contains a lot of noise. The standard deviation for N_E, the number of photoelectrons generated on a pixel, is [36]

$$\sigma_E = \sqrt{N_E} \qquad (3.71)$$

and if we take N_D as the number of electrons generated by dark current (due to thermal effects), the SNR will be

$$SNR_{eff} = \frac{N_E}{N_D + \sqrt{N_E}} \qquad (3.72)$$

so that, in the best case of a very short exposure (where we can neglect N_D) and pixel saturation (so that $N_E = N_{max}$), our best attainable SNR is

$$SNR_{max} = \sqrt{N_{max}} \qquad (3.73)$$

where N_{max} is between 25 000 and 50 000 for CCD pixels, and around 160 000 for CMOS pixels [36]. Thus, the limits on the achievable SNR by photon shot noise alone are between 160 and 225 (7.8 bits) for CCDs,[5] and around 400 (8.6 bits) for CMOS sensors (which, however, are slightly nonlinear).

Therefore, a digital resolution of 10 or even 12 bits (1024 or even 4096 gray levels) is not physically justified in most cases; indeed, it has been stated and demonstrated that sometimes "resolution of 10 or 12 bit only provides a better resolution of the noise" [36], and also takes up a lot of storage space unnecessarily. More by coincidence than anything else, the 8-bit digital format is almost ideal for storing irradiance readings. Note also that, because of Eq. (3.73), adding one significant bit of resolution requires increasing N_{max} by a factor of 4! On the other hand, when the irradiance I is decreased, the SNR decreases only as $\sqrt{I}$, so that even at a mean gray level of 128 (for 8-bit digitization), we still have 70% of the maximal SNR left.

Photon shot noise is by far the largest influence on the irradiance reading noise, and we can safely neglect the readout noise in practice (because long exposure times are best avoided anyway).

3.5.5
Digitization Noise

As stated above, we can safely assume that our true irradiance resolution corresponds to approximately eight bits. Digitization noise means the deviation of the integer readout from the true irradiance; just like rounding numbers (or simply truncating the unwanted digits), the error will be sometimes negative, sometimes positive.

If we look at irradiance alone and utilize the entire dynamic range for the fringe pattern, we would have a resolution of 256 gray levels for half a fringe (remember

---

5) Cooled CCD sensors with large pixels can store more photoelectrons and thereby exceed these specifications, but they are usually very expensive, specialized equipment for astronomy.

that one full fringe, $\Delta\varphi = 2\pi$, would go from white to black and back to white). But because $dI/d\varphi$ is zero at the fringe extrema, and maximal at the half-way points, the sensitivity of the irradiance signal to the phase varies across the fringe, and so does the influence of digitization noise.

Phase shifting removes this problem, because each point will go through offset phases with high and low $dI/d\varphi$ during the phase-shifting sequence. The problem has been treated analytically [39], and for four 90° phase steps the resulting standard deviation is

$$\sigma_\varphi = \frac{1}{\sqrt{3} \times 2M_I} \tag{3.74}$$

where $2M_I$ gives us the full range of gray levels actually used. Let us estimate roughly just how important the digitization noise is. For $2M_I = 224$ gray levels, we find $\sigma\varphi = 0.15°$ – assuming 548 nm light and one fringe per wave of path difference, this converts to 0.22 nm, which is safely below the level of other error sources. The digitization noise will decrease further (roughly as $\sqrt{N}$) if we use more phase steps per irradiance cycle. Consequently, digitization noise is not a large source of error in practice, and will exceed the 1 nm level only when $2M_I$ drops to about 50 gray levels.

3.6
Phase Unwrapping

We have now spent a great deal of effort on finding a way to get a good phase measurement with low systematic errors and low noise – the final step is to turn the wrapped phase map into a continuous surface map (or a discontinuous one with the correct step height). Unwrapping low-noise sawtooth maps from smooth fringes, for example, from reflective surfaces or fringe projection, can be very easy; if, on the other hand, the fringes are very noisy (typically in speckle interferometry), unwrapping can become very complicated, and there are volumes of research [40] on the best way to unwrap noisy phase maps. We start with the spatial unwrapping technique, and then discuss temporal unwrapping techniques for tracking high-speed phase measurements, as well as techniques for discontinuous phase distributions.

3.6.1
Spatial Phase Unwrapping

As depicted in Figure 3.19, the unwrapping is done by looking for the $0 \leftrightarrow 2\pi$ jumps in the wrapped phase map, where the criterion is

$$\varphi(x_{n+1}, y_n) - \varphi(x_n, y_n) < -\pi \Rightarrow \text{add } 2\pi \text{ to } \varphi(x_{n+1}, y_n); \text{ increment } k$$
$$\varphi(x_{n+1}, y_n) - \varphi(x_n, y_n) > \pi \Rightarrow \text{subtract } 2\pi \text{ from } \varphi(x_{n+1}, y_n); \text{ decrement } k.$$
$$\tag{3.75}$$

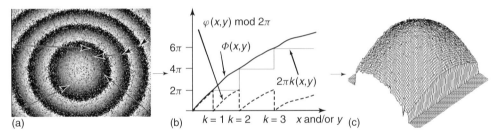

Figure 3.19 Sketch showing the principle of spatial phase unwrapping. (a) $\varphi(x,y)$ in sawtooth representation; arrows: examples of unwrapping paths leading to an unwrapping history as in (b). (b) Schematic of unwrapping along a line (but not necessarily along just one co-ordinate). (c) Final measurement result: the height map after conversion according to Eq. (3.1).

In other words, phase jumps of more than half a fringe are not allowed and are taken to indicate that the next-order fringe begins. Note that this is for one dimension only, and a similar operation must be carried out in the vertical direction to unwrap the entire image to a continuous distribution $\Phi(x, y)$.

What happens if we swap the x and y unwrapping steps? Will the results always be the same? Unfortunately, this is not always the case; in the presence of noise or spurious fringes, the criterion Eq. (3.75) is not particularly good. We must use 2-D information, that is, look at all eight neighbors of a given pixel (on the four sides and along the two diagonals), to decide whether or not, and where, the fringe order changes. Particularly in sawtooth images such as Figure 3.19a, which has speckle noise, this can become quite difficult because of the presence of so-called singularities in the phase map. These are points where the phase is undefined, and can be artifacts due to noise and other errors, or actual physical features (e.g., in irregular wave fields such as speckle patterns). Wherever there are points of zero field amplitude in a speckle pattern, the phase becomes undefined, any straight path through that point will result in a 180° phase jump, and successive wavefronts are not separated sheets but a continuous surface interconnected by a helix [41]. Figure 3.20 shows how this affects the fine detail of a speckle pattern.

Regardless of how cleverly we try to do it, the wavefront in Figure 3.20 cannot be transformed into a smooth wavefront, because it actually is not one: the phase dislocations are real, and are analogous to screw dislocations in crystals, where the atom layers can do the same as our wavefront sheets are doing here.

The problem can become worse when we subtract two such maps, as is done in speckle interferometry, because this will double singularities that have moved between the initial and final object states. So, to extract useful information from speckle phase maps, and to correct errors in smooth-surface phase maps, we must find a way to deal with singularities. This process starts by creating a map of singularities, as illustrated in Figure 3.21.

To detect singularities, the phase difference is accumulated over a closed path around (or including) the point of interest; the simplest version of this is

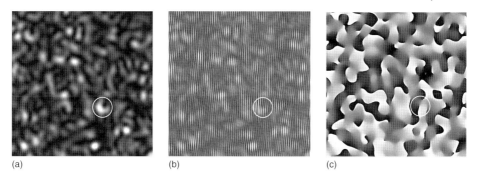

Figure 3.20 (a) Field of well-resolved speckles. (b) The same speckle field in a spatially phase-biased interferogram: phase singularities are recognizable by bifurcations of the fringes. (c) Sawtooth map calculated from (b); singularities abound. The white circle marks the same region in each image. (Images from Ref. [42].)

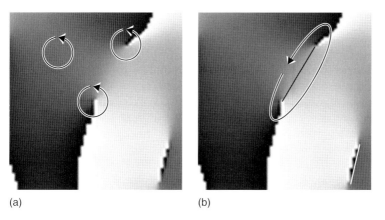

Figure 3.21 Magnified version of singularity region in Figure 3.20. (a) Singularities are detected by closed path summations. (b) "Branch cuts" are set for the unwrapping path and singularities are removed in opposite pairs. Here, we have two pairs in the field of view, one in the center and one at the lower right.

$$k = \begin{pmatrix} \{\varphi(x_l, y_j) - \varphi(x_{l+1}, y_j)\} + \{\varphi(x_{l+1}, y_j) - \varphi(x_{l+1}, y_{j+1})\} \\ + \{\varphi(x_l, y_{j+1}) - \varphi(x_l, y_j)\} + \{\varphi(x_{l+1}, y_{j+1}) - \varphi(x_l, y_{j+1})\} \end{pmatrix} \quad (3.76)$$

where $\{\bullet\}$ denotes a counting operator as in Eq. (3.75), l and j denote the pixel index in x and y, and the terms are spatially arranged to reflect the pixel positions in the counterclockwise accumulation loop expressed by the sum. If $S = 0$, the phase surface is smooth, as in the upper left circle in Figure 3.21a; if $S = 1$, the path includes a left-handed singularity (top circle in Figure 3.21a), and if $S = -1$, the singularity is right-handed (bottom circle in Figure 3.21a). Phase singularities always lie on lines (filaments) of zero amplitude within a wave field. These dark filaments do not suddenly begin and end, but are closed loops [41], and phase singularities are seen to appear and disappear in pairs of opposite handedness, depending on how a dark filament touches or pierces the plane of observation.

This phenomenon gives the physical justification and the recipe to remove as much ambiguity as possible even in the presence of singularities: we must find a way to force the unwrapping path to go around two opposite singularities at a time. To achieve this, the nearest neighboring opposite singularities are paired up and a "branch cut" is set between these two points. The term *"branch cut"* has been introduced to indicate that, as the pixels are processed in two dimensions, the unwrapping path can branch out over the image (and should, to make the unwrapping process faster), but is only allowed to enclose pairs of singularities, not single ones, so that the accumulated phase remains free from inconsistencies. A popular way to guide the development of the unwrapping path or "tree" into which it will branch out is to determine the reliability of all pixels, for example, according to Eq. (3.15), or the phase difference to neighboring pixels, to go along the very best paths first to obtain global consistency and then to deal with smaller area elements in later steps. This will also keep errors confined if and where they occur.

The branch-cut method alleviates the problem with speckle interferometry mentioned above, because if a dislocation moves in the second phase map, which is then subtracted from the first, the artifact will be a pair of opposite singularities in close proximity, which will disappear together (and of course we are hoping for other noisy features in a speckle phase-difference map to do the same). In general, however, any singularity can be paired up with any other singularity of opposite handedness besides the nearest neighbor, and defining the best combination with the smallest total length of branch cuts is a difficult optimization problem. Also, there is not always a way to match all singularities; if the total number of singularities is odd, at least one of them will remain unpaired, but more typically there will be several singles near the edges of the image. These must be assigned branch cuts as well, which are then continued toward a virtual singularity outside the image borders. This case will occur predominantly in measurements of smooth surfaces, where there are few if any singularities, and sometimes just one.

Another way to make the task of unwrapping much more robust is to filter the sawtooth images before unwrapping. Since they always contain $0 \leftrightarrow 2\pi$ edges, simple low-pass filtering cannot be used, as that will blur these edges. Using a median filter is a better choice, which at least maintains the sharpness of the edge, but it will be shifted a little. The very best choice is to start with the "sine" and "cosine" images (the numerator and denominator inputs to the arctan formula), filter those separately with an ordinary or apodized low-pass filter, and then calculate the arctangent. We can create "sine" and "cosine" images even after the original measurements have been discarded, because the phase is represented in the sawtooth image, and that is of course enough information to create the quadrature images. These can be filtered again and processed to a sawtooth image, and the process can be repeated as many times as desired, obviously giving up some

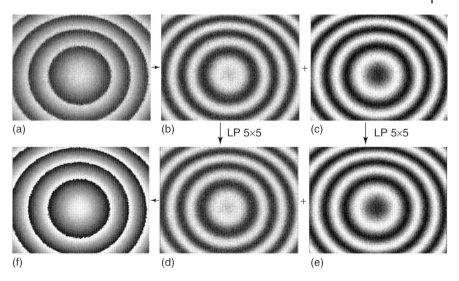

Figure 3.22 Sine–cosine filtering of Figure 3.19a. Left: original sawtooth image; center left: decomposition into sine and cosine fringes, center right: after 5 × 5 low-pass filtering; right: smoothed phase map calculated from the smoothed sine/cosine fringes.

information but preserving the sawtooth edges [43]. Applying a small filter mask (e.g., 5 × 5) several times is very often better than trying to do all the necessary smoothing in one step. Phase maps should be filtered only just enough to allow error-free unwrapping. Figure 3.22 gives an example of the filtering procedure.

3.6.2
Temporal Phase Unwrapping

The complications of spatial unwrapping can be circumvented if the phase on each pixel can be tracked in time rather than space, as depicted in Figure 3.23. If Eq. (3.75)

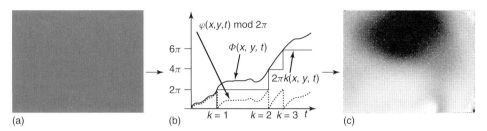

Figure 3.23 Principle sketch of temporal phase unwrapping. (a) Empty/reset phase map. (b) Time evolution of the phase at a given pixel, with unwrapping steps triggered at the $0 \leftrightarrow 2\pi$ phase jumps. (c) Unwrapped result, immediately available at the end of the measurement.

is to remain applicable, we must ensure that the phase jumps found indicate an increment or decrement of the fringe order, rather than an actual change in the object phase by more than half a fringe. In other words, phase measurements must be taken fast enough to ensure sampling of $\varphi\,(x,\,y,\,t)$ according to the sampling theorem. This can be done with TPS systems and dedicated electronics in quasi-real time, but is easier to implement in SPS [44].

In practice, the data processing is stabilized by median filtering the phase maps before determining $k(x,\,y,\,t)$, specifically in speckle interferometry where speckle patterns decorrelate over time. However, this opens the door to spatial spreading of errors from faulty pixels, and can be seen in Figure 3.23c as a slight grainy noise on the height map.

One way to suppress this problem is to store the measured phase distributions as a temporal stack, and then to apply 3-D phase unwrapping to the generated volume of phase data [45]. In this case, the structures to be avoided by the unwrapping algorithm are not points in a plane, but lines in a volume; more specifically, these lines are loops, just like singularities in speckle patterns form loops in space. The problem is then to define not branch-cut lines, but branch-cut surfaces that the (now 3-D) unwrapping tree is not allowed to cross. The algorithm can become quite complex, but draws on the same quality parameters as the 2-D techniques, and helps to suppress errors in any given phase slice $\varphi\,(x,\,y,\,t)$ by utilizing information from data after and before t.[6]

3.6.3
Multiwavelength Unwrapping Techniques

When measuring objects with step discontinuities, as can occur frequently in fringe projection, these are natural "branch cuts" in the image that can under no circumstances be crossed by a spatial unwrapping routine. However, the height information can still be retrieved with a temporal or spatial multiwavelength technique. The principle is the same as in heterodyne interferometry: we use two slightly different fringe periods (now in space rather than time) and get a much lower beat frequency.

The two wavelengths are then chosen so that the number of waves in the required height range Δh differs by (at most) one:

$$\left|\frac{\Delta h}{\lambda_1} - \frac{\Delta h}{\lambda_2}\right| = 1, \text{ or } \left|\frac{\Delta \varphi_1}{\lambda_1} - \frac{\Delta \varphi_2}{\lambda_2}\right| = 2\pi \qquad (3.77)$$

where we assume a sensitivity of one fringe per wavelength, and will account for the sensitivity below. This gives us the synthetic wavelength

---

6) This principle is also used in digital video encoding, where data are compressed in the spatial coordinates as well as in the time domain. This is the reason why the image has large errors and visibly settles down within a few frames after a discontinuous change, such as a scene cut.

$$\Lambda = \Delta h = \frac{\lambda_1 \lambda_2}{|\lambda_2 - \lambda_1|} \tag{3.78}$$

and with

$$\varphi_1(x,y) = \frac{2\pi}{\lambda_1} h(x,y) \quad \text{and} \quad \varphi_2(x,y) = \frac{2\pi}{\lambda_2} h(x,y) \tag{3.79}$$

we can evaluate the beat phase

$$\varphi_1(x,y) - \varphi_2(x,y) = 2\pi h(x,y) \left(\frac{1}{\lambda_1} - \frac{1}{\lambda_2} \right)$$

$$\Phi(x,y) = \frac{2\pi}{\Lambda} h(x,y), \tag{3.80}$$

which is nothing but Eq. (3.1) with a new wavelength, and can now be scaled for sensitivity as given by the geometry.

This method is known as *contouring*, and sometimes must be done with more than two wavelengths, because λ_1 and λ_2 cannot be spaced arbitrarily closely to make Λ larger: the absolute noise in the phase measurements will stay roughly the same as the wavelength is varied from λ_1 to λ_2, but the noise in $h(x,y)$ scales as Λ. It is therefore better to use three wavelengths and synthesize two phase measurements $\Phi_{12}(x,y)$ and $\Phi_{23}(x,y)$ with lower relative noise first, and then to evaluate the beat frequency between the two new signals to generate the final $h(x,y)$.

The contouring approach works well even when the phases fluctuate between the two measurements, because $\Phi(x,y)$ will in that case only acquire an offset, which is irrelevant.

In fringe projection, the relative phases of the projected fringe patterns are very well known, and are not subject to fluctuations. This enables another approach, which relies on number theory and is often called the *method of excess fractions*. Let us rewrite Eq. (3.77) for fringe projection as

$$\left| \frac{\Delta z}{P_1} - \frac{\Delta z}{P_2} \right| C = 1 \tag{3.81}$$

where the P_k are the fringe periods as in Section 3.3.3, and we introduce a sensitivity factor C. Condition (3.81) can be fulfilled by any pair of fractions whose difference is 1, but the fractions themselves can be nonintegers. As an example, if we take $P_1 = 5$ pixels and $P_2 = 7$ pixels and increase Δz, the first time that Eq. (3.81) is fulfilled is at $\Delta z C = 17.5$ pixels, where the fractions take the values 3.5 and 2.5, respectively, so that our Λ would be 17.5 pixels and could be calculated with Eq. (3.78). However, there is extra information in the remainders of the integer divisions (which we abbreviate $R(\bullet)$): we can keep tracking the remainders from zero until both become zero again together – and only after that will the combination of remainders repeat. The reason why this works is that $R(\Delta z/P_1)$ is a sawtooth function mod 5 (where we partition one fringe into five parts) and $R(\Delta z/P_2)$ is a sawtooth function mod 7 (where we partition one fringe into seven parts). Because the numbers are relative primes, the sawtooth functions do not coincide again until their product is reached. For our example, this would happen after $\Delta zC = 35$ pixels, so that the measurement range is extended by a factor of 2, without increasing the noise.

▪ Tutorial Exercise 3.11

Write down $R(\Delta z/P_1)$ and $R(\Delta z/P_2)$ for ΔzC from 0 to 35 and convince yourself that no combination occurs twice.

Solution:

There are two periodicities, one with 5 pixels, the other with 7 pixels. Written side by side

ΔzC	0 1 2 3 4 5 6 7 8 9 10 11 12 13 14 15 16 17 18 19 20 21 22 23 24 25 26 27 28 29 30 31 32 33 34 35
$R(\Delta z/P_1)$	0 1 2 3 4 0 1 2 3 4 0 1 2 3 4 0 1 2 3 4 0 1 2 3 4 0 1 2 3 4 0 1 2 3 4 0
$R(\Delta z/P_2)$	0 1 2 3 4 5 6 0 1 2 3 4 5 6 0 1 2 3 4 5 6 0 1 2 3 4 5 6 0 1 2 3 4 5 6 0

What this means is that, by projecting and phase-shifting two fringe patterns of slightly different pitch, we can infer the absolute fringe order and unambiguously assign surface heights over a range of several times their spacing. The phase-measurement accuracy needed to unwrap the measurement properly is only 1/7th and 1/5th of a fringe in the respective patterns, but this is only to determine the correct $k(x, y)$. The full sensitivity of the phase-shifting measurements is maintained, and φ_1 and φ_2 could even be averaged after appropriate scaling, but this is risky since periodic errors in the two patterns would undergo a beat between cancellation and doubling.

The unambiguous unwrapping range can be made larger by using larger numbers, such as $P_1 = 29$ pixels and $P_2 = 31$ pixels, for a measurement range of $\Delta zC = 899$ pixels (29×31); but in this case, the larger fringe spacing results in less accurate height measurement, and also the required phase resolution is 1/30th of a fringe, which can become difficult if the object cannot be illuminated well or its surface is dark.

Another solution using three fringe spacings is $P_1 = 9$ pixels, $P_2 = 10$ pixels, and $P_3 = 11$ pixels [46], which takes advantage of the fact that the numbers need to be only relative primes, and covers an unambiguous unwrapping range of $\Delta zC = 990$ pixels ($9 \times 10 \times 11$) with finely spaced fringe patterns, which falls only slightly short of a horizontal pixel count of 1024, as offered by a medium-resolution projector.

Unwrapping errors in this scheme (faulty assignments of one or more remainders) have drastic consequences, in that they can produce a vastly different fringe-order assignment. For our example with $P_k = \{9, 10, 11\}$, a remainder set of $\{5, 6, 7\}$ puts us on pixel 986, whereas a remainder set of $\{5, 5, 7\}$ belongs to pixel 95. High reliability is therefore very important, and the method of excess fractions has been used in a four-wavelength interferometer system to measure absolute distance with 0.4 ppm accuracy [46].

3.6 Phase Unwrapping

Another method to perform absolute unwrapping is called the *hierarchical approach*, where the idea is to construct a geometric sequence of fringe pattern spacings which records the smallest amount of data needed for successful unwrapping, and therefore optimizes the measurement time [47]. It starts with the smallest fringe period yielding a useful measurement P_0, and then increases the fringe period in a geometric sequence by a factor F:

$$P_{k+1} = F \times P_k, 0 \leq k \leq K \tag{3.82}$$

so that K denotes the index of the broadest projected fringe pattern (which must contain one fringe or less). Figure 3.24 presents an example for a hierarchical phase measurement.

The unwrapping is then carried out as follows: the phase values of each individual phase map φ_k are multiplied by F^k, which converts them to phase maps Φ_k, all of which have the same phase slope: that of the first densest one with the most fringes. Starting from Φ_K, which is free of $0 \leftrightarrow 2\pi$ transitions, step functions S_k are then determined to generate the absolute phase maps $F^k \varphi_k + S_k = \Phi_k + S_k$ that should equal Φ_K but have higher precision due to the higher sensitivity afforded by the smaller P_k. This will work only if the measurement errors from step to step are smaller than $180°$, which is again the familiar unwrapping criterion, and constitutes an upper bound to F. If the phase-measurement noise is denoted by ε, F is found by the condition

$$\varepsilon F + \varepsilon \leq 180° \Leftrightarrow F \leq \frac{180°}{\varepsilon} - 1 \tag{3.83}$$

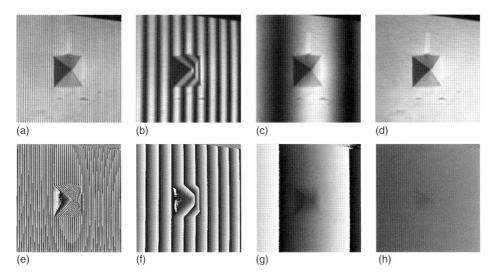

Figure 3.24 Hierarchical fringe-projection sequence on demonstration object, here with $F = 6$. (a–d) Fringe patterns. (e–h) Wrapped phase maps.

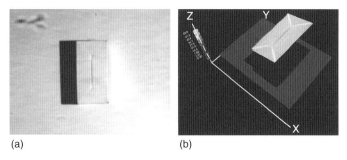

Figure 3.25 (a) Image of step object (cardboard box) on flat panel. (b) Unwrapping result; height of box is reconstructed correctly even though the bottom of the box is not connected to the flat panel by any measurement data. Fine detail of the box is also resolved.

because errors in φ_{k+1} will be multiplied by F to enable unwrapping, and will compound the errors already present in φ_k. If X is the number of projector pixels normal to the projected fringes, the number of measurement steps is determined by

$$P_0 F^K \geq X \tag{3.84}$$

which means that we have to find the first K that gives a fringe period equal to or larger than X, that is, we have one fringe or less on the projector and Φ_K is therefore unique:

$$K \geq \frac{\log X - \log P_0}{\log F} \tag{3.85}$$

Assuming $\varepsilon = 20°$ (which is a 3σ value for very high unwrapping reliability), $P_0 = 8$ pixels, and $X = 1024$, we get $K \geq 2.33$, so that we need to project and phase-shift four patterns: $P_0 = 8$ pixels, $P_1 = 64$ pixels, $P_2 = 512$ pixels, $P_1 = 4096$ pixels. Note that the ultimate precision is determined by P_0; but still, F could be reduced a little in this case to decrease the noise in Φ_K.

Figure 3.25 shows an unwrapping result of a discontinuous surface that was measured with the hierarchical unwrapping method.

Problems

3.1 Letting φ run from 0 to 2π, plot the I_n, and the numerator and denominator terms composed from them, for Eq. (3.22) for $N = 3$ and 4. Do this in different diagrams to keep it simple. Why do the b_n go as $-\sin\alpha$, but the numerator comes out as $+\sin\varphi$?

3.2 Again letting φ to run from 0 to 2π, plot the I_n, and the numerator and denominator terms composed from them, for Eq. (3.63), and demonstrate that it measures $\varphi - 3\pi/4$.

3.3 Convert Eq. (3.66) into Eqs. (3.11) and (3.24) by using phase factors in the characteristic polynomial. Hint: these factors are multiples of 45° so the rotations are a bit easier to do than in Tutorial Exercise 3.9.

3.4 Again letting φ run from 0 to 2π, plot a Lissajous figure of $\sin\varphi$ against $\cos\varphi$ in one graph, and $\varphi_{\text{calc}} = \mathtt{atan2}(\sin\varphi/\cos\varphi)$ against φ in another graph. Then vary the amplitude or phase offset of one of the terms a little. You will see how the Lissajous figure deforms to an ellipse and how φ_{calc}, previously linear in φ, starts to oscillate, with two full cycles in the $[0, 2\pi]$ range of φ. Plot $\varphi_{\text{calc}} - \varphi_i$ to see this more clearly, and watch the ordinate as you do this. Compare the output error to the input error.

3.5 Print Figure 3.20c magnified, pair up the dislocations, and draw a consistent unwrapping path through the image. You can do this in multiple ways, so make a few copies.

4
Experimental Stress Analysis – An Overview
Krishnamurthi Ramesh

4.1
Introduction

Experimental stress analysis is a vast field, and several experimental methods have been developed and several new ones are being developed to suit various situations. From a mechanics point of view, what quantities one expects to measure from experimental techniques is brought out in this chapter. A basic understanding of what is stress and strain is important, and these are discussed first. Material characterization is necessary to apply the principles of mechanics, and the rudiments of a simple tension test are discussed to highlight the extent of elastic and plastic regions in terms of strains. The concept of stress concentration and the role of inherent flaws affecting the performance of structures, as well as the need for determining a new set of information from experimental methods, are then presented. Although many experimental techniques exist, a question may arise in the minds of readers as to why there are so many of them, and also whether they are redundant. To answer this, the fundamental difference between analytical, numerical, and experimental methods is highlighted. In contrast to analytical or numerical methods, each of the experimental techniques can provide only a particular type of information, and these are then summarized. A brief discussion of how to go about selecting an experimental technique is discussed next. Finally, two case studies are presented: the first one shows how innovative use of simple techniques can solve a complex industrial problem and the second one focuses on identifying a comprehensive experimental facility for microelectromechanical systems (MEMS) application.

4.2
Concept of Stress and Strain

Understanding the load–elongation behavior of materials has attracted the attention of many scientists. Consider, for example, a rod made of a homogeneous material

Optical Methods for Solid Mechanics: A Full-Field Approach, First Edition. Edited by Pramod Rastogi and Erwin Hack.
© 2012 Wiley-VCH Verlag GmbH & Co. KGaA. Published 2012 by Wiley-VCH Verlag GmbH & Co. KGaA.

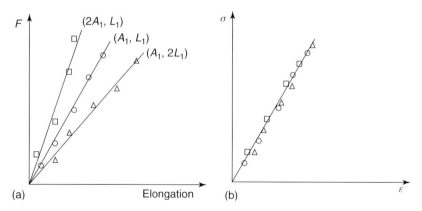

Figure 4.1 (a) Load–elongation curves for various rods of the material. (b) Single curve summarizing the stress–strain behavior of the material.

of cross section A_1 and length L_1. If it is pulled, and if sufficient provisions are made for noting its elongation, the load–elongation curve will be obtained, as shown in Figure 4.1a. Suppose the length is doubled ($L_2 = 2L_1$); then the load–elongation behavior will also be different. If the experiment is repeated with a rod of twice the cross-sectional area ($A_3 = 2A_1$) and length L_1, then also the load–elongation behavior will be different. These graphs are shown in Figure 4.1a and many such graphs will be obtained for different cross-sectional areas and lengths. If for a single material one gets an infinite number of load–elongation curves, then for the purpose of design the information required is mind boggling. This has forced scientists to come out with a different way of representing the collected information. If the graph is not drawn between load and elongation but between (load/area) and [(change in length)/(original length)], a great simplification can be achieved. The multiple load–elongation graphs merge into just one graph (Figure 4.1b), which greatly simplifies the representation of the material behavior.

The quantity (load/area) is termed as *stress* with units of megapascals. The quantity [(change in length)/(original length)] is known as *strain*. From the foregoing discussion, it would appear that these are scalar quantities, which is not so. Further developments in engineering have taught us that these are indeed tensorial quantities.

Consider a simple example of breaking a chalk, available as a cylindrical piece. Let the chalk (Figure 4.2a) be pulled until it breaks. Figure 4.2b shows the nature of separation of the chalk when it is pulled. Now, let the chalk piece be twisted as shown in Figure 4.2c. At a particular load, the chalk will break. The broken chalk is shown in Figure 4.2d. It is strikingly different from the case when the chalk was simply pulled (Figure 4.2b). The failure plane is now at 45°. It is common knowledge that, when the chalk is subjected to excessive load, it will break. However, simple but careful experimentation has shown that the failure plane depends on the nature of the load applied.

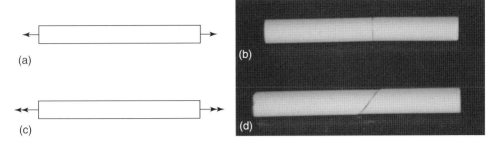

Figure 4.2 Failure planes in a brittle material. (a) Cylindrical chalk subjected to tension. (b) Broken pieces of chalk subjected to tension. (c) Cylindrical chalk subjected to torsion. (d) Broken pieces of chalk subjected to torsion.

If we have understood the material behavior correctly and also developed suitable mathematical tools, then we must be able to predict its behavior under different types of loading.

For us to proceed further, we need to understand the concept of stress and strain better. One needs to develop the concepts of "state of stress at a point" and "state of strain at a point." The above experiment has brought out a key information that the failure plane is a function of loading at the point of interest. On a given plane, the stress can have a component parallel to the plane and another perpendicular to the plane. The component parallel to the plane is labeled *shear stress* and the component perpendicular to the plane is labeled *normal stress*. The resultant of these two defines the stress vector acting on the plane of interest. At a point of interest, many planes can pass through. From a failure analysis point of view, one needs to know the stress vector on all the possible planes passing through the point of interest. Developments in mechanics have taught us that it is enough if one knows the stress vectors on any three mutually perpendicular planes (say, e.g., the Cartesian frame of reference x, y, and z) passing through the point of interest; then the stress vector on any arbitrary plane can be found out. The three stress vectors are bundled together as a mathematical entity called the *stress tensor*. Mathematically, stress is a tensor of rank 2 and is defined as [1]

$$[\tau_{ij}] = \begin{bmatrix} \sigma_{xx} & \tau_{xy} & \tau_{xz} \\ \tau_{yx} & \sigma_{yy} & \tau_{yz} \\ \tau_{zx} & \tau_{zy} & \sigma_{zz} \end{bmatrix} \quad (4.1)$$

In the above equation, the first subscript defines the plane on which the stress component acts and the second subscript denotes the direction in which it acts. If we know the stress tensor at the point of interest, the stress vector on all the possible planes can be easily obtained by Cauchy's formula (Eq. (4.2)). A plane is defined by its outward normal **n** and, for any generic plane, the stress vector can be obtained as

$$\begin{pmatrix} \overset{n}{T_x} \\ \overset{n}{T_y} \\ \overset{n}{T_z} \end{pmatrix} = \begin{bmatrix} \sigma_{xx} & \tau_{xy} & \tau_{xz} \\ \tau_{yx} & \sigma_{yy} & \tau_{yz} \\ \tau_{zx} & \tau_{zy} & \sigma_{zz} \end{bmatrix} \begin{pmatrix} n_x \\ n_y \\ n_z \end{pmatrix} \qquad (4.2)$$

where the superscript n defines the plane for which stress vector is obtained. $n_x, n_y,$ and n_z are the direction cosines of the plane $\mathbf{n}$.

The stress tensor has nine independent components and from the moment equilibrium considerations it is possible to show that the cross-shears are equal. Thus $\tau_{xy} = \tau_{yx}$, and so on.

Thus for most materials, one will have six independent quantities of stress that need to be determined.

Tutorial Exercise 4.1

The stress tensor acting at a point of interest is given as

$$[\tau_{ij}] = \begin{bmatrix} 50 & 10 & -5 \\ 10 & 20 & 30 \\ -5 & 30 & 30 \end{bmatrix} \text{ MPa}$$

Find the stress vector acting on a plane ($\mathbf{n}$) with direction cosines (0.866, 0.5, 0). Also determine the normal and shear stresses acting on that plane.

Solution:

By applying Cauchy's formula, the stress vector can be found.

$$\begin{pmatrix} \overset{n}{T_x} \\ \overset{n}{T_y} \\ \overset{n}{T_z} \end{pmatrix} = \begin{bmatrix} 50 & 10 & -5 \\ 10 & 20 & 30 \\ -5 & 30 & 30 \end{bmatrix} \begin{pmatrix} 0.866 \\ 0.5 \\ 0 \end{pmatrix}$$

$$\overset{n}{T} = 48.3\mathbf{i} + 18.66\mathbf{j} + 10.67\mathbf{k}$$

$$\sigma_n = \overset{n}{T} \cdot \mathbf{n} = (48.3\mathbf{i} + 18.66\mathbf{j} + 10.67\mathbf{k}) \cdot (0.866\mathbf{i} + 0.5\mathbf{j}) = 51.16 \text{ MPa}$$

$$\tau_n = \sqrt{((48.3^2 + 18.66^2 + 10.67^2) - 51.16^2)} = 13.33 \text{ MPa}$$

Strain is also a tensor of rank 2 and is given as

$$[\varepsilon_{ij}] = \begin{bmatrix} \varepsilon_{xx} & \varepsilon_{xy} & \varepsilon_{xz} \\ \varepsilon_{yx} & \varepsilon_{yy} & \varepsilon_{yz} \\ \varepsilon_{zx} & \varepsilon_{zy} & \varepsilon_{zz} \end{bmatrix} \qquad (4.3)$$

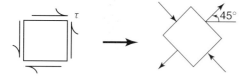

Figure 4.3 Decomposition of pure shear as combination of tension and compression at 45°.

where $\varepsilon_{xx}, \varepsilon_{yy}$, and ε_{zz} are the linear strains at a point in x, y, and z directions. The components $\varepsilon_{xy}, \varepsilon_{yz}$, and ε_{zx} are the tensorial components of shear strains in xy, yz, and zx planes, respectively. The tensorial components of shear strain obey the laws of tensorial transformation and they are one-half of the respective engineering components of shear strain which are usually labeled as γ_{xy}, γ_{yz}, and γ_{zx}. Similar to the stress tensor, the strain tensor is also symmetric. Knowing the strain tensor at a point, it is possible to calculate the linear strains in any direction at that point. The totality of all linear strains in every possible direction defines the state of strain at a point.

We know that chalk can be classified as a brittle material and its failure can be predicted by the *maximum stress theory* or *maximum strain theory* (MST). When the chalk is pulled axially, the maximum stress/strain plane is perpendicular to the loading direction and hence the chalk has failed as shown in Figure 4.2b. On the other hand, when the chalk is subjected to torsion, one induces a pure shear stress state on the chalk, which could be thought of as a combination of tension and compression at 45° (Figure 4.3). Thus, when the torsion or twist is increased, the local stresses developed reach a critical value and the chalk breaks at 45°!

It simply fails as dictated by its material behavior, and as engineers we have been able to correctly conceptualize the material behavior by developing the concept of stress and the associated mathematics to predict its failure behavior.

Tutorial Exercise 4.2

Consider a cylinder made of a ductile material and predict its failure planes in tension and torsion.

Solution:

A ductile material fails by shear and hence the failure planes would be different from that of a brittle material. One can invoke maximum shear stress theory such as the *Tresca yield criteria* or maximum distortional energy theory such as *von Mises yield criteria* (Box 4.1).

In torsional load, the failure planes will be perpendicular, and in tension the failure planes would be at 45°. Try it out with a soft material and verify.

4.3
Stress-Strain Relations

Both stress and strain obey the tensorial transformation laws. The interrelationship between stress and strain is dictated by the material behavior. The law that relates these two is generalized Hooke's law which states that stress is linearly proportional to strain.

$$\sigma_{ij} = E_{ijkl}\varepsilon_{kl} \tag{4.4}$$

where E_{ijkl} is the fourth-order elasticity tensor containing the elastic constants [2]. Fortunately, this tensor exhibits certain symmetry properties that reduce the total number of independent components to 21 for a material that does not have any axes of symmetry. Such a material is known as *anisotropic*. The number of elastic constants will reduce further when the number of planes of symmetry increases. For materials having two axes of symmetry (typically fiber-reinforced composites), the number of elastic constants reduces to nine. Three-dimensional (3-D) orthotropy requires nine independent elastic constants and for two-dimensional (2-D) orthotropy, four independent constants. The number of elastic constants required for an isotropic material is just two for both 2-D and 3-D stress states.

> **Tutorial Exercise 4.3**
>
> Write the strain–stress relations for in-plane stress for an isotropic material and for a specially orthotropic lamina with L and T as the reference axes. What do you infer from these relations?
>
> **Solution:**
>
> Isotropic material:
>
> $$\varepsilon_{xx} = \frac{1}{E}\left(\sigma_{xx} - \nu\sigma_{yy}\right)$$
>
> $$\varepsilon_{yy} = \frac{1}{E}\left(\sigma_{yy} - \nu\sigma_{xx}\right)$$
>
> $$\gamma_{xy} = \frac{\tau_{xy}}{G}; \quad G = \frac{E}{2(1+\nu)}$$
>
> These are the well-known strain–stress relations, and from the above expressions it is enough if one knows two elastic constants such as E and ν to characterize the material. An interrelationship exists between the elastic modulus and the shear modulus. For a specially orthotropic lamina, the relations are
>
> $$\varepsilon_{LL} = \left(\frac{\sigma_{LL}}{E_{LL}} - \nu_{TL}\frac{\sigma_{TT}}{E_{TT}}\right)$$
>
> $$\varepsilon_{TT} = \left(\frac{\sigma_{TT}}{E_{TT}} - \nu_{LT}\frac{\sigma_{LL}}{E_{LL}}\right)$$
>
> $$\gamma_{LT} = \frac{\tau_{LT}}{G_{LT}}$$

The inspection of the equations shows that one requires four independent elastic constants to characterize the material. Similar to the case of an isotropic material, it is possible to establish a relationship between the elastic and shear moduli.

4.4
Rudiments of a Tension Test

Although the response of actual structures under service loads is quite complicated, from a conceptual understanding the material behavior can be captured well by performing a simple tension test. Detailed codes exist on what the size of the specimen is, and so on, for a given material. The result of a typical tension test on mild steel is shown in Figure 4.4.

Mild steel is a ductile material and the stress–strain behavior is linear up to a point (upper yield strength) when it becomes nonlinear and reaches a peak stress at ultimate tensile strength, and finally the specimen breaks when fracture strength is reached. The figure shows two distinct regions. The elastic region is very small compared to the plastic region. In the elastic region, the material returns to its original shape when the loads are removed. However, in the plastic region, the material develops permanent deformation because of plastic flow and only the elastic strain is recovered when the loads are removed.

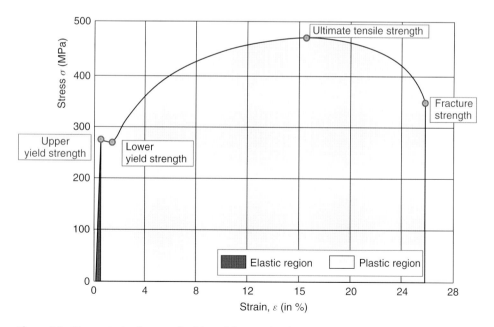

Figure 4.4 Stress–strain diagram of mild steel from a simple tension test (Courtesy [3]).

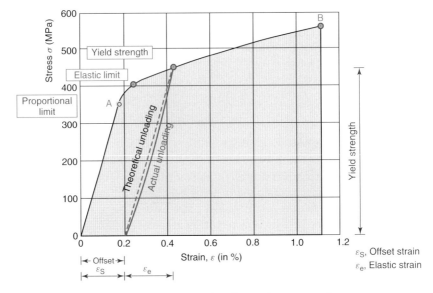

Figure 4.5 Closeup of stress–strain curve for alloy steel (Courtesy [4]).

Most mechanical, aerospace, and civil structures operate well within the elastic limit, and it is clear from Figure 4.4 that they need to operate at very small strain regions. From material characterization, one needs to know the yield strength and it is quite obvious to find out for mild steel as the demarcation from elastic to plastic region is quite sharp and distinguishable. However, this may not be so for a general material. Consider the stress–strain curve of an alloy steel (Figure 4.5). The point A is the proportional limit up to which stress–strain behavior is linear. Beyond this for a short while, the material behavior is elastic but the stress–strain behavior is nonlinear. It is customary to set an offset strain of 0.2% (2000 $\mu\varepsilon$) to identify the point of yield strength. The value of strain at the point when the material starts plastic deformation is quite important and, from the foregoing discussions for metals, it is quite small, on the order of only 0.2% strain.

From the point of view of selecting a suitable experimental technique for a given problem, the knowledge of the strain range in which the experiment needs to be conducted is important.

4.5
Principal Stress and Strain

It was noted earlier that, in general, on a plane there could be a tangential component of stress and a normal component of stress. As an infinite number of planes pass through a point of interest, it is possible to identify a set of planes on which stress is wholly normal. Such planes are known as *principal planes*. Three such planes that are mutually perpendicular can be identified at each point and the corresponding stresses are labeled as *principal stresses*. The algebraically largest principal stress is labeled as σ_1, the next highest is σ_2, and the least one is σ_3. The concept of principal stresses has greatly simplified the utility of stress analysis. Failure theories require only the value of the principal stresses to assess the onset of yielding. From a stress analysis point of view, one needs to know the principal stresses, and from material behavior the yield strength to assess the structural behavior at least for isotropic materials.

The stress tensor in Eq. (4.1) is written with respect to the generic coordinate system x, y, and z. In many instances, it is convenient to express the stress tensor in a different coordinate system such as with respect to the principal stress directions. The advantage is that the stress tensor is simply a diagonal matrix, which is given as follows:

$$[\tau_{ij}] = \begin{bmatrix} \sigma_1 & 0 & 0 \\ 0 & \sigma_2 & 0 \\ 0 & 0 & \sigma_3 \end{bmatrix} \quad (4.5)$$

Experimental techniques basically exploit suitable physical phenomena, and some can provide the difference or sum of principal stresses and their orientations rather than the individual stress components with respect to generic axes. Thus, knowledge of principal stresses is useful to an experimentalist.

On similar lines, one can also think of principal strains and their directions. The principal strains are the extremal values of the linear strain at a point, and the associated directions are the principal strain directions. The strain tensor is then given as

$$[\varepsilon_{ij}] = \begin{bmatrix} \varepsilon_1 & 0 & 0 \\ 0 & \varepsilon_2 & 0 \\ 0 & 0 & \varepsilon_3 \end{bmatrix} \quad (4.6)$$

■ **Tutorial Exercise 4.4**

The state of 2-D stress at a point is defined as follows:

$$(a) \begin{bmatrix} 80 & 60 \\ 60 & 30 \end{bmatrix} \text{MPa} \quad (b) \begin{bmatrix} 110 & 40 \\ 40 & 50 \end{bmatrix} \text{MPa}$$

Determine the principal stresses and label them as σ_1, σ_2, and σ_3.

Solution:

The matrices must be brought into diagonal form:

$$(a)\ \sigma_{1,2} = \frac{1}{2}(\sigma_x + \sigma_y) \pm \frac{1}{2}\sqrt{(\sigma_x - \sigma_y)^2 + 4\tau_{xy}^2}$$

$$= \frac{1}{2}(80 + 30) \pm \frac{1}{2}\sqrt{(80 - 30)^2 + 4 \times 60^2}$$

$$= 55 \pm 65$$

$$\sigma_1 = 120\ \text{MPa and}\ \sigma_2 = -10\ \text{MPa}$$

However, the stresses have to be algebraically labeled and in such a case one will have to label them as

$$\sigma_1 = 120 \, \text{MPa}, \sigma_2 = 0 \, \text{MPa} \quad \text{and} \quad \sigma_3 = -10 \, \text{MPa}$$

(b) $\sigma_{1,2} = \dfrac{1}{2}(\sigma_x + \sigma_y) \pm \dfrac{1}{2}\sqrt{(\sigma_x - \sigma_y)^2 + 4\tau_{xy}^2}$

$= \dfrac{1}{2}(110 + 50) \pm \dfrac{1}{2}\sqrt{(110 - 50)^2 + 4 \times 40^2}$

$= 80 \pm 50$

$\sigma_1 = 130 \, \text{MPa} \quad \text{and} \quad \sigma_2 = 30 \, \text{MPa}$

In this case, the stresses have to be labeled as

$$\sigma_1 = 130 \, \text{MPa}, \quad \sigma_2 = 30 \, \text{MPa} \quad \text{and} \quad \sigma_3 = 0 \, \text{MPa}$$

Note the way the principal stress σ_3 is labeled in both cases. The stresses have to be labeled algebraically and, while doing so, the stress of zero value has to be properly accounted for. It is a subtle point not to be missed. Particularly, while applying the Tresca yield criteria, if both principal stresses are positive (as in case "b") or negative, then the zero stress has to be properly accounted to calculate the maximum shear stress.

Box 4.1: Yield Criteria

A *yield criterion* is a mathematical expression of the stress states that will cause yielding or plastic flow. The simplest yield criterion is the one first proposed by Tresca. It states that yielding will occur when the largest shear stress reaches a critical value. The largest shear stress is

$$\tau_{max} = \dfrac{(\sigma_{max} - \sigma_{min})}{2}$$

The critical value is found from a uniaxial tension test; let the yield stress determined be σ_{ys}, which is twice the maximum shear stress.

If the convention is maintained that $\sigma_1 \geq \sigma_2 \geq \sigma_3$, the Tresca yield criterion can be written as

$$(\sigma_1 - \sigma_3) = \sigma_{ys}$$

While employing Tresca yield criterion, the identification of the minimum principal stress needs to be done carefully. However, once done, the numerical calculation is relatively simple. The von Mises yield criterion involves the use of all the three principal stresses and is stated as

$$(\sigma_1 - \sigma_2)^2 + (\sigma_2 - \sigma_3)^2 + (\sigma_3 - \sigma_1)^2 = 2\sigma_{ys}^2$$

Box 4.2: Unknowns to Be Determined for a Generic Problem

From a stress analysis point of view, at a point of interest one needs to know the stress tensor, the strain tensor, and the displacement components. If these are referred to generic axes, then one needs to find out at every point 6 stress components, 6 strain components, and 3 displacement components that totals to 15 variables! In a complete sense, one may want to evaluate these 15 quantities at every point in the model/component and such a solution is known as a *closed-form* solution. For simple problems such as a uniform plate subjected to tension, slender members subjected to pure bending, or circular shafts subjected to torsion, analytical methods based on strength of materials itself can give closed-form solutions. The list can be extended for a few more planar problems if the theory of elasticity approach can be used. A closed-form solution is a luxury and is available for only a few problems. For many problems of practical interest, near-field solutions or even point solutions are sufficient.

4.6 Concept of Stress Concentration

Consider the problem of a finite plate with a small hole subjected to tension. The fringe patterns observed in a photoelastic test are shown in Figure 4.6a. If there were no hole, one would only see a uniform color fringe in a photoelastic test. Owing to the presence of the hole, one observes fringe loops with different colors. The fringe patterns are nevertheless impressive but, from a stress analysis point of view, they represent a complex stress field.

The uniform stress field σ_{xx}, which otherwise would exist, is disturbed by the presence of the hole and the stress field has become biaxial in the vicinity of the hole. Moreover, the magnitudes of the stress have significantly increased in the vicinity of the hole, which results in stress concentration. A closed-form solution exists from theory of elasticity if the plate is of infinite dimensions and the hole is very small. The variation of the boundary stress (Figure 4.6b) on the boundary of the hole ($r = a$) is given as

$$\sigma_{\theta\theta} = \sigma_{xx}(1 - 2\cos 2\theta) \qquad (4.7)$$

The interesting interpretation of this equation is that, at $\theta = 0°$, the stress is compressive and at $\theta = 90°$ it reaches a maximum of three times the average stress! (Figure 4.6b). The *stress concentration factor* (SCF) is defined as maximum stress divided by far-field average stress. For this problem it reduces to 3. A word of caution here: only for an infinite plate with a small hole does the SCF reach a value of 3, but for a finite plate it is much more and one has to resort to an experimental

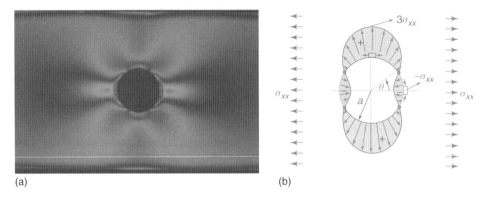

Figure 4.6 (a) Photoelastic fringe patterns in the vicinity of a hole in a finite plate subjected to axial tension. (b) Blowup of the nature of stress distribution in the boundary of a hole in an infinite plate subjected to uniaxial tension (Adapted from [3]). (Please find a color version of this figure on the color plates.)

technique or a numerical technique to evaluate. From a design point of view, one may be interested only in a point solution, that is, the maximum boundary stress which can be easily determined by employing the photoelastic technique!

What is highlighted here is that one should know what essential information needs to be determined for a given problem. Understanding the stress field near a hole is important from engineering practice, as bolted and riveted connections are material-joining techniques that are widely used.

Tutorial Exercise 4.5

The problem of a finite plate with a hole is studied, and from the measurements the maximum stress at the boundary of the hole is found to be 250 MPa. The ligament stress (the average stress in the net section) is found to be 120 MPa and the average far-field stress is estimated to be 70 MPa. Determine the value of SCF from the definition of theory of elasticity and how a designer would compute the same.

Solution:

The theory of elasticity defines SCF as

$$\text{SCF}_{\text{TE}} = \frac{\sigma_{\max}}{\sigma_{\text{farfield}}}$$

$$= \frac{250}{70} = 3.57$$

It is to be noted that it is greater than 3! Thus experimental or numerical evaluation of the SCF is a must from a design point of view.

The design handbooks use a different definition of SCF, which is related to the ligament stress as

$$\text{SCF}_{DE} = \frac{\sigma_{max}}{\sigma_{ligament}}$$

$$= \frac{250}{120} = 2.08$$

By looking at this value which is less than 3, one should not jump to the conclusion that the SCF is lower than for a small hole in an infinite plate – in fact, it is quite high and the difference in value comes from a modified definition of the SCF. In all finite-body problems, the SCF is higher than 3 if the theory of elasticity definition of SCF is used.

4.7
Birth of Fracture Mechanics

In the beginning of the nineteenth century, although structures were designed based on the best design practices of that time, they failed in service. From simple design methodologies, improvements were made by looking at what happened if there were repeated loads and how this affected the service behavior. From such an understanding, the concept of *fatigue* emerged and got incorporated into design methodologies. Even this was found lacking as more and more complex service conditions such as working at extreme temperatures were demanded from structures. During World War II, several Liberty ships that were made by welding broke into two in the North Atlantic Sea. This and several other spectacular failures prompted scientists to look back and see what could have caused such failures [3]. A study by Inglis for a small elliptical hole in an infinite plate subjected to uniaxial tension showed that for an elliptical hole of semimajor axis a and semiminor axis b, the maximum stress developed is given by

$$\sigma_{max} = \sigma_{xx}\left(1 + \frac{2a}{b}\right) \tag{4.8}$$

Equation (4.8) indicates that, when one has a crack, that is, as b tends to zero, the maximum stress could theoretically become infinite! The result from this study was that cracks could be very much more dangerous than originally thought. Further, the simple representation of discontinuities contributing to increase in stress concentration was no longer sufficient, and a new field of mechanics known as *fracture mechanics* came into existence through research done by several investigators across the world [5]. The relative influence of a circular hole, an elliptical hole, and a crack can be well understood from Figure 4.7. For the same load and similar geometric features, the fringe density at the tip of the crack is much higher than for a circular hole. For the case of an elliptical hole, the fringe density lies in between these two.

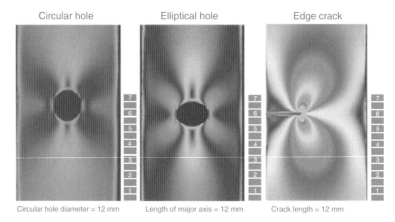

Figure 4.7 Isochromatics (contours of difference in principal stresses, see Chapter 9) observed near various stress raisers (Courtesy [3]). (Please find a color version of this figure on the color plates.)

With such new developments, the requirement of the experimentalist is to determine new parameters from experiments. One needs to find out for fracture mechanics studies what are known as *stress intensity factors* (SIFs) from experimental results. In fact, many methods exist for such evaluations [3, 6]. Thus it is not always the determination of the 15 quantities mentioned earlier but it could be just any information pertinent to a given problem such as SCF or SIF.

4.8
Peculiarities of Experimental Approach

There is a fundamental difference between experimental and other methods such as numerical or analytical approaches. In general, both numerical and analytical methods give complete information of either stress or displacements based on the formulation and, once these are determined through stress–strain or strain–displacement relations, it is possible to get the other components. On the other hand, experiments exploit a particular physical information and hence they may just give only a particular information for the problem studied [4, 7]. For example, moiré methods can give u or v displacements, but one may require special arrangements to get the w displacement (Chapter 7). On the other hand, photoelasticity exploits the phenomenon of temporary (stress or strain induced) birefringence and may give only the difference in principal stresses or strains. With suitable additional optical arrangement, it can also give the principal stress or strain directions (Chapter 9). Thus it is worthwhile to know what the basic quantities an experimental technique can give and what the basic physics that a technique employs. Further, the experimental techniques could be classified as whole-field or point-by-point methods. Most of the optical techniques discussed in this book come under the category of whole-field methods.

4.9
Information Directly Obtainable from Various Experimental Techniques and Their Typical Applications

4.9.1
Photoelasticity

Photoelasticity provides the principal stress/strain difference and their orientations (Chapter 9). The basic physical principle exploited is the phenomenon of *temporary birefringence*. The material is optically isotropic normally, but behaves like a crystal when stressed and becomes doubly refractive. Thus by understanding the crystal optics it is possible to relate the information of stress/strain to the observed fringes. It is to be noted that the refractive index is a tensor of rank 2 and stress and strain are also tensors of rank 2.

In view of the physical principle exploited, it can measure only the difference in principal stresses/strains which are the so-called isochromatic fringes (representing contours of $(\sigma_1 - \sigma_2)$ or $(\varepsilon_1 - \varepsilon_2)$) and their orientations (θ), the so-called isoclinic fringes.

Photoelasticity comes under the category of common-path interferometers, and vibrations do not have much effect on these. This is one of its greatest strengths and hence the method widely used in the industry for its simplicity in optical arrangements and ruggedness in data acquisition. Many variants of photoelasticity have been developed to handle different problems. Table 4.1 summarizes the variants of photoelasticity and their relevant use.

4.9.2
Holography

Holography provides the displacement vector (Chapter 6). If one has to determine specific components of displacement, then special efforts must be made to constrain the model as well as use suitable illumination and observation directions.

In contrast to conventional photography, holography records both the intensity and phase of the wavefronts. The phase information is recorded using the principles of interferometry. In view of the use of a reference beam, the vibration isolation requirements are quite stringent for fringe formation. However, the resolution obtainable is high and, of late, this technique has found wide application in the study of MEMS. It is also attractive for recording mode shapes of vibrating objects. Over the years, many variants of holography have been developed. Table 4.2 summarizes these variants and their typical applications.

For plane problems subjected to in-plane loading, the out-of-plane displacement can be related to the sums of principal stresses and such contours are labeled *isopachics* (representing contours of $(\sigma_1 + \sigma_2)$). If used in conjunction with photoelasticity, principal stresses can be separated over the domain.

Table 4.1 Variants of photoelasticity and their salient applications.

Variants of photoelasticity	Typical applications
Two-dimensional photoelasticity	Excellent tool to teach various principles of solid mechanics such as St Venant's principle, principle of superposition, concept of stress concentration, and so on
Three-dimensional photoelasticity	To study stresses interior to the model
(a) Stress freezing and slicing	Extends the use of principles of two-dimensional photoelasticity to solve three-dimensional problems. Has found wide acceptance in the industry for solving a variety of real-life complex problem situations
(b) Scattered light photoelasticity	Replaces mechanical slicing by optical slicing. Has found utility in finding residual stresses in plate glass
(c) Integrated photoelasticity	Mathematically demanding. Has been successful in evaluating stresses in axisymmetric problems. Has found wide use in estimating residual stresses in glass articles such as tumblers, fiber preform, CRT, and so on
Photoelastic coatings	Extends transmission photoelastic methods to opaque prototypes. Has found applications in various fields ranging from aerospace to biomechanics. Quite useful for finding assembly stresses and residual stresses
Dynamic photoelasticity	Useful for the study of time varying phenomena. Provided phenomenological understanding of stress wave propagation, crack propagation, and crack-stress wave interaction studies
Photo-orthotropic elasticity	Exclusively developed for studying fiber-reinforced plastics
Photoplasticity	Extension of photoelasticity to the plasticity domain. Has been used to evaluate the SCFs under plastic loading conditions and in modeling large deformations in metal-forming operations
Digital photoelasticity	Intensity data over the model domain is processed to find photoelastic parameters at every point in the model domain through the use of digital image processing hardware and software. The methodology is applicable to all the variants of photoelasticity listed above

4.9.3
Grid Methods

In grid methods, very coarse rulings are applied to an object. Grids are not always rulings of parallel lines. The deformation of the grid lines are directly measured for finding the displacements. In metal-forming studies, circles are marked on the specimens and the deformation of circles into ellipses helps in identifying the principal strain directions at a point conveniently. Coarse grids find applications in the study of plastic flow and for the qualitative study of displacements in rubber models simulating elastic members.

Table 4.2 Variants of holography and their salient applications.

Variants of holography	Typical applications
Real-time hologram interferometry	The advantage is that the fringes can be seen in real time. Useful to find displacement vector as in double-exposure hologram interferometry
Time-average hologram interferometry	Useful to record resonant mode shapes and frequencies of transverse vibration. Bessel fringes are seen that have unequal brightness. The bright and most visible fringes correspond to the stationary vibrational nodes of the object
Stroboscopic time-average hologram interferometry	Pulsed lasers are used and are useful for vibration studies. The fringes observed are similar to those recorded in double-exposure hologram interferometry – the intensity variation is cosinusoidal

4.9.4
Geometric Moiré

One can get in-plane displacements and, with different optical arrangements, the out-of-plane displacements (Chapter 7). The fringe formation is due to mechanical interference of gratings and the fringes are labeled *isothetics*. Thus one will get u-displacement isothetics, v-displacement isothetics, and so on. One of the advantages of this technique over the other methods of displacement measurement is that it is quite simple to associate the fringe patterns observed to a particular component of displacement.

If one wants to record the slope or curvature, special methods exist and for each of these one needs to employ different optical arrangements. Several variants of geometric moiré exist such as shadow moiré, reflection moiré, and projection moiré. Table 4.3 summarizes typical uses of these variants.

4.9.5
Moiré Interferometry

Moiré interferometry is widely used to obtain in-plane displacements. It uses gratings of very fine pitch and, in view of high fringe densities possible, these displacements can be differentiated to get the strain components [8]. Suitable optical arrangements have also been reported for obtaining out-of-plane displacements.

4.9.6
Speckle Interferometry

Speckle interferometry shares a commonality with holography but, in view of its modified experimental approach, it is easily amenable for digital recording and

Table 4.3 Variants of moiré and their salient applications.

Variants of Moiré	Typical applications
Shadow moiré	Gives out-of-plane displacement. Camera needs to resolve only the moiré fringes. Useful for buckling and post-buckling studies. Has also been found to be quite useful to study warpage of printed circuit boards, silicon chips, and MEMS
Reflection moiré	Gives the slope of out-of-plane displacements. Suitable modifications have also been proposed to record even the curvature
Projection moiré	Here the grating is projected onto the specimen. Hence, the recording system must be fine enough to record the grating line. Large structures such as aircraft can be analyzed. Essentially gives slope of out-of-plane displacements. Has found use in the study of electronic packaging applications

processing (Chapter 6). It provides the displacement vector. With suitable optical arrangements, one can find the in-plane and out-of-plane displacements.

When a specularly diffuse surface is illuminated with a coherent light source such as a laser, speckles are formed. Speckle pattern acts like a random marker, similar to gratings in geometric moiré, on the specimen and for small displacements of the specimen it moves with it. In speckle interferometry, a double exposure is made, one before and one after the deformation of the specimen. The two speckle patterns give raise to an apparent fringe pattern covering the component. These fringes are termed *speckle correlation fringes* and appear much grainier than a holographic interference fringe pattern.

Several variants such as electronic speckle interferometry, digital speckle interferometry, and so on, have been reported. These are basically procedures that greatly simplify data reduction by digital processing of intensity data. A modification of speckle interferometry known as *shearography* provides slope contours and finds immense use in nondestructive testing of composites.

4.9.7
Digital Image Correlation

Digital image correlation provides in-plane and out-of-plane displacements (Chapter 5). Unlike in speckle interferometry where the speckles are formed naturally, here the speckle patterns are artificially introduced on the specimen. The size of the speckles can be changed to accommodate different length scales from nanoscale to macroscale. Since white light is used for experimentation, the methodology is also known as *white light speckle correlation*. Extensive statistical processing of the intensity data of speckles is used to evaluate the displacements of the component under study. Thus the method is unique conceptually and is in tune with the computer era, as without extensive computation the experimental results cannot be interpreted.

At its current development, it reports an error in strain measurement [7] on the order of $\pm 200\ \mu\varepsilon$, which is high. The method is suitable for large deformation studies and also for analyzing exotic materials such as metallic foam. It has great potential, and development of the technique is under way to make it a general-purpose experimental tool.

4.9.8
Thermoelastic Stress Analysis

Thermoelastic stress analysis (TSA) provides the change in sum of principal stresses or strains under cyclic or random loading (Chapter 8). Under cyclic loading, the local temperature of the component varies on the order of $0.001\,°C$. This is measured using infrared cameras. Thermodynamic principles are invoked to correlate the variation of temperature to the stress information. The name of the technique derives from the physical principle exploited, and the technique is not to be confused with that for the measurement of thermally induced stresses.

4.9.9
Brittle Coating

In this, a thin brittle coating is fractured to provide the principal stress directions over the model domain. The contours thus observed are the isostatics. The experiment is performed for various load steps, and for each load step the isostatics are carefully noted for the whole model domain. In addition, for each of the loadings, the ends of the cracks are joined to get contours known as *isoentatics*. For quantitative extraction of data isoentatics are useful. However, from a practical standpoint isostatics are more useful. If used in conjunction with strain gauges, the method greatly simplifies experimental analysis of large structures.

4.9.10
Strain Gauge

Contrary to the general perception, a strain guage does not provide strain but only a component of the strain along the gauge length. As strain is a tensor, one needs multiple strain gauges to measure the strain at a point. On a free surface, one requires a minimum of three strain gauges to measure the strain tensor. Prealigned strain gauges in a single backing comes for easy installations and these are known as *strain rosettes*. A three-element strain rosette would require three channels of the measurement system.

In contrast to the techniques discussed so far, the method of strain gauge is a point-by-point technique. The measurement system used in general will have an upper limit in processing a specific number of channels. One of the important considerations in the implementation of the technique is how to minimize the number of channels per point for a given application. This requires a priori information on the nature of the strain field at various locations of the structure/component. In general, one would require a three-element rosette per point for determining the strain tensor. However, if the principal strain directions are known at a point, then it is enough to use just two strain gauges, one each along the principal strain directions. The luxury of using a single strain gauge is possible only if the strain field is uniaxial. Thus by judiciously selecting a single-element gauge or two-element or three-element strain rosettes, one can obtain pertinent information of the problem under consideration at more locations for the same upper limit on the number of channels that can be handled by the measurement system.

The technological aspects of the technique are very well-developed and, if used carefully, even 0.5 $\mu\varepsilon$ can be measured reliably.

4.9.11
Caustics

Caustics are particularly suited for studying high strain gradients such as stress concentration and fracture mechanics problems. The elegance of the approach is that the measurement of a single geometric parameter is sufficient to get the pertinent stress/strain gradient.

In optics, a caustic is the envelope of light rays reflected or refracted by a curved surface or object, or the projection of that envelope of rays on another surface. When a plane specimen under in-plane loading deforms, because of Poisson's effect, near the point of stress concentration the lateral thickness of the specimen changes significantly. In the transmission arrangement, the specimen behaves like a divergent lens. In view of it, parallel light rays incident on the model gets deflected and a caustic shadow is formed with a bounding silver line.

In the development of fracture mechanics, caustics was quite useful to experimentally demonstrate the existence of the (HRR) field (Box 4.3) near the crack tip.

Box 4.3: HRR Field

Hutchinson, Rice, and Rosengran have independently shown that the *J*-integral characterizes the crack-tip condition in a nonlinear elastic solid. The plastic behavior in the close vicinity of the crack tip could be well described by a nonlinear elastic solid and from an engineering approach; the *J*-dominated field is convenient for analysis. The *J*-dominated field near the crack tip is termed the *HRR field* or the *HRR–Cherepanov field*. The HRR field plays an important role in the extension of nonlinear fracture mechanics into the elastoplastic domain.

4.9.12
Coherent Gradient Sensor

A coherent gradient sensor provides the full-field sum of in-plane normal stress gradients and the out-of-plane displacement gradients. It is a whole-field, real-time, lateral shearing interferometric method. The method was originally developed to measure the crack-tip stress fields to offset the limitations of the method of caustics. With the development of MEMS, it is useful to find the slopes and curvature in thin (500 µm) silicon wafers.

4.10
Selection of an Experimental Technique

One of the crucial steps in experimental stress analysis is to identify a suitable technique or a combination of them to solve a given problem on hand [4, 9]. The selection of a technique depends on several factors:

- Time available for analysis
- Level of accuracy required
- The range of strain/stress to be measured
- Influence of extreme conditions such as high temperature, high strain rates, and so on.
- Thoroughness of the study required
- The type of study (whether on models or on prototypes)
- The cost permissible for the study.

It is to be noted that, although several experimental techniques are available to measure similar parameters, the inherent accuracy of each of them is different. The demand in accuracy can also dictate the selection of a particular technique over the others. There are no fixed guidelines in selecting a technique; however, certain strategies can be suggested. The first and foremost is to consider the techniques that one has expertise in or at least access to. This dictum will not apply if one has to consider a critical and complex scenario where any mistake in assessment can lead to serious consequences. In such instances, one may have to explore all possibilities and even think of devising a new technique for experimental evaluation. Obviously, this would be possible only if sufficient time and cost is permitted, or the administrators are impressed upon to relax such restrictions to carry out the study. In fact, if one looks at the development of newer techniques such as thermoelastic stress analysis, the cost has been secondary to its development.

For routine stress analysis, one may think of the use of general-purpose techniques such as photoelasticity and strain gauges to see whether they can yield the intended result. Photoelasticity is an optical whole-field technique and therefore can give information over the field instantly. Thus, it is a very convenient tool for identifying stress concentration zones, less stressed zones, and so on. For optimization studies, one may want to identify the less stressed zones to scoop out further material, but from failure analysis point of view one may want to know the zones of stress concentration. Being an optical technique, accessibility of the object for visual recording (at least from a remotely operated camera) is a must. On the other hand, although the strain gauge uses a point-by-point technique, it is suitable for the study of even invisible locations. With its recent developments, it can be conveniently employed with the facility of computer readout if the zone of interest is known a priori.

Although the focus of this book is mainly on the utility of optical methods, a discussion on strain gauges will not be out of place because of its versatility. In view of aging structures such as bridges, dams, and so on, the concept of *in situ* measurements is gaining importance and strain gauges are an ideal choice for such applications. One may employ electrical resistance strain gauges or optical fiber-based strain gauges. In fact, with fiber-reinforced composites replacing several metallic structures including the fuselage of aircrafts (e.g., Boeing 787), the use of structural health monitoring is gaining importance because of the multiplicity of failure modes. Optical fiber-based strain gauges are becoming important for such applications.

Electrical resistance gauges have several advantages. Remote locations can be studied through telemetry, and rotating components can be examined by the use of slip rings or with telemetry. In complex problems, one may not know whether the structure will be loaded beyond its plastic regions. As strain gauges can be conveniently employed from elastic to plastic range, they are the good candidates for such studies. Dynamic and transient studies can be conducted through the use of suitable data acquisition systems.

In many problems of practical interest, a single experimental method may not yield all pertinent information. Further, the time of study may also need to be optimized. From such a standpoint, the use of brittle coatings – a whole-field optical technique – and strain gauges – a point-by-point technique – make a good combination. If the structure under analysis is very large and one does not know which zones of the structure need refined analysis, then the first step could be to conduct a brittle coating test on the structure. This will reveal zones of stress concentration, less stressed zones, as well as the direction of principal stresses at various locations including the information on whether the stress field is uniaxial, biaxial, or isotropic. This is a wealth of information, which not only helps the analyst to focus on specific zones but also reduces the number of channels required per point for strain gauge instrumentation. Even reducing one channel per point is a great saving for large structures, as there is always an upper limit on the number of channels and it may not be possible to repeat the experiments.

For whole-field evaluation, optical methods are the best. Here again, one should know whether for the problem on hand one needs to obtain displacements, slope, curvature, strain, or stress information to decide the technique. As mentioned before, photoelasticity is a versatile technique and the optical arrangement required is quite simple and thus it is an industry-friendly technique.

Photoelasticity directly provides the information on the principal stress difference and the principal stress direction at the point of interest. Using these, one can find the normal stress difference and in-plane shear stress by invoking equations in mechanics of solids or Mohr's circle. For comparative evaluation of various designs, even a qualitative study of isochromatic fringe patterns can provide sufficient information [4]. Shape optimization can be effected by altering the boundary of the component such that the fringes on the boundary follow the boundary contour rather than cutting it. Such fillets are known as *streamline fillets* [4]. The evaluation of SCF, contact stress parameters, and SIFs require the knowledge of only the isochromatic fringe order at selected points in the field. Further, in 2-D problems, if the principal stresses have opposite signs, then the isochromatic fringe contour is equivalent to the Tresca yield contour for zones where the yield condition is satisfied. Thus, for a wide variety of design problems, photoelasticity provides direct information.

Special techniques are required to address special situations. For example, moiré interferometry has found wide acceptance in electronic packaging industry to measure thermally induced stresses on tiny components. Speckle and holographic methods are found to be suitable for the analysis of MEMS. Thermoelastic stress analysis is useful to measure stresses developed because of fatigue or random loadings. Recently, the method has been found suitable for unconventional applications such as stresses introduced by the slamming of automobile doors! Holography was developed to study the mode shapes of vibrating objects. Speckle method in its variation as shearography is an ideal nondestructive testing (NDT) tool for the study of delaminations in the aerospace industry.

The recent development of digital image correlation (DIC) is promising as a general-purpose technique such as strain gauges or photoelasticity. However, in its current development, the accuracy is not very high and further improvements in the techniques are needed for its wider use. A broad classification of the selection of various techniques for different strain ranges is shown in Figure 4.8.

Rapid technological developments have simplified experimentation, ensured higher accuracies, and removed the drudgery in carrying out the experiments. If time and cost are taken care of, the accuracy can be improved until it reaches sufficient limit to give better information of the problem on hand.

The above guidelines are not exhaustive. The experimentalist's acumen is needed in deciding an appropriate technique or intelligently using an existing technique to unconventional situations. The case study discussed next illustrates this point. Further, the practicing experimentalist should also know the current literature to learn which techniques have been used for related problems.

Tutorial Exercise 4.6

Select a suitable technique(s) to determine the stresses developed because of assembly of different components. After assembly of the components, the surface of interest is (i) visible and (ii) invisible. For each case, indicate whether the analysis has to be carried out on a model or a prototype.

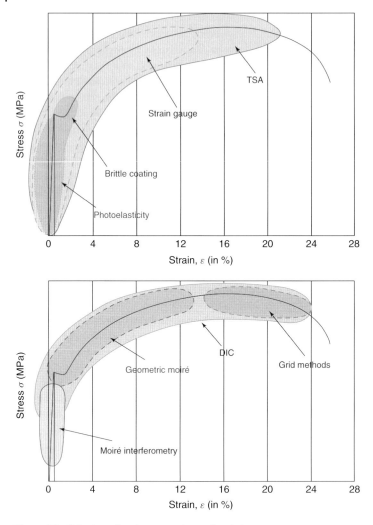

Figure 4.8 Selection of various experimental techniques based on the strain range (Courtesy [4]).

Solution:

The problem is one of assembly stress and obviously one has to perform the experiment on the prototype.

Optical methods can be used only when the surface of interest is visible or accessible through a remotely operated camera. For the case where the surface is visible, the photoelastic coating technique is ideal.

When the surface of interest is not visible, then one has to explore the possibility of using the strain gauge technique. Suitable precautions may be taken during assembly so that the strain gauge installation is not disturbed.

4.11
Case Studies

One of the most challenging aspects in solving an industrial problem is identifying a technique or a combination of suitable techniques to solve the problem on hand. In recent times, experimental mechanics is witnessing the development of newer and automated measurement techniques. To solve a problem, it is not necessary to employ the newer methods but even existing methods can be used ingeniously. In this context, the first case study explains how transmission and reflection photoelastic techniques are innovatively used to investigate the root cause for random failures of chain plates.

Investigation of objects over a small field of view is of particular importance in the electronics industry where characterization of components is required to determine their reliability. MEMS are the result of the integration of mechanical elements, sensors, actuators, and electronics on a common silicon substrate through microfabrication technology. Experimental methods for characterizing the mechanics of silicon chips and wafers are of great interest. Silicon wafers contain residual stresses because of a variety of reasons, which may eventually induce warpage on the chips. Silicon is transparent to infrared radiation over the wavelength range of 1100–1500 nm. In view of this, photoelasticity has proved to be useful for stress measurement in silicon wafers [10]. Warpage of silicon wafers is of increased concern because of the continuing increase in the wafer diameter and decrease in wafer thickness for improved design of sensors. Coherent gradient sensing (CGS) is found to be useful for such studies [11]. Equally important is the knowledge of the response of the sensors made out of silicon wafers to external loads. The second case study focuses on identifying a comprehensive experimental facility to make the requisite measurements on a MEMS pressure sensor.

4.12
Experimental Study on Investigation of Random Failure of Chain Plates

Most automotive and industrial chains are an assembly of five parts, as shown in Figure 4.9. They are the outer plate, the inner plate, the bush, the pin, and the roller [12]. Two inner plates are press-fitted with two bushes to form an inner block assembly. The outer plates are press-fitted with pins after keeping the pins through the assembled bushes of the inner block. The roller is a rotating member that is placed over the bush during the inner block assembly. Inner block assembly is the load-transfer member from sprocket tooth, and the outer block assembly helps in holding and also pulling the inner block over the sprocket teeth. Thus interference fits are a necessary feature in a chain assembly to keep the parts together with sufficient holding force. The bushes shown in Figure 4.9 can be made by cold-forming, cold-extrusion, or cold-drawing processes.

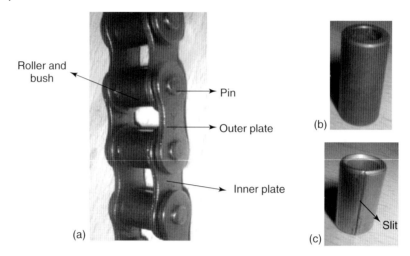

Figure 4.9 Typical chain and its parts. (a) General chain assembly with five parts. (b) Extruded bush without slit. (c) Formed bush with fine slit along its length. (Courtesy [13]).

In a cold metal-forming process, a steel strip of specified width is cut into smaller lengths and rolled into the shape of a cylinder using a forming die set. There is a fine slit along the length of the formed bush (Figure 4.9c). Cold extrusion is a process where extrusion punches are pushed against a solid pin, which is held firmly, to remove the material to make it hollow. In the cold-drawing process, a tube of higher diameter is pulled through a die set to make it a smaller diameter tube with increased wall thickness. In both cold-extrusion and cold-drawing processes, the geometric shape is well controlled (Figure 4.9b) and no fine slit exists along its length such as in a formed bush. The chain plates are made by blanking and piercing operations.

The holding force of an interference fit is checked by the pullout force test on the shop floor. In this test, the bushes and pins are pulled out using special punches and the maximum force is recorded. The recorded pullout force values are checked with the customer-specified values to decide whether the assembly of chain with bushes and pins is acceptable or not. The pullout force test is an industrial quality control tool to check the pullout force at regular intervals during mass production to assess and correct the process parameters.

In spite of rigorous testing, there have been random fatigue failures of chain assemblies in the field. Most of the failures were observed with the inner plate in the problem under study, and the nature of the failure is shown in Figure 4.10. The reason for the random failure could not be found out through regular quality checks. Use of strain gauges could not give any satisfactory results. Preliminary studies by numerical methods only shifted the focus and suggested a true 3-D photoelastic analysis for problem identification.

Figure 4.10 Fatigue failure of inner chain plate. (Courtesy [13]).

4.12.1
Possible Investigation Methodologies

In view of the interference fit, a true simulation of the actual service load due to the combination of axial pull and the interference fit would be cumbersome if one resorts to stress freezing. Hence alternative methods of analysis need to be thought of [13].

4.12.1.1 Analysis of Combined Stress Fields

In this, all elements of the chain can be made out of photoelastic material and a few plates assembled to form a short chain assembly. The chain assembly can be loaded in a test rig and viewed in a transmission polariscope as shown in Figure 4.11. The applied load has to be equivalent to the application load, which can be calculated through model to prototype relations. It can be seen from Figure 4.11 that all the four plates (two inner and two outer) are loaded equally and the fringes observed are an integrated effect of the fringe fields of the plates involved. For data interpretation, principles of integrated photoelasticity need to be invoked [4, 14], which is quite tedious. One can also think of utilizing scattered light photoelasticity for the analysis. However, such a facility was not available.

Alternatively, reflection photoelastic techniques can be used on the assembly to selectively reveal the stress field, which is illustrated in Figure 4.12. The inner plate can be coated with a reflective backing as shown in Figure 4.12a, b and loaded in a

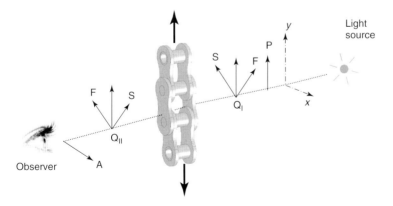

Figure 4.11 Transmission polariscope setup for integrated stress analysis. For the explanation of the working principle, see Chapter 9. (Courtesy [13]).

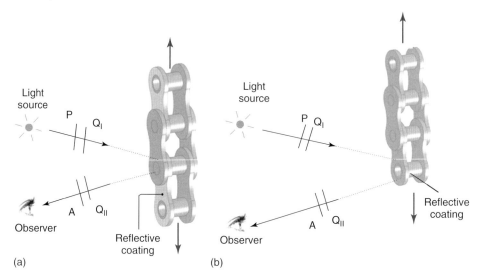

Figure 4.12 Reflection polariscope setup to study combined stresses selectively. (a) Reflective coating on inner plate outer surface to view the fringes in outer plate and pin assembly when it is loaded. (b) Reflective coating on inner plate inner surface to view fringes in inner plate because of bush interference and external axial load. (Adapted from [13]).

test rig. The loaded specimens can be seen through a reflection polariscope and the stress fields analyzed. The arrangement in Figure 4.12a can provide the combined result of the axial pull and the stresses because of the interference fit and provide an integrated effect of both inner and outer plates. On the other hand, arrangement of Figure 4.12b can be used to get the stresses on the inner plate because of the interference fit and also because of the combination of both interference and axial load. Though these two methodologies offer scope for revealing the combined stress, the data interpretation can be tedious.

4.12.1.2 Decoupled Analysis of Assembly and Application Stress Fields

As the whole chain assembly needs to be only elastically loaded, it would be prudent to decouple the study of the assembly stress and tensile loads simulating the actual condition. The principle of superposition can then be invoked to know the final state of stress. Further, since only the inner plate has failed, more focus is needed to study that. An approach of this kind greatly simplifies the analysis of the problem on hand.

4.12.2
Analysis of Assembly Stress due to Interference Fit between Bush and Inner Plate Using Transmission Photoelasticity

While studying the stresses due to assembly, it was initially decided to check whether the form of the chain plate had any major influence on the nature of

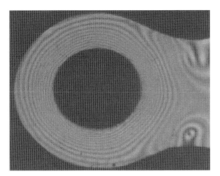

Figure 4.13 Fringe pattern in an epoxy chain plate model press-fitted with machined pin (Courtesy [13]).

Figure 4.14 Fringe pattern observed in an epoxy chain plate model press-fitted with slit bush. (Courtesy [13]).

the stresses developed. A model of the inner plate was made with epoxy sheet of thickness 6.0 mm. A cylindrical pin was machined with an appropriate outer diameter such that it created interference equal to the interference of the metallic bush and the chain plate combination. The fringes observed in a transmission polariscope setup after pressing the pin are shown in Figure 4.13. The fringes in the figure are concentric and depict the uniform pressure exerted by the cylindrical pin on the inner periphery of the specimen. Since the fringes are concentric, the outer form of the chain has no major influence on the stress field developed. In order to truly simulate shop-floor conditions, bushes obtained from the production unit were inserted into another epoxy model and the fringes observed in a transmission polariscope. The fringes were expected to be of the same kind as observed in the case of a cylindrical pin; however, they were not concentric and a "flower pattern" was observed, as shown in Figure 4.14. The flowery fringe pattern suggests that the formed bush is not a true cylinder and the outer surface of the formed bush has many undulations that resulted in a nonuniform pressure/interference with the inner plate. In zones where more fringes appear, the pressure exerted by the formed bush is higher. This observation threw some light on the behavior of the formed bush and chain plate under interference fit. It is to be noted that information of this nature could not have been obtained either by a numerical simulation or a true model study, as no prior information is available on the possible geometric undulations of the formed bush. Thus the information obtained in the current study is quite important and only a whole-field experimental approach could reveal this.

Though the diameter of the hole in the epoxy sheet and the outer diameter of the bush combination were carefully selected to simulate the interference as in the actual condition, the current study can be only qualitative in nature as there could

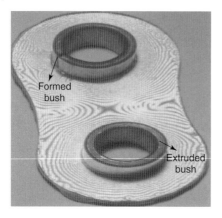

Figure 4.15 Fringes observed on a photoelastic sheet pasted on a chain plate and press-fitted with both extruded bush and formed bush. (Courtesy [13]).

be small deviations in the interference level because of the difference in elastic properties.

4.12.3
Analysis of a Chain Plate and Bush Assembly Using Reflection Photoelasticity

In order to understand the assembly stresses more realistically, a new experiment was planned to press-fit the bush in an actual chain plate for instant appreciation of the stress distribution using reflection photoelasticity.

A Photoelastic sheet PS 1A of thickness 2.97 mm and strain coefficient K of 0.15 (Vishay Micro-Measurements) was selected as the coating material. The photoelastic coating was cut to the same profile as the chain plate using a routing machine and holes of required sizes were made in steps using different size drills. The alignment of the holes in the coating and the steel plate are important for the success of the test. A fixture was made for aligning both the chain plate and the photoelastic coating properly, and positioning pins were also made to maintain the hole concentricity between chain plate and the photoelastic coating. Using standard procedures, the coating was pasted on to the chain plate with proper alignment.

The sample was press-fitted with a formed bush in one of the holes and the extruded bush in the other. The appearance of fringes when the specimen was viewed in a reflection polariscope is shown in Figure 4.15. In the case of the formed bush, the fringe order is maximum near the zone where the bush has its end joining together. When press-fitted with the extruded bush, the fringes near the hole are by and large uniform. The procedure to evaluate quantitative information is discussed next.

The strain-optic law of reflection photoelasticity relates the fringe order and the difference in principal strain (Section 4.8 and [6, 14]) as

$$\varepsilon_1 - \varepsilon_2 = R_f \frac{N}{2h_c} \frac{\lambda}{K} \tag{4.9}$$

where R_f is the reinforcement factor, N is the fringe order, h_c is the photoelastic coating thickness, λ is the wavelength of light, and K is the strain coefficient of photoelastic coating. Since the chain plate is loaded in-plane, the correction factor for plane stress conditions can be used, which is given as

$$R_f = 1 + \frac{h_c E_c (1 + \nu_s)}{h_s E_s (1 + \nu_c)} \qquad (4.10)$$

where E_c is the Young's modulus of the photoelastic coating, E_s is the Young's modulus of the chain plate material, ν_c is the Poisson's ratio of the photoelastic coating, ν_s is the Poisson's ratio of the chain plate material, and h_s is the thickness of the chain plate.

The edges of the plate are far away from the hole and, as the fringes are concentric as in a Lame's problem for a perfectly cylindrical bush, as a first approximation the plate can be considered as a thick cylinder subjected to internal pressure. The maximum values of the radial stress (σ_r) and circumferential stress (σ_θ) components are given by [1]

$$\sigma_{\theta\,max} = \frac{p_c(a^2 + b^2)}{(b^2 - a^2)} \qquad (4.11)$$

$$\sigma_{r\,max} = -p_c \qquad (4.12)$$

where a is the inner diameter of the chain plate hole, b is the outer diameter of the chain plate, and p_c is the contact pressure between the bush and the chain plate. Since the radial and circumferential components themselves are principal stresses, the principal strain difference can be obtained from the stress–strain relations as

$$(\varepsilon_\theta - \varepsilon_r) = \frac{1 + \nu}{E_s} (\sigma_\theta - \sigma_r) \qquad (4.13)$$

Based on photoelastic data, it is possible to find the contact pressure using Eqs. (4.9) and (4.13). Once the contact pressure is known, the radial and tangential stress components can easily be determined from Eqs. (4.11) and (4.12).

4.12.4
Stress Concentration Factor due to Applied Load

In order to evaluate the stress field due to the application of the load, a photoelastic model of the inner plate was pulled with a load equivalent to the operating load on a test rig. Figure 4.16 shows the fringe pattern. The experiment revealed that the stress concentration in the vicinity of the hole was 3.5 times the far-field stresses. To arrive at the maximum stress at the interface, the tangential stress due to the interference fit needs to be added to the direct stresses due to the application of the load. The combined stresses have led to failure in the field. Fatigue test of actual chain assembly by varying the bush slit orientation of the formed bush demonstrated that the chain plate with the bush slit oriented perpendicular to the applied load direction had a lower fatigue life. The fatigue life improved substantially when an extruded bush was used.

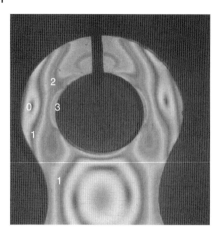

Figure 4.16 Fringe pattern observed in an epoxy chain plate model under tensile load. (Courtesy [13]).

4.13
Comprehensive Experimental Study on a MEMS Pressure Sensor

A firm involved in the manufacture of MEMS pressure sensors needs to set up an experimental facility to calibrate its pressure sensor as well as to inspect for defects in the fabricated sensors. Figure 4.17 shows the design of the pressure sensor. It is to be checked whether the deflection of the silicon wafer is linearly related to the external pressure applied. In view of its size, one has to look at selecting an

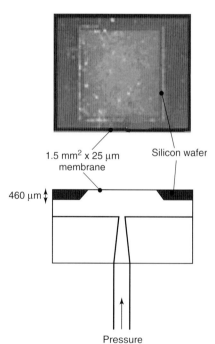

Figure 4.17 Silicon-based MEMS pressure sensor along with the rigidly mounted leak-proof pressure housing unit. (Adapted from [17]).

experimental approach that is capable of microscale measurements. One possibility is to use micro-moiré interferometry [15]. Although it is convenient to measure the deflection due to pressure, it is not quite convenient to inspect for defects.

Speckle interferometry can be used for out-of-plane measurements. A variation of speckle interferometry known as *shearography* measures the slope of the out-of-plane displacement, which facilitates easy detection of defects. Rapid strides have been achieved in shearography and it has matured into identifying defects in a variety of problems. With recent developments, it is possible to obtain even quantitative full-surface strain measurement [16]. Thus microscale speckle interferometry could be used for deflection measurement and, with suitable modification of the equipment, it is possible to conduct microscale shearography.

4.13.1
Microscale Speckle Interferometry and Shearography

Speckle interferometry with the use of a CCD camera and the digital processing of the correlation speckle patterns is also referred to as *TV holography*. A specialized microscopic TV holographic system is required for the study. The use of ordinary microscope objectives to magnify microelements restricts the working distance available for imaging the object. The appropriate microscopic scheme has to meet the needs of imaging the microelement with high resolution and magnification, and at the same time ensure enough working distance so as to enable good illumination of the object.

A schematic representation of the microscopic TV holographic system [18] to meet some of the requirements cited above is shown in Figure 4.18. The microscopic imaging system consists of a Thales-Optem Zoom 125C LDM microscope with extended zoom range and a Sony 2/3″ CCD camera (XC-ST70CE). The resolution of the CCD is 752(H) × 582(V) pixels and the size of each pixel is 11.6 µm(H) × 11.2 µm(V). The CCD is interfaced to a PC with an NI1409 frame grabber card. The zoom LDM provides a 12.5 : 1 zoom ratio, at a working distance of 89 mm with a 1.0× objective. The zoom ratio in the system can be further increased by using a higher magnification objective lens. However, the working distance will reduce for a higher zoom ratio. In the present setup, the magnification can be varied in steps (1.0–12.5×) and the specimen dimensions of 8.4 mm × 6.3 mm at low magnification (1.0×) and 0.68 mm × 0.51 mm at high magnification (12.5×) cover the full area of the 2/3″ CCD.

In the microscopic TV holography system, a narrow 50 mW laser beam from a diode-pumped solid-state mini 532 nm continuous-wave (CW) Nd:YAG laser (CompassTM315M) is expanded using a spatial-filtering (SF) setup and collimated with a 150 mm focal length collimating lens (CL). A variable neutral density (VND) filter in front of the laser controls the power of the laser beam, and an iris diaphragm in front of the lens is used to adjust the size of the collimated beam. The collimated laser beam illuminates the object and a reference mirror via a cube beam splitter (BS). A neutral density filter (NDF) in front of the reference mirror allows controlling the intensity ratio between the object and reference wave. The

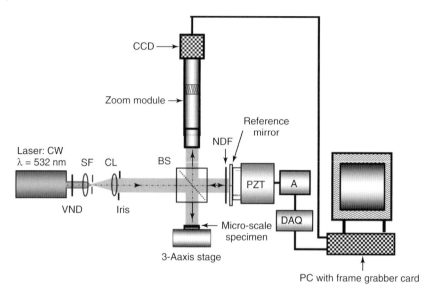

Figure 4.18 Schematic representation of a microscopic TV holographic system: VND, variable neutral density filter; SF, spatial filter; CL, collimating lens; BS, cube beam splitter; NDF, neutral density filter; A, amplifier; DAQ, data acquisition system. (Courtesy [17]).

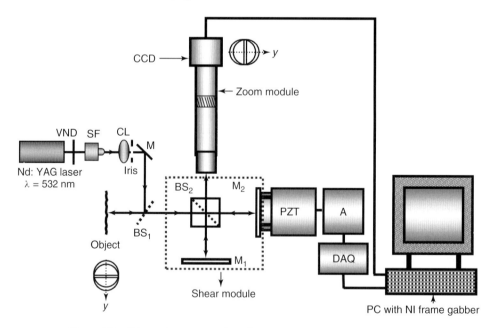

Figure 4.19 Schematic representation of a microscopic TV shearographic arrangement. (Courtesy [17]).

specimen is mounted on a three-axis stage for alignment. The system is fitted with a fiber-based white light illumination (VSI 220 Illuminator, 220 V/150 W) for initial alignment and focusing of the test specimen onto the CCD camera.

The optical arrangement provides enough space between the test object and the LDM, and hence allows the optimization of the sensitivity vector of the system, so that the deformation components can be determined with the highest sensitivity. In the present arrangement, the collimated illumination and the observation beams are in line and hence the sensitivity vector is perpendicular to the test object. Therefore, the system is predominantly sensitive for the measurement of out-of-plane deformation.

By including a conventional Michelson shearing interferometer, Paul Kumar *et al.* [19] have shown that their microscopic TV holography system can function as a microscopic TV shearography system (Figure 4.19).

4.13.2
Deflection Measurement of the Pressure Sensor

For deflection measurement, the specimen is coated with a white spray to act as a diffusely scattering specimen. The pressure sensor is connected to a pressure

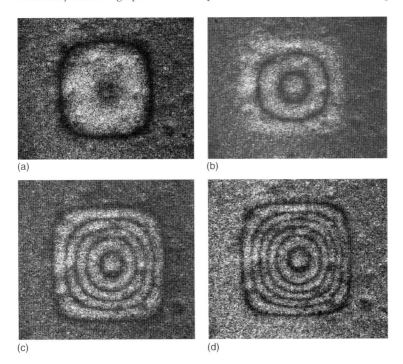

Figure 4.20 Real-time speckle correlation fringe patterns obtained from the pressure sensor at pressures (a) 10 kPa, (b) 20 kPa, (c) 40 kPa, and (d) 50 kPa. (Courtesy [18]).

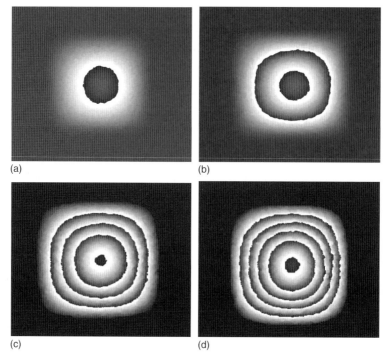

Figure 4.21 Filtered phase maps at pressures (a) 10 kPa, (b) 20 kPa, (c) 40 kPa, and (d) 50 kPa. (Courtesy [17]).

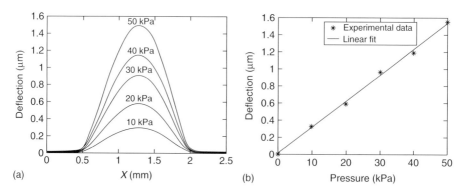

Figure 4.22 (a) Deflection profiles for various pressure loads and (b) maximum deflection as a function of pressure. (Courtesy [18]).

pump, which is monitored using a precision pressure gauge of 0–60 kPa with an accuracy of ±0.5 kPa. The sensor area is suitably focused on to the CCD with white light illumination. The test surface and the reference mirror are then illuminated by a collimated laser beam via the BS. The scattered object wave and the smooth reference wave are combined at the CCD plane and, by using image subtraction, the real-time correlation fringes (Figure 4.20) are obtained for the pressure values of 10, 20, 40, and 50 kPa, respectively.

In order to enhance the quality of data reduction, a seven-step phase shifting was adopted, and the resultant filtered phase maps are shown in Figure 4.21.

Figure 4.22a shows the deflection profile obtained by processing the filtered phase maps for different pressure loadings. From this graph, the maximum deflection of the membrane is plotted as a function of the pressure. The graph obtained establishes that a linear relationship exists between the two and the design of the pressure sensor is satisfactory.

The same pressure sensor, which is of good quality without any defects, is analyzed to obtain the slope fringes. The real-time slope fringes and filtered phase maps for two different pressure loads (40 and 60 kPa) obtained by microscopic TV shearography [19] are shown in Figure 4.23.

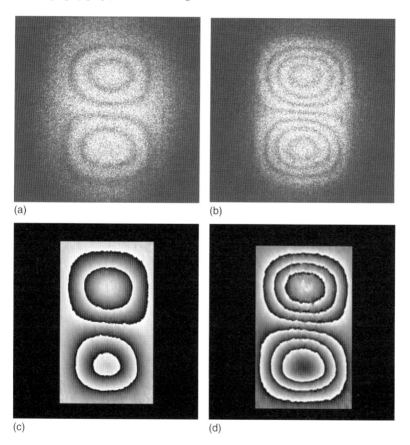

Figure 4.23 Slope fringes by shearography: real-time fringes (a) 40 kPa, (b) 60 kPa; filtered phase maps: (c) 40 kPa, (d) 60 kPa. (Courtesy [17]).

The fringes thus obtained serve as benchmark fringes to accept or reject a pressure sensor fabricated. In view of the successful demonstration of the capability of the microscopic speckle interferometry and shearography to measure relevant parameters of the MEMS pressure sensor, the firm in question was advised to invest in such a system. It may be worthwhile to point out that the possible extensions of the basic system to function as microinterferometry [20] as well as for time-average holography [17] have already been demonstrated.

4.14
Conclusions

In this chapter, rudiments of solid mechanics followed by a brief overview of experimental methods were given. Initially, the genesis of stress and strain was discussed. It was shown how looking at new ways to summarize the results of simple load–elongation relations has helped in arriving at the quantities such as stress and strain. The concept of stress vector, stress tensor, and strain tensor were then discussed. With the success of mathematical development of stress as stress tensor and from it the stress vector, the possibility of predicting the failure planes correctly was illustrated.

The essential difference in an experimental approach compared to numerical or analytical approaches, as well as what dictates the determination of direct information in an experimental technique, was then brought out. The basic information one would need to determine in a general stress analysis scenario and the need for determining new parameters such as SCF and SIF through experimental means were indicated. The steps involved in selecting a suitable technique were briefly discussed.

Finally, two case studies were presented: one that involves a simple mechanical system and the other one deals with the latest MEMS application. The focal issues were different in these two case studies. The first case study was on determining the cause of random failures in a chain assembly. Initial analysis by finite elements demanded a complete three-dimensional experimental stress analysis to identify the root cause. However, it was shown how the innovative use of transmission and reflection photoelastic analysis helped in identifying the root cause of the problem. The study revealed that the stresses developed due to interference fitting are quite significant. Further, the geometric undulation of the formed bush had an influence on the pressure distribution on the inner hole periphery of the inner plate. The geometric undulation of the formed bush could not have been easily modeled either in a numerical study such as finite elements or if only a model study using experiments was undertaken. The use of analytical solution of the Lame's problem was judiciously used to interpret the experimental results. The case study chosen also illustrated that the industry wanted their practical problem to be solved through the use of experimental stress analysis rather than the determination of stresses as a routine exercise.

The second case study focused on identifying a suitable experimental facility to determine the out-of-plane displacements of a MEMS pressure sensor for calibration purposes and also to monitor the quality of the fabricated pressure sensor. The use of microscopic speckle interferometry for calibration purposes as well as of microscopic shearography for quality inspection was suggested. The novelty of the chosen system for future expansion was also indicated.

Acknowledgments

The author gratefully acknowledges the discussions he had with Prof. M. P. Kothiyal, Dr N. Krishna Mohan, and Dr Paul Kumar in writing the case study related to the MEMS pressure sensor. Special thanks are due to Dr Paul Kumar for having provided results and figures, which are part of his Ph.D. dissertation.

Problems

4.1 Several advancements in engineering and science were possible through proper use of selecting the variables for plotting a graph. Can you explain with neat sketches how the concept of stress and strain came about from plotting the load–displacement relations for a material?

4.2 Understanding the relative values of the elastic strain and plastic strain is important. Using the result of the tension test on a mild steel sample, indicate the elastic and plastic regions. What is the procedure adopted for finding the yield stress of a material if the stress–strain curve is smooth.

4.3 The figure shows a specimen subjected to tension. For a general point P shown in the figure

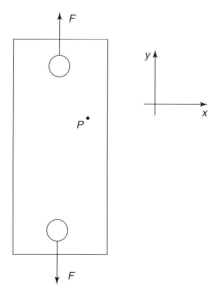

a) Sketch of stress vector acting on all the planes passing through the point as a function of θ. Represent it as a polar plot.
b) Polar plot of normal stress acting on all the planes.
c) Polar plot of shear stress acting on all the planes.

4.4 At a point in a stressed material, the stress tensor is given as

$$\tau_{ij} = \begin{bmatrix} -40 & 72 & 32 \\ 72 & 80 & 46 \\ 32 & 46 & 120 \end{bmatrix} \text{ MPa}$$

a) Calculate the stress vector on the plane whose normal makes an angle 48° with the x-axis and 61° with y-axis.
b) Calculate the component of traction perpendicular to the surface and tangential to the surface.
c) Calculate the angle between the stress vector and the normal to the surface.

4.5 Define principal stress and principal strain at a point. Can stress or strain tensors be written based on these? If so, mention their forms.

4.6 Find the principal stresses for the following cases of plane stress:

(a) $\sigma_x = 40$ MPa
$\sigma_y = 0$ MPa
$\tau_{xy} = 80$ MPa

(b) $\sigma_x = 140$ MPa
$\sigma_y = 20$ MPa
$\tau_{xy} = -60$ MPa

(c) $\sigma_x = -120$ MPa
$\sigma_y = 50$ MPa
$\tau_{xy} = 100$ MPa

4.7 A batch of aluminum alloy yields in uniaxial tension at the stress $\sigma_{ys} = 330$ MPa. If this material is subjected to the following state of stress, will it yield according to (a) von Mises criterion or (b) Tresca yield criterion?

$$\sigma_x = 138 \text{MPa} \quad \tau_{xy} = 138 \text{MPa}$$
$$\sigma_y = -69 \text{MPa} \quad \tau_{yz} = 0$$
$$\sigma_z = 0 \quad \tau_{zx} = 0$$

4.8 What do you understand from a closed-form solution? Do you need closed-form solutions for all problems on hand?

4.9 What does a stress raiser do? Capture all its features and explain your answer with a suitable example, and if possible support your reasoning with fringe patterns.

4.10 What was the key finding of Inglis? How did it prompt researchers to look for new approaches in mechanics?

4.11 Highlight the essential difference between the analytical, numerical, and experimental methods. Mention in which classes of problems these methods are useful.

4.12 List out the information directly obtainable from various experimental methods.

4.13 What are the various fringes that can be seen in different experiments? Summarize in a table the name of the contour and the physical parameter it represents.
4.14 List the various factors in choosing an experimental technique.
4.15 Based on the strain range, can you guide how the various experimental techniques can be selected?
4.16 When do you think the strain gauge technique is the ideal choice?
4.17 How do you go about in selecting a whole-field technique versus a point-by-point technique?
4.18 Do you always need newer techniques to solve a problem? Comment.
4.19 Selection of an appropriate experimental technique is the most crucial step in the analysis of practical problems. In most of the cases, one may have to resort to a combination of several techniques. Owing to the complexity of the problem, in some cases only model analysis can be carried out, while in others only prototype analysis is simpler to do. Further, the accuracy needed also guides the choice of an experimental technique.

For the following problems, mention the appropriate technique/combination of techniques. For each case, indicate whether analysis on model/prototype or both is required. If several techniques are possible, grade them according to accuracy.

a) Structural optimization of fillets and cutouts.
b) *In situ* measurements on TV towers over prolonged period of time.
c) Evaluation of residual stresses in heat-treated components.
d) Stress analysis of truck tyres.
e) SIF evaluation of flaws in pressure vessels. (i) Surface flaws. (ii) Subsurface flaws.
f) Weight reduction of aircraft landing gears.
g) Thermal stress analysis of IC chips.

5
Digital Image Correlation

François Hild and Stéphane Roux

5.1
Introduction

The basics of optics, imaging and image processing have been treated in previous chapters, and the widespread and ever-growing fields of digital image correlation (DIC) are covered in this chapter. DIC consists in registering two (or more) images of the same scene (e.g., the surface of a loaded sample or structure) and extracting the displacement fields that enable obtaining the best match. DIC, which is used in solid mechanics, is the counterpart of particle image velocimetry (PIV) used in fluid mechanics. DIC [1–4] was initiated a few years later than PIV [5–9]. However, because the sought displacement and strain resolutions were smaller than the velocity resolutions, its development was slower because the algorithmic challenges were more difficult to address. Today, though, it can be stated that the performance of DIC techniques allows the experimentalist to utilize them in most practical cases.

The reasons for the growing interest in DIC include the following:

- The cost of charge-coupled device (CCD) and complementary metal-oxide-semiconductor (CMOS) sensor-based cameras have decreased substantially, whereas their definition has improved (e.g., 10 Mpixel cameras are common). The dynamic range (or bit depth) generally exceeds 8 bits; 12-bit and even 16-bit devices are available nowadays.
- It provides magnifications leading to a physical size of 1 pixel ranging from 1 cm (or even more) to 1 µm. For instance, far-field (or long-distance) microscopy enables magnifications equivalent to those of optical microscopy [10, 11], although the working distance varies between 15 and 35 cm.
- The development of analog and, more recently, digital high-speed cameras provide pictures at a rate of 10^4 to 10^6 frames per second, with full resolution, which can be analyzed by resorting to DIC [12–14].
- The natural texture of samples (e.g., concrete, raw surface of a ceramic) or those patterned by spraying B/W paints to create a random texture can be used with

no influence on the mechanical response, and no irreversible alteration of the surface if needed for further analysis.
- The analysis can be performed on digital images provided by other imaging techniques (e.g., atomic force microscopy (AFM) [15–17], and scanning electron microscopy (SEM) [18, 19]) when distortions and artifacts are controlled or corrected. Special attention should be paid to minimize the measurement errors. The physical size of one pixel may reach the nanometer, and even subnanometer, range.
- When two (or more) cameras are used, 3-D shape and displacement fields can be measured by stereo-correlation [20]. The latter allows one to monitor curved surfaces of samples and structures. Moreover, it is possible to register more than one pair of pictures at a time by resorting to spatiotemporal analyses [21].
- X-ray computed microtomography or magnetic resonance imaging (MRI) can be utilized to image opaque (to the eye) materials [22], and their microstructure is registered when *in situ* experiments are carried out by resorting to digital volume correlation (DVC [23–27]). The physical size of one voxel ranges from submicrometer levels to 100 μm.

The listed items show that measurements can be performed at different scales with essentially the same processing technique. To be completely quantitative, noise, distortions, artifacts, and drifts should be assessed very carefully since they are dependent on the imaging device from which the pictures are obtained.

The main output of a DIC analysis is a displacement field that contains a very large number of degrees of freedom (typically 10^3 to 10^4 of kinematic unknowns for 2-D pictures, and 10^4 to 10^6 of kinematic unknowns for reconstructed volumes), which may be compared with classical measurement devices of universal testing machines (e.g., extensometers, strain gauges, linear variable displacement transducers (LVDTs)) providing a single quantity. Advanced DIC codes also provide additional information, such as residual maps, a priori uncertainty evaluation, or noise sensitivity. The latter ones are important to set or adjust analysis parameters to make the best out of a set of images for the projected use of the kinematic fields. It is even possible to control an experiment with DIC by prescribing average strains [28] or, say, a history of stress intensity factors (SIFs) [29].

Finally, let us note that identification techniques of material parameters using full-field measurements (e.g., provided by DIC analyses) call for specific developments [30]. The next step then consists in validating material models and numerical models [31]. Some simple examples will be discussed thereafter.

The chapter is organized as follows. Section 5.2 introduces the basic principles of DIC. Simple one-dimensional examples are used to illustrate what type of analysis can be performed. 2D-DIC is discussed in Section 5.3. Section 5.4 is devoted to the 3-D extension (i.e., stereo-correlation or 3D-DIC) of 2D-DIC. DVC is then introduced in Section 5.5. After a brief conclusion in Section 5.6, some unsolved problems (Section 5.7) with hints to answers allow the reader to make sure he/she has mastered the key concepts of this chapter.

5.2 Correlation Principles

5.2.1 Determination of the Optical Flow

Optical flow corresponds to the apparent motion of surfaces, objects, or even volumes in a scene with respect to the observer (e.g., human eyes or cameras). In the field of experimental mechanics, the displacement or velocity fields associated with flows or loaded samples and structures are sought. This is made possible by using pictures recorded at different instances of time. As mentioned previously, different acquisition devices can be used to generate digital pictures. Sequences of pictures are analyzed to estimate the motion (i.e., instantaneous velocities in fluid mechanics, displacements in solid mechanics). In most of the approaches followed up to now, the motion is determined by considering two image frames f and g at each pixel (or voxel) position $\mathbf{x}$ shot at times t and $t + \Delta t$. Pictures f and g are therefore matrices for which each elementary component (pixel in a 2-D case, voxel in a 3-D case) is a gray-level value. It is worth noting that, if color pictures are used, it is important to know the color filter array of the sensor (e.g., Bayer filter) when storing the raw data. In many instances, they lead to unreliable interpolations if the raw pixel information is used, but can be easily transformed into B/W images by averaging 2×2 pixels in the case of a Bayer filter.

For an n-D case, a pixel at location $\mathbf{x}$ with intensity $f(\mathbf{x})$ will have moved by $\mathbf{u}(\mathbf{x})$ between the two image frames, so that the equation

$$f(\mathbf{x}) = g[\mathbf{x} + \mathbf{u}(\mathbf{x})] \tag{5.1}$$

corresponds to the local (i.e., for any pixel) gray level conservation. When the movement is small (i.e., the Lagrangian and Eulerian configurations coincide), a first-order Taylor expansion of $g[\mathbf{x} + \mathbf{u}(\mathbf{x})]$ reads

$$g[\mathbf{x} + \mathbf{u}(\mathbf{x})] \approx g(\mathbf{x}) + \nabla g \cdot \mathbf{u}(\mathbf{x}) \tag{5.2}$$

or equivalently,

$$f[\mathbf{x} - \mathbf{u}(\mathbf{x})] \approx f(\mathbf{x}) - \nabla f \cdot \mathbf{u}(\mathbf{x}) \tag{5.3}$$

From these two equations, it follows that

$$f(\mathbf{x}) - g(\mathbf{x}) - \nabla f \cdot \mathbf{u}(\mathbf{x}) \approx 0 \quad \text{and} \quad f(\mathbf{x}) - g(\mathbf{x}) - \nabla g \cdot \mathbf{u}(\mathbf{x}) \approx 0 \tag{5.4}$$

This linearized equation in n unknowns cannot be solved as such. This is known as *the aperture problem of the optical flow algorithms*. Let us imagine a set of parallel white lines along a direction $\mathbf{i}$ in a black background. By definition, $\nabla f(\mathbf{x}) \cdot \mathbf{i} = 0$. If the motion occurs in a direction parallel to the lines (i.e., $\mathbf{u} = u_i \mathbf{i}$), it cannot be detected since $\nabla f \cdot \mathbf{u}(\mathbf{x}) = 0$. The motion direction of a contour is ambiguous, because the component parallel to the line cannot be inferred based on the visual input. This means that a variety of contours of different orientations moving at different speeds can cause identical responses. Conversely, the motion is only detectable along the direction of $\nabla f(\mathbf{x})$. To evaluate the optical flow, a set of additional hypotheses is needed.

At this stage, it is worth asking whether Eqs. (5.2) and (5.3) make any sense. Are motions always small in amplitude? The answer is clearly no. Even if they were small, is a picture gradient always well defined? Yet again, the answer is no! To circumvent these two paradoxes, pyramidal (or coarse-graining, or multiscale) approaches are considered (see Section 5.3.1). They consist in defining a series of coarsened images at different scales starting from the finest one, which is the picture itself. To coarsen an image, the mean gray level of each set of 2^n (where n is the dimensionality of the image) neighboring pixels is assigned to a "super-pixel". The series is built from the recursive application of the coarsening step to the reference picture. For each scale transition, the amplitude of the displacements (when expressed in terms of super-pixels) is divided by 2. Consequently, multiscale approaches are a logical way of determining the optical flow, starting from the coarsest scale down to the original one of the picture. As a by-product of this type of approaches, the problem size (and hence computation time) becomes smaller and smaller as coarser scales are considered.

It is worth noting that in all practical cases, Eq. (5.1) is not strictly satisfied because of acquisition noise or even reconstruction artifacts when dealing with AFM, SEM, or tomography pictures. This "simple" fact shows that, in addition to the aperture problem, a pixel determination of the motion is possible only with additional hypotheses (i.e., regularization). In some cases, the nonconservation may even carry additional information (e.g., an out-of-plane motion in AFM pictures [17], a temperature variation when infrared (IR) pictures are used to measure temperature and displacement fields [32]).

To measure $\mathbf{u}$, a "suitable" norm between the signal difference is chosen. The norm is then minimized with respect to $\mathbf{u}$

$$\min_{\mathbf{u}} \|f - g(. + \mathbf{u})\|^2_{\text{ZOI}} \tag{5.5}$$

Usually, the 2-norm is chosen. The minimization of the gray-level residual is therefore written on a subimage, or a zone of interest (ZOI)

$$\int_{\text{ZOI}} \left(f(\mathbf{x}) - g[\mathbf{x} + \mathbf{u}(\mathbf{x})]\right)^2 d\mathbf{x} \tag{5.6}$$

and corresponds to a local approach (also referred to as *block matching*) that consists in minimizing the sum-of-squared differences or maximizing the cross-correlation (see Sections 5.2.2 and 5.3.2.1). The whole region of interest (ROI) may also be considered so that the global residual is minimized

$$\mathcal{T}_{\text{DIC}} = \int_{\text{ROI}} \left(f(\mathbf{x}) - g[\mathbf{x} + \mathbf{u}(\mathbf{x})]\right)^2 d\mathbf{x} \tag{5.7}$$

This corresponds to a global (or variational) approach (see Sections 5.2.3 and 5.3.3.1). The following sections introduce the general concepts of image correlation. For the sake of simplicity, the considered "images" are 1-D signals.

5.2.2
Local DIC

Let us consider two signals $f(x)$ and $g(x) = f(x - u)$, the second being a mere translation by u of the first one. To determine u, the sum-of-squared differences

$$\int_{ZOI} [f(x) - g(x + u)]^2 \, dx \tag{5.8}$$

is minimized with respect to u. When u is a constant (as in the present case), the minimization is equivalent to maximizing

$$(f \star g)(u) = \int_{ZOI} f(x) g(x + u) \, dx \tag{5.9}$$

The symbol $\star$ represents the (cross)correlation of functions f and g. This is the reason why this type of registration is called a *correlation technique*. By extension, it is still called a *correlation technique* even if the minimization is based on the sum-of-squared differences.

The correlation function can be calculated in the real space. It can also be obtained in Fourier space. In practice, the signals are sampled. Consequently, the discrete Fourier transform of the signal

$$f_p = f(x_0 + p \delta x) \quad \text{with} \quad p = 0, 1, \ldots, P - 1 \tag{5.10}$$

reads

$$\text{FT}[f]_k = \frac{1}{\sqrt{P}} \sum_{p=0}^{P-1} f_p \exp(-2 i \pi k p / P) \tag{5.11}$$

The inverse Fourier transform is written as

$$f_p = \frac{1}{\sqrt{P}} \sum_{k=0}^{P-1} \text{FT}[f]_k \exp(+2 i \pi k p / P) \tag{5.12}$$

Continuous or discrete Fourier transformation can be tedious. Since 1965, fast algorithms have been proposed [33]: they are the so-called fast Fourier transforms (FFTs). Equation (5.9) is rewritten as

$$f \star g = \sqrt{P} \, \text{FT}^{-1} \left[\overline{\text{FT}[f]} \, \text{FT}[g] \right] \tag{5.13}$$

where $\overline{\bullet}$ is the complex conjugate of $\bullet$. Having determined the correlation function, the most likely estimate corresponds to the pixel location of the maximum of the former. When subpixel resolutions are sought, one way is to interpolate (e.g., by a parabolic polynomial) the correlation function around its maximum value. To iterate, shifted pictures are needed. This can be achieved by resorting to the shift/modulation property. Let us consider the translation operator defined by $[T_u f] = f(. - u)$ and calculate its Fourier transform

$$\text{FT}[T_u f](k) = \frac{1}{\sqrt{P}} \sum_{p=0}^{P-1} f_{p-u} \exp\left(\frac{-2 i \pi k p}{P}\right) \tag{5.14}$$

which can be written as $FT[T_u f](k) = \exp(-2i\pi ku)FT[f](k)$. By introducing the modulation operator E_u such that $[E_u f](k) = \exp(-2i\pi kp/P)FT[f](k)$, the previous result becomes $FT[T_u f] = E_u FT[f]$.

Up to this point, only elementary translations for each ZOI were considered. Most available commercial DIC codes are local but consider more sophisticated kinematics. A uniform displacement gradient, or even more complex, say a quadratic, displacement field, is used for each ZOI. However, the only output of the analysis is still the mean displacement at the middle of the ZOI. The term "local" refers to the fact that ZOIs are small subsets that are treated independently. In contrast, the following section introduces "global" approaches.

5.2.3
Global DIC

A different (global) route is followed in this part. Let us assume that the displacement field is approximated by

$$u(x) = \sum u_p \psi_p(x) \qquad (5.15)$$

where $\psi_p(x)$ are chosen, and u_p are the unknown degrees of freedom. This a priori hypothesis corresponds to the regularization that was alluded to earlier. The functional to minimize [see Eq. (5.7)] reads

$$\mathcal{T}_{DIC}(u_p) = \int_{ROI} \left[f(x) - g\left(x + \sum u_p \psi_p(x)\right) \right]^2 dx \qquad (5.16)$$

An iterative scheme is followed by resorting to its first-order approximation [see Eq. (5.4)], which becomes a quadratic function of the corrections δu_p

$$\mathcal{T}_{lin}(\delta u_p) = \int_{ROI} \left[f(x) - \tilde{g}(x) - \sum \delta u_p \psi_p(x) f'(x) \right]^2 dx \qquad (5.17)$$

At each iteration, a new picture $\tilde{g}$ is constructed by using the current estimate $\tilde{u}$ of the displacement field ($\tilde{g}(x) = g[x + \tilde{u}(x)]$). The previous reconstruction requires gray-level interpolations (e.g., linear, spline) since subpixel resolutions are sought. The simple form chosen for u yields a minimization condition that can be written in the matrix form as

$$\mathbf{Mu} = \mathbf{b} \qquad (5.18)$$

where $\mathbf{u}$ is a vector collecting all the kinematic unknowns u_p. The components of matrix $\mathbf{M}$ and vector $\mathbf{b}$ read

$$M_{ij} = \int_{ROI} \psi_i(x) f'^2(x) \psi_j(x) \, dx \qquad (5.19)$$

and

$$b_i = \int_{ROI} (f - \tilde{g})(x) f'(x) \psi_i(x) \, dx \qquad (5.20)$$

In the present case, the search direction is given by f'. It can also be g' or $(f' + g')/2$. The advantage of choosing f' is that matrix $\mathbf{M}$ is computed once and for all, and only vector **b** is updated for each new iteration. When large displacements occur, a multiscale approach (e.g., coarse-graining technique and spatial filtering) is desirable.

5.2.4
Gray-Level Interpolation

In each iteration of the previous analysis, a new picture $\tilde{g}$ is constructed by using the current estimate $\tilde{u}$ of the displacement field. The evaluation of $\tilde{g}$ requires gray-level interpolations (e.g., linear, spline) since subpixel resolutions are sought. To discuss this aspect, the treated case is experimental in order not to bias the analysis and thus deals with genuine difficulties associated with real images. The chosen example fulfills accurately the one-dimensional kinematics (i.e., it was checked by running a 2D-DIC code). A metallic plate is moved with a two-axis motorized stage. Twenty-two pictures were shot by a 12-bit camera (1024 × 1280 definition) and correspond to displacement increments of about 0.05 pixel along the horizontal direction. Figure 5.1 shows the reference picture and the extracted 1-D "picture" at $y = 512$ pixels. A random texture was created by spraying B/W paint.

The dynamic range of the picture is $3951 - 489 = 3462$ gray levels (i.e., 85% of the depth of the CCD sensor). Different gray-level interpolations labeled by a three-letter acronym will be tested in the present analysis:

- (lin) linear,
- (cub) cubic,
- (spl) spline.

This is one of the key parameter to choose in a correlation algorithm, be it local [34] or global in what follows. In all cases, N denotes the number of degrees of freedom chosen to describe the kinematics. Only one type of discretization is considered, namely, $N = 64$ degrees of freedom (or 63 elements with a linear interpolation of the displacement).

Figure 5.2a shows the mean displacement as a function of the picture number for the three different gray-level interpolations. From an affine interpolation of the results, it can be concluded that the displacement increment is equal to 0.051 pixel. Any displacement interpolation leads to the same result (this is true in particular for a piecewise constant displacement, mimicking a local approach). This effect will not be studied hereafter since it is of second-order influence in the present case. It will be addressed in Section 5.2.5.

The best result in terms of mean error (or bias) for the 21 pictures is obtained with the spline interpolation [Figure 5.2b]. The linear and cubic interpolations reach approximately the same level. The dependence of the root mean square (RMS) displacement with the picture number (or the prescribed displacement) is similar for all three interpolations [Figure 5.2c]. The spline interpolation leads to slightly lower RMS levels. The analysis of the correlation residuals is another way of studying the results. Figure 5.2d shows the dimensionless correlation residual for

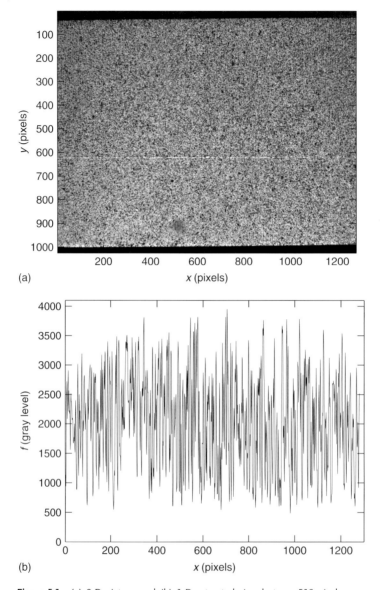

Figure 5.1 (a) 2-D picture and (b) 1-D extracted signal at y = 512 pixels.

the different gray-level interpolations. There is a clear advantage of using a higher order interpolation (i.e., cubic and spline) in comparison with the linear one. The spline interpolation outperforms again the cubic one.

From this analysis, it is concluded that the spline gray-level interpolation leads to the least biased results, the smallest RMS levels (maximum value less than 0.02 pixel), and the smallest correlation residuals (mean value of 0.011 times the dynamic range of the picture).

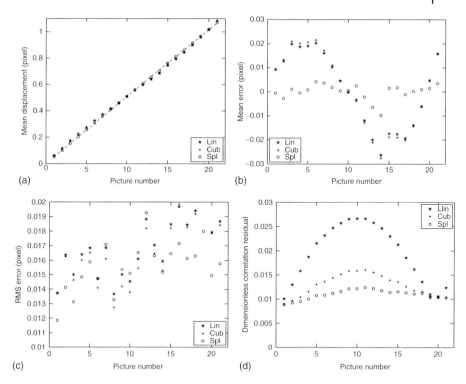

Figure 5.2 (a) Mean displacement, (b) mean displacement error, (c) root mean square (RMS) error, and (d) dimensionless correlation residual versus picture number for different gray-level interpolations when $N = 64$. The dashed line corresponds to an affine interpolation of all the mean displacements as a function of the picture number.

5.2.5
Relaxation of the Gray Level Conservation

In many cases, the gray level conservation is violated. Yet, it is possible to perform a DIC analysis after relaxing the gray level conservation if some information is available. In particular, a commonly encountered situation is that of an affine transformation of the gray-level scale

$$g[x + u(x)] = a(x) + [1 + b(x)]f(x) \qquad (5.21)$$

where a and b are either constant or spatially varying. The above strategy can be followed without much changes. Both fields $a(x)$ and $b(x)$ are decomposed over a suitable basis $a(x) = \sum a_p \varphi_p(x)$ and $b(x) = \sum b_p \varphi_p(x)$. The linearized functional to minimize becomes

$$\mathcal{T}_{\mathrm{lin}}(u_p, a_p, b_p) = \int_{\mathrm{ROI}} \Big[f(x) - g(x) - \sum u_p \psi_p(x) f'(x) \\ + \sum a_p \varphi_p(x) + \sum b_p \varphi_p(x) f(x) \Big]^2 \, \mathrm{d}x \qquad (5.22)$$

5 Digital Image Correlation

All the unknowns are gathered in a single vector

$$\mathbf{u}^T = \{u_p, -a_p, -b_p\} \tag{5.23}$$

and determined simultaneously from a linear system. In the particular case where the interpolation functions chosen for u, a, and b are the same (i.e., $\varphi_p = \psi_p$), it is convenient to introduce a three-component vector

$$\mathbf{G}^T(x) = \{f'(x), 1, f(x)\} \tag{5.24}$$

such that matrix $\mathbf{M}$ consists of elementary 3×3 blocks such that

$$\mathbf{M}_{ij} = \int \psi_i(x) \mathbf{G}(x) \otimes \mathbf{G}(x) \psi_j(x) \, dx \tag{5.25}$$

and, similarly, vector $\mathbf{b}$ consists of vector blocks of length 3

$$\mathbf{b}_i = \int (f - g)(x) \mathbf{G}(x) \psi_i(x) \, dx \tag{5.26}$$

Box 5.1: Various Analyses of a 1-D Signal

This subsection illustrates the above principles in a simple one-dimensional example. The chosen case fulfills quite accurately one-dimensional kinematics. A glass wool whose Poisson's ratio is close to 0 is considered (i.e., the transverse displacements are vanishingly small), and the loss of thickness recovery is studied after a severe compression. Figure 5.3 shows a transmission image of such a glass wool sample before and after the compression test along the vertical axis of the figure.

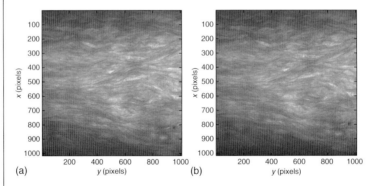

Figure 5.3 (a) Reference and (b) deformed image of light transmission through a thin glass wool specimen before and after compression along the vertical axis x.

A vertical line ($y = 500$ pixels) is extracted from each image and they appear to present a comparable texture as shown in Figure 5.4. However, a significant displacement has taken place, which corresponds to a permanent strain in the medium after a significant compression.

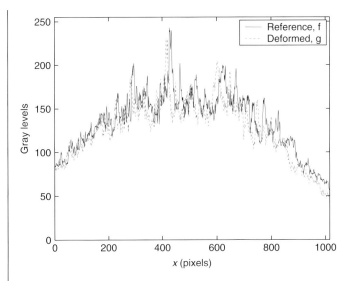

Figure 5.4 One-dimensional "image" taken from the pictures of Figure 5.3 for $y = 500$ pixels (the solid curve is the reference image, while the dashed one is after a uniaxial compression).

As the sample is deformed, the local density changes, and hence the local absorption and scattering are expected to increase. Consequently, the basic hypothesis of gray level conservation is not expected to be exactly satisfied, and gray-level corrections will have to be considered.

Different bases of functions ψ labeled by a three-letter acronym are considered:

- (con) piecewise constant functions,
- (lin) continuous piecewise linear functions,
- (pol) polynomial functions,
- (nsp) natural[1] spline functions (C_1 interpolation),
- (fou) Fourier functions, with a linear function to account for contraction.

In all cases, in order to compare the different methods, N denotes the number of degrees of freedom chosen to describe the kinematics. The same type of interpolation is chosen to analyze the displacement field and the gray-level corrections. However, the number of degrees of freedom was chosen independently for these fields. The gray-level interpolation is carried out with cubic splines.

1) "Natural" means that the curvature vanishes at the two ends of the spanned interval

A Priori Analysis

Prior to the determination of the displacement field, it is of importance to quantify the uncertainty level attached to a specific basis. This analysis is based on the reference image only. The latter is artificially translated by 0.5 pixel to probe the effect of subpixel interpolation errors. This translation is performed in the Fourier space through the "shift-modulation" property (Section 5.2.2). However, because of the lack of periodicity in the reference image, a linear end-to-end gray-level variation is subtracted from the original image prior to the phase shift and added back after translation. This artificially translated image is then treated as the deformed image and the DIC analysis is run. The resulting displacement field is typically close but not strictly equal to the prescribed rigid translation. The difference between the mean (measured) and the prescribed displacements is the systematic error (or *bias*), while the standard deviation of the displacement gives the *uncertainty* level. Note that this procedure involves no noise, and simply probes the quality of the subpixel gray-level interpolation, which is a function of the texture of the original image and of the parameters of the DIC analysis.

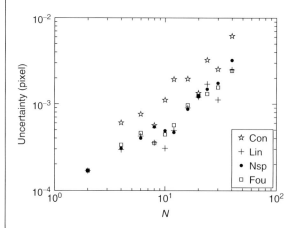

Figure 5.5 Standard displacement uncertainty as a function of the number of degrees of freedom for different displacement interpolations.

One key parameter in this uncertainty analysis is the number N of degrees of freedom chosen for the analysis. The larger the value of N, the fewer the pixels available to determine the value of the kinematic degrees of freedom, and hence the larger the uncertainty. Figure 5.5 shows the change of the uncertainty with N. In that case, the gray-level corrections were fixed to two degrees of freedom (i.e., linear spatial variation) for the gray-level offset (or brightness), and the same for the gray-level rescaling (or contrast). The uncertainty level grows roughly in proportion with N for all bases (note the log–log scale of the graph). All bases give a comparable level of uncertainty, apart from the piecewise constant basis, which is about 3–4 times higher.

The main difference between this basis and the other ones is the absence of coupling between elements (i.e., no continuity is prescribed for a piecewise constant field, as opposed to all other considered bases). Spatial coupling of other functions helps reduce the fluctuations in the displacement field. This trend is also observed in higher dimensions [35], and represents an extra bonus for global methods as compared to local ones. The remarkably low value of the uncertainty is to be emphasized. For a few degrees of freedom, displacement uncertainties less than 10^{-3} pixel can be achieved.

In this plot, the results of the polynomial basis are not shown. For the latter, when N is less than 10, the uncertainty level is comparable with, say, the natural splines basis. However, a large N tends to produce a rather badly conditioned system to invert. On the basis of the MATLAB implementation, the lower eigenvalue modes are discarded from the analysis in such a case, and hence, although the resulting apparent uncertainty level is quite low, this is due to an artificial truncation in the effective number of degrees of freedom used in the analysis. Such a basis may be very convenient, but its use is to be restricted to low-order polynomials. The change of the bias is quite comparable to that of the uncertainty, but much lower, and hence the standard uncertainty is systematically the most limiting parameter.

Effect of Basis Type

The actual deformed image (Figure 5.4) is now systematically used in the sequel. The sensitivity of the resulting displacement field with respect to the DIC parameters is investigated.

Figure 5.6 shows the effect of the number of kinematic degrees of freedom N on the displacement field and the resulting global residual for one particular choice of the basis (i.e., the natural spline). A similar conclusion would be obtained for other bases. The global correlation residual is the RMS of the difference between the reference and the corrected deformed image scaled by the dynamic range of the reference image (i.e., $\max_{ROI} f - \min_{ROI} f$). This global residual is less than 2%, which is very small. It is observed that it decreases with N since displacement fluctuations are better captured. However, the incremental gain decreases. The displacement field appears to be quite stable and only small wavelength features change when N increases.

Figure 5.7 shows the measured displacement field for different bases ($N = 48$ for the kinematic degrees of freedom, and $N' = 2$ for the gray-level corrections). The piecewise constant functions again provide the worst performance in terms of correlation residuals. The other bases, where all degrees of freedom are spatially coupled, give rise to smaller residuals. However, the displacement field appears to be quite close for all bases.

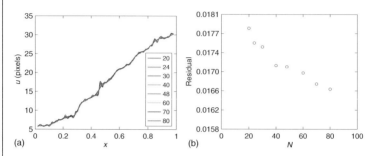

Figure 5.6 (a) Effect of the number of degrees of freedom N on the determined displacement field versus dimensionless abscissa. The natural splines basis is chosen and different discretizations N (given in the caption) are compared. The gray-level corrections both set to $N' = 2$. (b) Corresponding change of the global correlation residual. (Please find a color version of this figure on the color plates.)

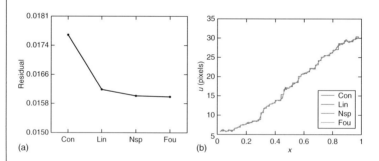

Figure 5.7 Comparison between (a) residuals and (b) displacement fields for different interpolation bases when $N = 48$ and $N' = 2$. (Please find a color version of this figure on the color plates.)

In all the previous examples, the gray-level corrections were reduced to a minimum. Figure 5.8 shows the determined gray-level corrections and kinematic fields for different basis types ($N = 48$ for the kinematic degrees of freedom, and $N' = 28$ for the gray level corrections). It is observed that the results are quite consistent, and the inclusion of a large number of additional parameters did not affect significantly the displacement field. Gray-level corrections appear to be less stable than the displacement, yet the main features are captured by all basis types.

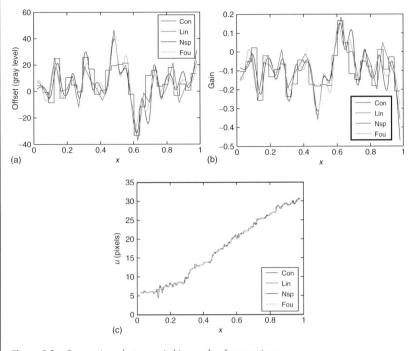

Figure 5.8 Comparison between (a,b) gray level corrections and (c) displacement fields for different bases when $N = 48$ and $N' = 28$. (Please find a color version of this figure on the color plates.)

Tutorial Exercise 5.1: Displacement Resolution

a. The resolution is defined as the "smallest change in the value of a quantity being measured by a measuring system that causes a perceptible change in the corresponding indication. The resolution of a measuring system may depend on, for example, noise (internal or external) [36]. A reference signal is considered and the sensitivity of the measured degrees of freedom u_p to noise associated with image acquisition is assessed. Both reference and deformed images are affected by the same noise. Devise a simple way of accounting for this effect.

b. Are matrix **M** and vector **b** affected by this noise?

c. Express the mean value of δu_p and its covariance $C(\delta u_p, \delta u_q)$.

Solution:

a. The noise-free reference being unknown, this is equivalent to considering a noise ζ in the difference $(g - f)$ of variance $2\sigma^2$ that can be conventionally attributed to g while f is considered as noiseless. ζ is therefore a random white noise of zero mean and covariance $2\sigma^2 \delta(v,w)$, where v and w denote the indices of the pixels, and δ the Kronecker delta.

b. Matrix **M** is unaffected by this noise, but **b** is modified by an increment

$$\delta b_m = \sum_v (f'\psi_m)(v)\zeta(v) \tag{5.27}$$

whose average is thus equal to 0, and its covariance reads

$$C(\delta b_m, \delta b_n) = 2\sigma^2 \sum_v \sum_w (f'\psi_m)(v)\delta(v,w)(f'\psi_n)(w) = 2\sigma^2 M_{mn} \tag{5.28}$$

where M_{mn} are the components of the matrix **M** to be inverted in the DIC problem [see Eq. (5.18)].

c. By linearity [see Eq. (5.18)], the mean value of δu_p vanishes, and its covariance becomes

$$C(\delta u_p, \delta u_q) = 2\sigma^2 M^{-1}_{pm} M_{mn} M^{-1}_{nq} = 2\sigma^2 M^{-1}_{pq} \tag{5.29}$$

To illustrate the result, the reference picture f of the previous analysis (Figure 5.4) is considered. The "deformed" picture g is constructed by adding a white Gaussian noise to f of standard deviation $\sqrt{2}\sigma$. Figure 5.12 shows the change of the standard displacement uncertainty as a function of N for various interpolations.

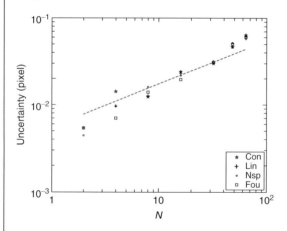

Figure 5.9 Standard displacement uncertainty as a function of the number of degrees of freedom for different displacement interpolations when $\sigma = 1$ gray level. The dashed line corresponds to a square root fit.

In the present case, very similar results are obtained. When compared with the results of the uncertainty analysis (Figure 5.5), it can be concluded that the limiting factor is associated with the level of noise and not the intrinsic performance of the correlation algorithm. This is due to the nature of the studied texture which has a very small correlation length. Therefore, it becomes very sensitive to noise, which is uncorrelated by hypothesis.

To understand the trend observed in Figure 5.9, let us note that each degree of freedom has a length of influence ℓ of the order of L/N, where L is the total length of the picture (i.e., the number of pixels). Consequently, each element of matrix $\mathbf{M}$ is proportional to $\ell \approx L/N$, and to the mean square gradient $\langle (f')^2 \rangle$. From Eq. (5.29), the variance of each measured degree of freedom $\sigma_{u_p}^2 = 2\sigma^2 M_{pp}^{-1} \propto \sigma^2 / \langle (f')^2 \rangle \ell$. The standard displacement uncertainty is thus inversely proportional to $\sqrt{\ell}$ or proportional to $\sqrt{N}$, which corresponds to the interpolation shown in Figure 5.9.

5.3
2-D Digital Image Correlation

5.3.1
Multiscale Analyses and Sequences of Pictures

To increase the maximum measurable displacement between a couple of pictures, two major steps are added [37–39]. First, different scales are introduced, namely, the finer scale is that of the original picture (i.e., typically 1 Mpixels or more with classical CCD cameras). It is referred to as *scale 0* (Figure 5.10). The ROI is a smaller part of this image; it defines the region over which the multiscale procedure will be applied as mentioned in Section 5.2.1. This corresponds to scale 1 (Figure 5.10). From scale 2 on, a coarse-graining procedure is followed to calculate the gray level of each super-pixel (i.e., the average of four neighboring pixels). It is worth noting that at each scale transition s to $s + 1$ (when $s \geq 2$) the displacement amplitude, when expressed in the new frame (i.e., super-pixels), is divided by 2. Consequently, the larger the total number of scales S, the larger the displacement that can be detected by DIC [37] when starting at the highest scale.

When starting at the coarser (i.e., highest) scale S, a first correlation is performed until convergence. From this first computation, the displacement field is interpolated by using the chosen kinematic basis at the given scale for global approaches (i.e., the number of measured degrees of freedom decreases with s). For local approaches, the user has the choice, and in the following section a linear interpolation will be used. To proceed to scale $S - 1$, a first estimate of the displacement field is that determined at scale S. It is used to position the points (e.g., centers of ZOIs for a local approach, nodes of a finite-element (FE) mesh for a global approach) on scale $S - 1$ according to the transformation rules inverted from those described above. The same approach is followed from scales $S - 1$ to 1.

Second, when a sequence of more than two images is analyzed, two routes can be followed. The first one consists in considering the same reference image. It follows that the errors are not acumulated but there exists a maximum strain level above which the method fails [37]. The second one considers that the reference image is the deformed image of the previous step (i.e., updating procedure). Under these hypotheses, there is no real limitation, apart from the fact that the errors are

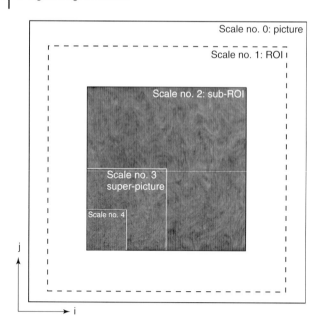

Figure 5.10 Different notations used in the multiscale approach. Sub-ROI of a reference image (scale 2) and corresponding superimages on scales 3, and 4.

now cumulated. Strains of the order of 100% and more are routinely observed in a full-field assessment even with a conventional DIC technique [40].

5.3.2
Local Approach

5.3.2.1 FFT-Based DIC

To determine the displacement field of one image with respect to a reference image, a series of subimages (e.g., a square region or ZOI) is considered. The aim of correlation method is to register the ZOI in the two images. It consists in maximizing the correlation product (see Section 5.2.2)

$$(f \star g)(\mathbf{u}) = \int_{ZOI} f(\mathbf{x})g(\mathbf{x} + \mathbf{u})\,d\mathbf{x} \tag{5.30}$$

The computation of a cross-correlation product can be performed either in the original space or in the Fourier space by using an FFT

$$f \star g = P\,\text{FFT}^{-1}\left[\overline{\text{FFT}\left[f\right]}\,\text{FFT}\left[g\right]\right] \tag{5.31}$$

where the complex conjugate is overlined and P^2 is the number of pixels. To determine the average displacement, the largest value R of an inscribed ROI of size $2^R \times 2^R$ centered in the reference image is extracted. The same ROI is considered in the deformed image. A first FFT correlation is performed to determine the average displacement $\mathbf{u}_0$ of the deformed image with respect to the reference image. This displacement vector is expressed by an integer number of pixels and is obtained as the maximum of the cross-correlation function evaluated for each pixel of the ROI. This first prediction enables the determination of the maximum number of pixels that belong to both images. The ROI in the deformed image is now centered at a point corresponding to the displaced center of the ROI in the reference image by an amount $\mathbf{u}_0$.

The user usually chooses the size L_{ZOI} of the ZOIs by setting the value of $p < P$ so that its size is $L_{ZOI} = 2^p$ pixels. To map the whole image, the second parameter to choose is the shift $\delta = \|\delta \mathbf{x}\| = \|\delta \mathbf{y}\|$ between two consecutive ZOIs: $1 \leq \delta \leq 2^p$ pixels (or even larger, see Section 5.3.2.2). This parameter defines the mesh formed by the centers of each ZOI used to analyze the displacement field. The analysis is then performed for each ZOI *independently*. A first FFT correlation is carried out and a first value of the in-plane displacement correction $\Delta \mathbf{u}$ is obtained. The components of $\Delta \mathbf{u}$ are again integer numbers so that the ZOI in the deformed image can be displaced by an additional amount $\Delta \mathbf{u}$. The displacement residuals are now less than $1/2$ pixel in each direction.

It can be noted that small displacement levels can be accurately measured by using a subpixel algorithm. An additional cross-correlation is performed. A subpixel correction of the displacement $\delta \mathbf{u}$ is obtained by determining the maximum of a parabolic interpolation of the correlation function. The interpolation is performed by considering the maximum pixel and its eight nearest neighbors. Therefore, a subpixel value is evaluated. By using the "shift-modulation" property of Fourier transforms (Section 5.2.2), the deformed ZOI can be moved by an amount $-\delta \mathbf{u}$. Since an interpolation is used, some (small) errors are induced, thereby requiring to re-iterate by considering the new "deformed" ZOI until convergence. The used procedure checks whether the maximum of the interpolated correlation function increases as the number of iterations increases; otherwise, the iteration scheme is stopped.

To measure large displacements as well, a multiscale procedure is applied as discussed in Section 5.3.1. The multiscale correlation algorithm is summarized in the flowchart of Figure 5.11. Starting on the highest (i.e., coarser) scale, a first correlation is performed. From this first computation, the displacement field is interpolated as

$$\mathbf{u}_S(\mathbf{x}_S) = \mathbf{B}_S \mathbf{x}_S + \mathbf{b}_S \tag{5.32}$$

where $\mathbf{B}_S$ and $\mathbf{b}_S$ are first estimates of the mean displacement gradient and rigid body translation at scale S, respectively, and $\mathbf{x}_S$ is the location of the considered ZOI in the reference picture at scale S. To check this first estimate, the center of each ROI of the "deformed" image is relocated according to these first estimates. A new correlation is performed and corrections to the previous displacement field are evaluated. The iterations are stopped as soon as there is no new correction between two iterations for *any* analyzed ZOI. This iterative procedure is implemented to make the displacement evaluation more robust. This robustness is the key to the success of the procedure. If ever the displacement is not properly evaluated on higher scales, there is no chance of getting a good final result [37]. Consequently, the first evaluations have to be performed very carefully.

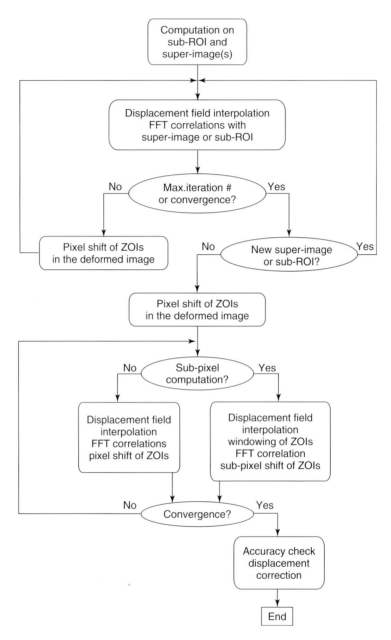

Figure 5.11 Flowchart of the principal steps in the multiscale correlation algorithm [37].

In practice, the user can choose between a first evaluation that is very fast since it remains on the pixel (or super-pixel) level, and an accurate evaluation requiring several iterations. On scale 1, the user selects either a pixel or a subpixel resolution of the displacement by following the above-described subpixel procedure, thereby allowing for the measurement of small levels of displacements as well as large ones.

5.3.2.2 Measurement Uncertainties

DIC aims at *measuring* a displacement field. Hence, it is of key importance to provide a quantitative estimate of the uncertainty attached to a given measurement [36]. The basic ideas have been presented in Section 5.2.5. Many factors contribute to the uncertainty. Bias and noise in the image acquisition are key factors, which should be investigated separately, as they are specific to the chosen imaging device technology and also to the operating conditions in the case of interest. DIC also introduces intrinsic uncertainties that are dependent on the texture of the analyzed image, and hence the resulting uncertainties and systematic errors can (and should) be evaluated prior to any analysis. The proposed principle of analysis is to generate an artificial image from the reference picture with a *known* displacement field, and to blindly measure it by DIC (see Section 5.2.5). The reference image can be itself an artificial one [41] (for validating a specific methodology), but as a common usage, the actual reference image of a test should be chosen [42, 43]. The comparison between the prescribed $\mathbf{u}_p$ and measured $\mathbf{u}_m$ displacement fields provides useful pieces of information.

One choice for $\mathbf{u}_p$ is a uniform translation. The spatial average of $\Delta \mathbf{u} = \mathbf{u}_m - \mathbf{u}_p$ gives an estimate of the systematic error, while its standard deviation σ_u provides the uncertainty [43] for a given $\mathbf{u}_p$. A translation by an integer number of pixels is an ideal case that should not introduce any error. However, a fraction-of-pixel translation necessarily calls for a subpixel interpolation scheme (see Section 5.2.4), which is fundamental for most applications. It is important to realize that the subpixel interpolation is again an ill-posed problem. Hence, the chosen scheme exploits a priori assumptions on the texture at hand. For instance, a continuous linear gray-level interpolation is a possible choice, which is easy to implement and efficient to compute. However, it is only a poor approximation, and more regular C_1, C_2, or C_∞ interpolation schemes have been tested over the years, which have been found to perform better [34]. Spline or Fourier interpolations are convenient and accurate ways to interpolate images. For a uniform translation, the shift/modulation property of Fourier transforms (Section 5.2.2) is very well suited to such artificially "deformed" image generation.

The displacement error δ_u and uncertainty σ_u are to be evaluated as averages over the displacements $\mathbf{u}_p(u_p, u_p)$ such that $|u_p| < 1$ pixel. A half-pixel translation $u_p = 1/2$ is typically the worst case, and its error and uncertainty are typically equal to $2\delta_u$ and $2\sigma_u$, respectively [42]. Figure 5.12 shows the change of σ_u with the ZOI size. A power-law decrease is generally observed so that

$$\sigma_u = \frac{A^{\alpha+1}}{L_{ZOI}^\alpha} \tag{5.33}$$

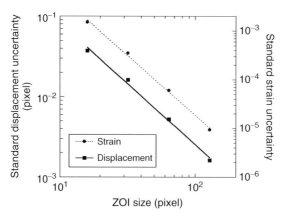

Figure 5.12 Standard displacement and strain uncertainties as functions of ZOI size L_{ZOI}. The solid symbols are correlation results. The solid line is the best fit by a power law of exponent -1.5, and the dashed line is the best fit by a power law of exponent -2.5. In the present case, the strain gauge length is equal to twice the ZOI size.

with $A = 1.43$ pixel (and $\alpha = 1.5$). The larger the ZOI size, the more numerous the pixels, and the smaller the uncertainty. The main limitation is that the spatial resolution decreases, and hence, if the analyzed displacement field is not smooth, it cannot be captured by a coarse description. It is, however, to be emphasized that a displacement uncertainty as small as 5×10^{-3} pixel can be obtained with a ZOI size of 64 pixels.

The strain uncertainty is evaluated with the same approach. A uniform strain field can be chosen to generate an artificially deformed image from the reference picture. However, it is to be noted that the previous analysis can also be used to evaluate the strain uncertainty. The strains are typically evaluated by using a finite differences scheme over a strain gauge length L_ϵ. The statistical independence of the fluctuation in displacement for nonoverlapping ZOIs allows us to express the strain uncertainty σ_ϵ as

$$\sigma_\epsilon = B \frac{\sigma_u}{L_\epsilon} \qquad (5.34)$$

Figure 5.12 shows that such a relationship is very well obeyed for $L_\epsilon = 2L_{ZOI}$, with $B \approx 1.3$ and an exponent $\beta \approx 2.5$, whereas the above argument would lead to $B = \sqrt{2}$ and $\beta = 1 + \alpha = 2.5$. For instance, a ZOI size of 64 pixels gives a strain uncertainty of 6×10^{-5}.

Box 5.2: Controlling a Testing Machine

DIC has the potential to provide a very efficient way of controlling a testing machine not only to follow a specific displacement or strain history but also a more sophisticated kinematic condition. As a simple demonstration of this ability, a strain-controlled experiment was designed where DIC was used to follow a prescribed strain history [28].

For the close-loop system to operate in an efficient way, it is of utmost importance to manage the associated uncertainty level of the measured quantity over which control is set. In the present case, L_ϵ was chosen to be greater than L_{ZOI} to lower the strain uncertainty while keeping small zones of interest, so that the computation time remains small. For the sake of robustness, 3×3 ZOIs are considered with an identical separation $L_\epsilon/2$ between neighboring points. Equation (5.34) holds with a reduced constant $B \approx 0.8$ (see Section 5.3.3.2). The corresponding strain uncertainty 10^{-5} is reached when L_ϵ exceeds 400 pixels for $L_{ZOI} = 64$ pixels. This level was validated a posteriori. In the case of interest, the 550×1000 pixel ROI covers most of the central part of the specimen.

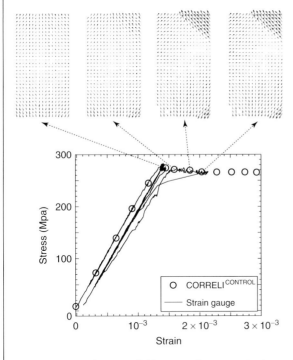

Figure 5.13 Displacement fields (magnification: ×38) determined by DIC ($L_{ZOI} = 32$ pixels, $\delta = 32$ pixels) for different longitudinal strain levels of a tensile test controlled by DIC.

A strain-control mode based on DIC is chosen instead of strain gauges. One major challenge has to be faced, namely, the CCD camera provides the images, and the DIC algorithm has been run to process these data and obtain the strain evaluation over the selected region. Image storage and processing are time-consuming steps that result in a low sampling rate as compared to the testing machine characteristic time. Hence a specific feedback architecture, named CORRELI^{CONTROL}, had to be designed. A tensile test was carried out on a conventional steel, operated by such a DIC-based control. During the test, full images were recorded to allow for post analyses. Figure 5.13 shows different displacement fields at various stages of loading obtained with a classical DIC software, with a uniform coverage of the ROI, and with smaller ZOIs (32 pixels) than the ones used to control the test. In the elastic domain, the a posteriori analysis revealed a uniform strain, with good agreement between the strain gauges and the DIC-control estimates.

However, in the elastoplastic regime, this agreement begins to fail. Post analysis reveals in the elastoplastic region clear shear bands in the lower and upper right corners of the ROI. From the location of the rosette, it can be guessed that the shear band traverses the strain gauge, and hence the measured signal is no longer representative of the actual mean strain. The same conclusion would have been drawn if the strain gauge had been located outside of the shear band. Had the experiment been controlled with the strain gauges, it would not have been possible to carry out the test up to this level of mean strain. A full-field measurement is the only way to handle such cases with the richness of information that allows for both the control and the subsequent interpretation of the data.

Tutorial Exercise 5.2: Spatial Resolution

A simple case is considered to illustrate the fact that the finite extension of the ZOI introduces a filtering of the displacement field. The study is recast as a one-dimensional problem since all directions may be treated in a similar manner. The space dimensionality will come into play only in the evaluation of the effect of noise. The reference picture is characterized by an auto-correlation function $C_0(r)$, which is assumed to be Gaussian:

$$C_0(r) = \exp\left(-r^2/2\xi^2\right) \tag{5.35}$$

where ξ is the correlation length, considered here to be small as compared to the ZOI size ℓ. A displacement field varying as

$$u(x) = A\sin\left(\frac{2\pi x}{\lambda}\right) \tag{5.36}$$

is chosen to exemplify the effect of probing the displacement over a finite sized window. The wavelength λ is chosen to be large compared to the correlation length of the image texture, but it may be comparable to the ZOI size ℓ.

Let x_0 denote the ZOI center. Mere translations are first considered. Hence the correlation $C_1(u)$ is computed over the ZOI as a function of the translation v between the two zones. Assuming a statistically homogeneous texture and a small correlation length, evaluate the cross-correlation.

Solution:

The cross-correlation reads

$$C_1(v) = \frac{1}{\ell} \int_{x_0-\ell/2}^{x_0+\ell/2} C_0(v - u(x)) \, dx \qquad (5.37)$$

The fact that the displacement changes over the ZOI induces a bias in the argument that maximizes the cross-correlation. Unfortunately, it is not possible to obtain a closed-form expression for an arbitrary displacement field or even the chosen harmonic expression. Compute the above expression and identify the argument that maximizes C_1.

The key parameter is the ratio $a = \ell/\lambda$. For a nonzero value of this parameter, the displacement can no longer be considered as constant over the ZOI and hence a bias is expected. Figure 5.14a shows that the estimated displacement departs significantly from the actual local value.

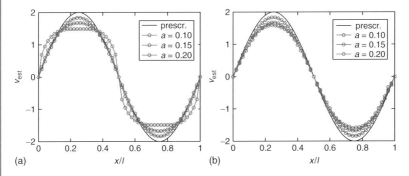

Figure 5.14 Prescribed (prescr.) and estimated displacements as functions of the ZOI center coordinate for different ratios $a = \ell/\lambda$ when (a) a piecewise constant and (b) linear interpolation is used. (Please find a color version of this figure on the color plates.)

Introducing a possible displacement gradient over the ZOI allows one to account much more precisely for the displacement profile. The above expression (5.37) holds, with v varying linearly with x. Figure 5.14b shows that the induced error is smaller near the inflexion point. Close to the maximum displacement (since the mean strain vanishes), the error level is comparable in both cases. To reduce it, a quadratic variation of the displacement would be needed. This observation is the rationale behind the common choice of having an at least linear variation of the displacement over the ZOI.

5.3.3
Global Approach

5.3.3.1 General Formulation
A global approach to DIC consists in minimizing the sum-of-squared differences

$$\mathcal{T}_{\text{DIC}} = \int_{\text{ROI}} (f(\mathbf{x}) - g[\mathbf{x} + \mathbf{u}(\mathbf{x})])^2 \, d\mathbf{x} \tag{5.38}$$

over the considered ROI in which the displacement field is expressed in a *chosen* basis

$$\mathbf{u}(\mathbf{x}) = \sum_n u_n \boldsymbol{\psi}_n(\mathbf{x}) \tag{5.39}$$

where $\boldsymbol{\psi}_n$ are (chosen) vector functions, and u_n the associated degrees of freedom. The measurement problem then consists in minimizing $\mathcal{T}_{\text{DIC}}$ with respect to the unknowns u_n. A Newton iterative procedure is followed to circumvent the nonlinear aspect of the minimization problem. Let $\mathbf{u}^i$ denote the displacement at iteration i, and $\mathbf{u}^i$ the vector containing all the unknown degrees of freedom. By assuming small increments $d\mathbf{u} = \mathbf{u}^{i+1} - \mathbf{u}^i$ of the solution, a Taylor expansion is used to linearize $g(\mathbf{x} + \mathbf{v}(\mathbf{x})) \approx g(\mathbf{x}) + \mathbf{v} \cdot \nabla g(\mathbf{x}) \approx g(\mathbf{x}) + \mathbf{v} \cdot \nabla f(\mathbf{x})$, and then $\partial \Phi_c^2 / \partial \mathbf{u}^i$ is recast in a matrix-vector product as

$$\frac{\partial \Phi_c^2}{\partial \mathbf{u}^i} = \mathbf{M} d\mathbf{u} - \mathbf{b}^i = 0 \tag{5.40}$$

with

$$M_{mn} = \int_\Omega \nabla f(\mathbf{x}) \cdot \boldsymbol{\psi}_m(\mathbf{x}) \nabla f(\mathbf{x}) \cdot \boldsymbol{\psi}_n(\mathbf{x}) d\mathbf{x} \tag{5.41}$$

and

$$b_m^i = \int_\Omega [f(\mathbf{x}) - g(\mathbf{x} + \mathbf{u}^i(\mathbf{x}))] \nabla f(\mathbf{x}) \cdot \boldsymbol{\psi}_m(\mathbf{x}) \, d\mathbf{x} \tag{5.42}$$

To capture large-scale displacements, it is important to perform an initial determination of the displacement field based on low-pass-filtered (i.e., coarsened) images where small-scale details are erased and to progressively restore finer details of images in the determination of subsequent corrections to the displacement field (see Section 5.2.5). This above-discussed multiscale or pyramidal scheme is essential for the robustness and accuracy of DIC.

At this level of generality, many choices can be made to measure displacement fields. In the following, two different routes are followed. The first one consists in using a standard finite element (FE) description of the displacements (Section 5.3.3.2). It is called *FE-DIC*. In that case, no mechanical statement is made except that the displacement field be continuous. Closed-form solutions are used in the second case. The measured degrees of freedom are not necessarily displacements or rotations, but mechanical quantities such as SIFs or elastic parameters. This second approach is referred to as *integrated DIC* (Section 5.3.3.2) in the sense that the measured displacement field satisfies the constitutive equation and equilibrium.

5.3.3.2 Q4-DIC: A Global Approach Using Q4 Elements

FEs are widely used in solid mechanics to perform numerical simulations. It is therefore natural to consider kinematic bases that can be compared directly with FE simulations for identification and validation purposes. In both cases, measured and simulated displacement fields are compared, and therefore it is desirable to utilize exactly the same discretizations so that no interpolation errors are induced at that stage.

Since pictures are sampled in pixels, it is logical to use square or rectangular elements to create a mesh. In the following, only square elements will be considered. The simplest element is a four-noded quadrilateral (Q4) with a bilinear displacement interpolation (i.e., 1, x, y, xy functions, where x and y denote the local coordinates of any point $M(x, y)$). The displacement field is therefore rewritten to account for the shape functions of the chosen Q4 element. Each component of the displacement field is treated in the same way (i.e., only scalar functions $N_n(\mathbf{x})$ are considered) so that the displacement $\mathbf{u}^e(\mathbf{x})$ in each element Ω_e reads

$$\mathbf{u}^e(\mathbf{x}) = \sum_{n=1}^{n_e} \sum_{\alpha} a^e_{\alpha n} N_n(\mathbf{x}) \mathbf{e}_\alpha \tag{5.43}$$

where n_e denotes the number of nodes (here $n_e = 4$), $\mathbf{e}_\alpha$ the unit vector associated with direction $\alpha = 1, 2$, and $a^e_{\alpha n}$ the unknown nodal displacements. Matrix $\mathbf{M}$ (see Section 5.3.3.1) is assembled by considering elementary matrices $\mathbf{M}^e$ whose components read

$$M^e_{\alpha m \beta n} = \int_{\Omega_e} \partial_\alpha f(\mathbf{x}) N_m(\mathbf{x}) \partial_\beta f(\mathbf{x}) N_n(\mathbf{x}) d\mathbf{x} \tag{5.44}$$

and vector $\mathbf{b}$ is assembled by calculating the elementary vectors $\mathbf{b}^e$

$$b^e_{\alpha m} = \int \left[f(\mathbf{x}) - \tilde{g}(\mathbf{x}) \right] \partial_\alpha f(\mathbf{x}) N_m(\mathbf{x}) d\mathbf{x} \tag{5.45}$$

where $\partial_\alpha f = \nabla f \cdot \mathbf{e}_\alpha$, $\partial_\beta f = \nabla f \cdot \mathbf{e}_\beta$, and $\tilde{g}$ is the corrected image.

Matrix $\mathbf{M}$ is symmetric, positive (when invertible), and sparse as in any FE computation, its mechanical equivalent being a mass matrix. However, contrary to classical FE procedures, quadrature formulas are not used because of the irregularity of the picture texture. Instead, a pixel summation is implemented [38].

Box 5.3: Integrated DIC Applied to a Cracked Sample

When dealing with cracks in, say, brittle materials, closed-form solutions are available in plane stress and plane strain conditions when analyzing the kinematics in the vicinity of the crack tip provided linear elasticity still holds [44]. The presence of a crack induces singular stress fields and discontinuous displacements (ψ_5 and ψ_6) across the crack mouth. Their amplitude is proportional to the mode I and II SIFs. The next displacement field (ψ_4) is associated with the so-called T-stress contribution (uniaxial tension along the crack direction). Two other subsingular fields ψ_7 and ψ_8 are added to better capture the kinematics close to the boundary of the ROI. Finally, three rigid body motions (i.e., two translations ψ_1, ψ_2, and one in-plane rotation ψ_3) are considered to account for average motions of the ROI with respect to the camera.

For the sake of simplicity, complex-valued fields are expressed in the crack frame, namely, the origin corresponds to the crack tip, and the crack face is described by the negative real axis. Any point M is represented by a complex number $\mathbf{z} = re^{i\theta}$, and any in-plane displacement $\mathbf{u} = u_x + iu_y$. The eight displacement fields read

$$\begin{aligned}
\psi_1(\mathbf{z}) &= 1 \\
\psi_2(\mathbf{z}) &= i \\
\psi_3(\mathbf{z}) &= i\mathbf{z} \\
\psi_4(\mathbf{z}) &= (\kappa - 1)\mathbf{z} + 2\bar{\mathbf{z}} \\
\psi_5(\mathbf{z}) &= \sqrt{r}[2\kappa e^{i\theta/2} - e^{3i\theta/2} - e^{-i\theta/2}] \\
\psi_6(\mathbf{z}) &= i\sqrt{r}[2\kappa e^{i\theta/2} + e^{3i\theta/2} - 3e^{-i\theta/2}] \\
\psi_7(\mathbf{z}) &= \sqrt{r^3}[2\kappa e^{3i\theta/2} - 3e^{i\theta/2} + e^{-3i\theta/2}] \\
\psi_8(\mathbf{z}) &= i\sqrt{r^3}[2\kappa e^{3i\theta/2} + 3e^{i\theta/2} - 5e^{-3i\theta/2}]
\end{aligned} \quad (5.46)$$

where

$$\kappa = \frac{(3-\nu)}{(1+\nu)} \quad (5.47)$$

in plane stress condition, as expected along the free observation surface, with ν being Poisson's ratio. The generalized degrees of freedom are related to the amplitude of rigid body translations (u_1, u_2) and rotation (u_3), or proportional to the T-stress component (u_4), mode I and II SIFs (u_5, u_6), and subsingular mode I and II contributions (u_7, u_8). The eight degrees of freedom are real-valued, and thus the decomposition (5.39) applies. The global approach discussed in Section 5.3.3.1 is particularized to account for these eight fields, some of which directly yield (i.e., with no additional postprocessing) the sought mechanical parameters (e.g., SIFs). This approach is therefore referred to as *integrated DIC* or *I-DIC* [35, 45], in the sense that the measured displacement field satisfies equilibrium and obeys a given material behavior (i.e., isotropic elasticity in the present case). The correlation residuals allow the user to check whether this type of hypothesis is valid.

The chosen test is a sandwiched beam (SB) [11] whose inner beam is the silicon carbide sample to be studied, and the steel beams provide the necessary stiffness for the crack propagation to be controlled in a three-point flexural test (Figure 5.15a). A notch is machined along the tensile loaded face of the SiC sample, and a preliminary loading is applied so that a crack is initiated from the notch. Once a sharp crack is formed, the steel beams are removed and a classical three-point flexural test is subsequently carried out on a cracked sample.

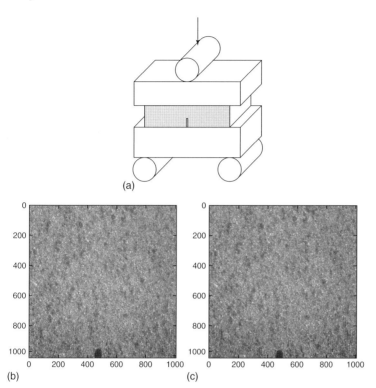

Figure 5.15 (a) Schematic view of the bending test. The observed face of the SiC sample is shown in gray. Eight-bit pictures of silicon carbide sample surface (b) before and (c) during loading. The notch on the lower edge of the sample is clearly visible (1 pixel corresponds to 1.85 μm).

Digital images of the sample face are taken with a long-distance microscope that allows a high magnification (i.e., 1.85 μm per pixel) in comfortable conditions (i.e., a working distance of about 20 cm) with respect to the monitored surface. A bright-field illumination is used. No specific surface preparation is performed since the natural surface roughness provides the necessary speckle inhomogeneity of the images. An example of one stage of cracking is shown in Figure 5.15(b,c). Bare eye examination of the picture does not allow the detection of the presence of a crack.

By using I-DIC, the amplitudes of the singular fields provide an estimate for the SIFs. An ROI of size 512×512 pixels, approximately centered on the crack tip position, is used in the analysis.

Figure 5.16(a,b) shows the displacement field in which the presence of the crack is clearly visible. Note that the maximum crack opening is of the order of 0.3 pixel (or about 500 nm), which explains why the crack is not visible with bare eyes but by I-DIC [45], Q4-DIC [45] and local DIC [42]. The correlation residuals provide a map of local contributions to the total residual $\mathcal{T}_{\mathrm{DIC}}$ (see Section 5.3.3.1 and Figure 5.16(c)). The units are in gray levels. Even though the error level is locally as large as 30 gray levels (for an 8-bit digitization, or 256 gray levels, of the initial images), the mean value of the correlation residuals is about 1.5 gray level. The error is highly concentrated along the crack path. In quantitative terms, the SIFs amount to

$$K_{\mathrm{I}} = 2.8 \pm 0.15\,\mathrm{MPa}\sqrt{\mathrm{m}}, \quad K_{\mathrm{II}} = 0.05 \pm 0.05\,\mathrm{MPa}\sqrt{\mathrm{m}} \tag{5.48}$$

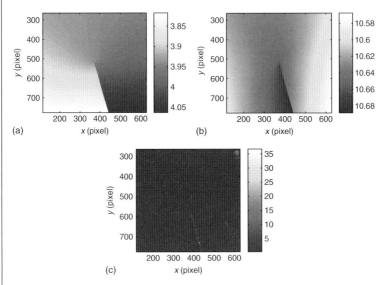

Figure 5.16 (a) Horizontal and (b) vertical components of the displacement field expressed in pixels and identified by using the integrated approach (1 pixel ↔ 1.85 μm). (c) Residual error map. The spatial integral of the square of this field gives the objective function to be minimized. The mean value of correlation residuals is about 1.5 gray level. (Please find a color version of this figure on the color plates.)

Note that the position of the crack tip is not known when analyzing experiments. One way of determining the location is to move the crack tip and keep the position that minimizes the global correlation residuals [45]. An alternative approach is given by canceling out the contribution of the first super-singular field. A mispositioned crack tip gives rise to nonzero amplitudes for this field, which vanish when the tip is correctly positioned [46].

This type of integrated analysis can also be carried out to analyze experiments on silica imaged with AFM. Contrary to classical configurations, the gray level value carries a physical information, namely, the local height. Consequently, because of out-of-plane motions and AFM artifacts, the gray level conservation is not satisfied. It can be relaxed as discussed in Section 5.2.5 to account for these effects [17] to evaluate SIFs *and* out-of-plane motions.

Integrated approaches, as proposed in the present section, can be used in any situation for which closed-form solutions are available. This is, for instance, the case when beams are tested and Euler–Bernoulli kinematics is used (see Section 5.7.2). When optical distortions are described by a parametric model, the present approach is also applicable and does not require any additional postprocessing of the measured displacements, that is, the signature of distortions. For nonparametric models, though, Q4-DIC can be used. Up to now, closed-form solutions were considered. Numerical solutions, say given by FE simulations, can also be chosen. Few changes are needed to use these fields [47]. This last case shows that the integrated approach may become more versatile in terms of constitutive law.

Box 5.4: DIC Regularization

It is a common practice in the field of inverse problems to regularize the sought field in order to compensate for the ill-posedness of the problem. Typically, gradients in the field are penalized by adding to the DIC objective functional to be minimized $\mathcal{T}_{\mathrm{DIC}}$ [see Eq. (5.7)] a regularization term $\mathcal{T}_R$ such that

$$\mathcal{T}_R(\mathbf{u}) = \alpha \int_{\mathrm{ROI}} (\nabla \otimes \mathbf{u})^2 d\mathbf{x} \tag{5.49}$$

This is the principle of the so-called Tikhonov regularization [48]. A similar approach can be followed for global DIC. However, the above form of regularization is not suited to the physics of the problem at hand, namely, it is not invariant under a rotation for instance. This property can be restored by introducing instead a norm that would be a quadratic form of the strain tensor. If, in addition, it is required that this norm be objective (i.e., based on invariants of the strain tensor), then the norm has to take precisely the form of an elastic energy. A natural consequence is to identify the expression of this regularization kernel with the actual, say elastic, behavior of the considered solid, thus providing an answer to the proper choice of the relative weight of the deviatoric and volumetric parts of this energy.

If an FE decomposition of the displacement field is considered, then it is natural to introduce the stiffness matrix $\mathbf{K}$ over the same mesh. The regularization kernel thus takes the following form

$$\mathcal{T}_R^{(1)}(\mathbf{u}) = \frac{\alpha}{2} \mathbf{u}^T \mathbf{K} \mathbf{u} \tag{5.50}$$

where **u** is the vector gathering all nodal displacements, and α a weighting prefactor. Higher order regularization may be considered. One could also consider the unbalanced part of the stress, and hence minimize the quadratic norm of the divergence of the stress (in the absence of body forces). Within the language of FEs, this would be written as

$$T_R^{(2)}(\mathbf{u}) = \frac{\alpha}{2} \mathbf{u}^T \mathbf{K}^T \mathbf{K} \mathbf{u} \tag{5.51}$$

One difficulty related to the use of such a regularization is the proper choice of the prefactor α. It gives the relative weight of the information coming from DIC and that given by the regularization. Various heuristic rules have been proposed in the literature (based on discrepancy principle, cross-validation, L-curve method, restricted maximum likelihood, or unbiased predictive risk estimator [49]).

The physical meaning of the regularization is obtained from some simple argument. Let us consider a trial displacement field in the form of a plane wave of small amplitude, $\mathbf{u} = \mathbf{a} \exp(i\mathbf{k} \cdot \mathbf{x})$. The DIC objective functional computed for such a displacement field will be quadratic in its amplitude **a** and independent of its wavevector **k**. The above two regularizations will be quadratic in **a**, but they will have a different dependence with respect to the wavevector

$$\begin{aligned} T_R^{(1)}(\mathbf{u}) &\propto |\mathbf{k}|^2 \\ T_R^{(2)}(\mathbf{u}) &\propto |\mathbf{k}|^4 \end{aligned} \tag{5.52}$$

Thus, balancing the relative contributions of the DIC and regularization kernel will select a specific cross-over spatial frequency, $\mathbf{k}_R$, such that for large wavevectors, that is, $|\mathbf{k}| \geq |\mathbf{k}_R|$, the regularization term dominates, while for long wavelengths, $|\mathbf{k}| < |\mathbf{k}_R|$, DIC functional will be larger. Hence, the inverse cross-over wavevector $2\pi/|\mathbf{k}_R|$ gives a characteristic scale comparable to the element size of a global DIC. However, below this length scale, rather than using an artificial FE shape function, here a more flexible and more mechanically meaningful interpolation is offered. Further developments and application of this technique can be found in Ref. [50].

Tutorial Exercise 5.3: Strain Resolution

a. In Section 5.3.2.2, the standard strain uncertainty σ_ϵ was evaluated when the standard displacement uncertainty σ_u was known. Nine points (i.e., 3×3) equally spaced are considered (separation: $L_\epsilon/2$). Let us interpolate the x-component of the measured displacement field $u_m(x, y)$ by an affine field, namely, $u_i(x, y, u_0, \epsilon_{xx}, \gamma_{xy}) = u_0 + \epsilon_{xx} x + \gamma_{xy} y$. In this example, u refers to the x component of the displacement. By resorting to least-squares minimization, evaluate the mean normal strain.

b. In Section 5.3.3.2, a Q4 discretization was considered. The following derivation aims at determining the strain resolution associated with each Q4 element. Let us compute the mean strain per element Ω_e, say the normal strain ϵ_{xx} along the x-direction. By using the divergence theorem, express the mean strain ϵ_{xx} over an element of area ℓ^2.

c. In Q4 elements, the displacement interpolation along any edge is a linear function of the position. Calculate ϵ_{xx}, and its standard uncertainty.

Solution:

a) The mean normal strain is assessed by minimizing

$$\sum_{i=1,3}\sum_{j=1,3} (u_m(x_i, y_j) - u_i(x_i, y_j, u_0, \epsilon_{xx}, \gamma_{xy}))^2 \quad (5.53)$$

with respect to u_0, ϵ_{xx}, and γ_{xy}. The mean strain ϵ_{xx} reads

$$\epsilon_{xx} = \frac{\sum_{j=1,3} u_m(x_3, y_j) - \sum_{j=1,3} u_m(x_1, y_j)}{3 L_\epsilon} \quad (5.54)$$

and its standard uncertainty

$$\sigma_\epsilon = \sqrt{\frac{2}{3}} \frac{\sigma_u}{L_\epsilon} \quad (5.55)$$

by assuming that the measurements are independent (i.e., $L_\epsilon/2 \geq L_{ZOI}$). Consequently, $B = \sqrt{2/3} \approx 0.8$ as stated in Section 5.3.2.2. As an additional exercise, the interested reader may want to evaluate the standard uncertainty of the shear strain.

(b) The mean strain ϵ_{xx} reads

$$\epsilon_{xx} = \frac{1}{\ell^2} \int_{\partial \Omega_e} u(\mathbf{n} \cdot \mathbf{x}) ds \quad (5.56)$$

where s is the curvilinear abscissa, and $\mathbf{n}$ is the outward normal. In the present case, only two edges of the element yield nonzero terms (i.e., $\mathbf{n} \cdot \mathbf{x} \neq 0$), namely, $\partial \Omega_e^+$ (with an outward normal $\mathbf{n}$) and $\partial \Omega_e^-$ (with an outward normal $-\mathbf{n}$).

(c) The mean displacement for any edge is equal to the half-sum of the nodal displacements (u_1, u_3, and u_2, u_4, respectively); therefore

$$\epsilon_{xx} = \frac{u_2 + u_4 - u_1 - u_3}{2\ell} \quad (5.57)$$

The result given in Eq. (5.29) still applies in the present case. It shows that the four degrees of freedom are *correlated* so that the strain uncertainty becomes

$$\sigma_\epsilon = \frac{\sqrt{2}\sigma}{2\ell} \sqrt{\sum_{i=1,4}\sum_{j=1,4} (-1)^{i+j} M_{ij}^{-1}} \quad (5.58)$$

If the cross-correlation terms are neglected and if the diagonal terms are identical, i.e., $M_{ij} \approx M\delta_{ij}$, the standard displacement uncertainty $\sigma_u = \sqrt{2}\sigma/\sqrt{M}$, and the corresponding standard strain uncertainty is such that $\sigma_\epsilon = \sigma_u/\ell$. This value coincides with the case of four independent displacement measurements for which a linear interpolation is sought by following a least-squares minimization (as performed above for nine points, i.e., Eq. (5.57) still applies).

5.4
3-D Digital Image Correlation

5.4.1
Basic Principles

So far, all the examples had to "live in flatland." Planar specimen surfaces (normal to the optical axis) were considered, and only in-plane displacements could be resolved. Furthermore, unless telecentric lenses are used, a displacement of the sample along the camera axis would produce an apparent dilatational strain that would be completely artifactual [51, 52]. This is a very strong limitation to 2D-DIC. It would be more than desirable to be able to evaluate the three components of the displacement of the observed surface, and, ideally, the latter should not be restricted to a planar geometry. *Stereo-correlation* is a technique that has the potential to meet these challenging demands [20].

Since the planar limitation comes from the two-dimensional nature of the images shot by the camera, the solution is to use more than one camera (usually two but the technique can be generalized to an arbitrary number). From two images taken from two different angles of the same object, it is possible to estimate its 3-D shape, through a technique known as *stereovision*. Then a second pair of images taken after deformation is matched to the first pair using the above-presented DIC technique. The two 2-D displacement fields can be further recombined in the same way as the shape reconstruction to produce a 3-D displacement vector field. This is the spirit of stereo-correlation as presented in this section.

5.4.2
Camera Calibration

The observed object is represented by a set of coordinates $\mathbf{x} = (x_1, x_2, x_3)$ in the lab frame, $\mathcal{R}$. Each camera (say l and r for "left" and "right," respectively) captures a 2-D image, where it is convenient to use the natural coordinates $\mathbf{x}^l = (x_1^l, x_2^l)$ for frame $\mathcal{R}^l$ (and $\mathbf{x}^r = (x_1^r, x_2^r)$ for frame $\mathcal{R}^r$). The translation and rotation transformations that relate $\mathcal{R}$ to $\mathcal{R}^l$ are denoted by $\mathbf{T}^l$ (and $\mathbf{T}^r$ for the right frame), and the projection from the 3-D space to the 2-D image frame is described by matrix $\mathbf{P}^l$ (and $\mathbf{P}^r$).

The first task consists in relating the three frames. This makes use of a specific optical model to relate $\mathbf{x}^l$ (or $\mathbf{x}^r$) to $\mathbf{x}$. It is based on the acquisition parameters (e.g., pixel size, camera position), but also contains the camera lens imperfections. These transformations are usually determined from a calibration stage where an object with a known geometry and surface pattern is used.

A simple optical model is the *pinhole camera model* (or "perspective projection"). Optical distortions can also be taken into account through a nonlinear correction step. More details can be found in Refs. [20, 53, 54]. This is of high importance for an accurate positioning of the 3-D points, but not covered in this presentation. The pinhole camera model consists in considering light rays passing through the focal point ω^l of, say, the left camera and hitting the observed scene. The intersection of the ray $\omega^l\mathbf{x}$ with the image plane of the camera provides the image point $\mathbf{x}^l$. The linearity of these transformations allows us to write the relationship between $\mathbf{x}^l$ and $\mathbf{x}$, which is conveniently rephrased by using homogeneous coordinates (i.e., $\tilde{\mathbf{x}}^T = \{x_1, x_2, x_3, 1\}$) as

$$\mathbf{x}^l = \mathbf{P}^l \mathbf{T}^l \tilde{\mathbf{x}} \tag{5.59}$$

where $\mathbf{P}^l \mathbf{T}^l$ is a 2×4 matrix. A similar relationship exists for $\mathbf{x}^r$ so that $\mathbf{x}$ is deduced from the pair $(\mathbf{x}^l, \mathbf{x}^r)$. This problem now becomes overdetermined.

The first operation to perform is the determination of matrices $\mathbf{P}$ and $\mathbf{T}$. This step requires that the two points $\mathbf{x}^l$ and $\mathbf{x}^r$ are identified to image the same physical point $\mathbf{x}$. Thus a calibration object whose geometry is known precisely is first imaged by the two cameras, and specific points are tracked on both images to provide enough information to identify the best transformation matrices. It is to be noted that, once a first determination of the transformation matrices is made, (local or global) DIC can be used to perform a final registration between, say, the original right image, and the right image that would be computed from the left one and the approximate left–right transformation. This procedure allows for an excellent accuracy on $\mathbf{P}$ and $\mathbf{T}$.

Stereo-correlation is then almost straightforward once the calibration step has been performed. A series of images is to be taken from the left and the right cameras simultaneously. Each left or right series can be treated independently using any 2D-DIC algorithm to evaluate the displacement vector field. Finally, these two displacement fields are recombined using $\mathbf{P}$ and $\mathbf{T}$ matrices to relocate the full 3-D displacement vector in space.

Box 5.5: Ultrahigh Speed Experiment

An example of application of the above method is to follow a dynamic experiment of bodies subjected to a rapid expansion (i.e., less than 100 μs) due to explosives located inside them [14]. This application requires extremely short exposure times (less than 1 μs). In order to follow such a fast expansion, a single mechano-optical camera is used, as it would not be possible to synchronize two of them. Moreover, for security reasons, the camera has to be located at a large distance from the scene (i.e., optical path of about 15 m). In order to use stereo-correlation, it is necessary to get two different views of the same scene. A pair of mirrors located close to the specimen has been used. An intense illumination is required because of the very short exposure time. It is provided by light emission of a plasma, itself produced by another explosive material.

The stereovision and stereo-correlation techniques have been implemented according to the above-presented principles. DIC revealed also to be very efficient to correct for a significant rigid body translation between pictures, due to the camera technology (different lenses are used for each image). Moreover, DIC was helped by initializing the displacement field at the start of the calculation by a computed displacement field. This procedure helped in gaining some robustness in the computation, as the displacement field is quite significant between subsequent images [14].

Figure 5.17 shows reconstructed geometries for two different instants of time. In particular, this analysis was able to reveal the onset of necking, where the axisymmetry of the test breaks down. Full-field measurements turned out to be very powerful in this extremely demanding context, in spite of the numerous experimental challenges that necessarily come along such setups.

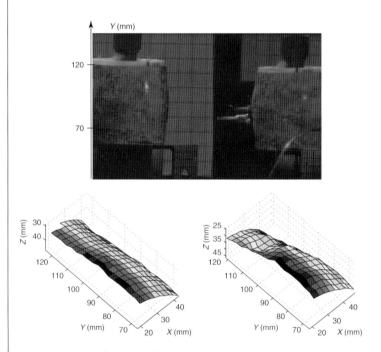

Figure 5.17 Pair of views used in the stereo-correlation analysis. Reconstructed geometry at two different instants of the experiment. Necking is observed in the second case.

■ Tutorial Exercise 5.4: Accuracy of Shape Reconstruction

a. *As an illustration of the stereovision principle, a very simple case is considered. The geometry is shown in Figure 5.18. A rough surface is observed in two directions forming an angle θ_i with the vertical axis. The dimension perpendicular to the figure is omitted for simplicity. The cameras are equipped with telecentric lenses, so that the focal point is ideally at infinity. The two images are formed along x_i-axes.*

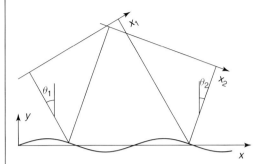

Figure 5.18 An undulating surface is observed with two telecentric lenses in the directions θ_1 and θ_2 so as to form two images along the x_1- and x_2-axes.

b. Considering the notations of Section 5.4.2, the transformation $\mathbf{T}_i$ is a rotation of angle θ_i. The perspective projection $\mathbf{P}_i$ is here simply an orthogonal projection. Therefore, a point M of coordinates (x, y) is imaged on a point of coordinate x_i in image i. Write the transformation in this particular case.
c. Why is it possible to invert the previous relationship?
d. A special case of the above geometry is when the two observation directions are symmetric with respect to the vertical axis. Let us consider $\theta_2 = -\theta_1 = -\theta$. Recast the previous relationship.
e. It is of interest to consider the change of gains $G_x \equiv \partial x/\partial x_1 = 1/(2\cos\theta)$ and $G_y \equiv \partial y/\partial x_1 = 1/(2\sin\theta)$ with θ. Plot these gains and comment on the results.

Solutions:

(a) The transformation reads

$$\begin{Bmatrix} x_1 \\ x_2 \end{Bmatrix} = \begin{bmatrix} \cos(\theta_1) & \sin(\theta_1) \\ \cos(\theta_2) & \sin(\theta_2) \end{bmatrix} \begin{Bmatrix} x \\ y \end{Bmatrix} \quad (5.60)$$

(b) Because two images are available, it is possible to invert the above relationship to obtain

$$\begin{Bmatrix} x \\ y \end{Bmatrix} = \frac{1}{\sin(\theta_1 - \theta_2)} \begin{bmatrix} -\sin(\theta_2) & \sin(\theta_1) \\ \cos(\theta_2) & -\cos(\theta_1) \end{bmatrix} \begin{Bmatrix} x_1 \\ x_2 \end{Bmatrix} \quad (5.61)$$

As soon as the two coordinates x_1 and x_2 have been ascribed to the same physical point through image matching, the absolute position (x, y) of this point is deduced.

(c) The latter becomes

$$x = \frac{x_1 + x_2}{2\cos\theta}$$
$$y = \frac{x_1 - x_2}{2\sin\theta} \quad (5.62)$$

(d) The result is shown in Figure 5.19. When θ is small, G_y diverges, and hence no accuracy is obtained on the out-of-plane position. The reverse holds for the in-plane coordinate when the observation angle is wide. Here again, a compromise has to be found between the sought x or y gains. An equal gain is obtained for $\theta = \pi/4$.

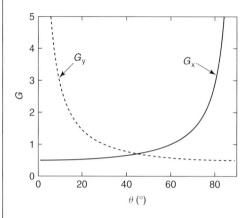

Figure 5.19 Change of the gains G in the x and y directions as functions of the observation angle for a symmetric configuration.

5.5
Digital Volume Correlation

5.5.1
3-D Extensions of 2-D Approaches

Local [23–25] or global [26, 55] correlation procedures are used to measure 3-D displacements in the bulk of materials imaged by, say, computed microtomography or MRI. For a global approach, the 3-D counterpart of Q4-DIC (Section 5.3.3.2) is C8-DVC [26] in which eight-noded cubes with trilinear shape functions are considered. Many of the implementations are direct extensions of 2-D algorithms. One of the main difference is related to the size of the reconstructed volumes which can contain more than one billion voxels. The correlation procedures are therefore more challenging in terms of computation time and memory storage, especially for global approaches. To minimize the computation time, the following adjustments are useful:

- Matrix **M** is estimated once and for all for the reference volume f and only vector **b** or equivalently $\tilde{g}$ has to be updated for each iteration (see Section 5.3.3.1).
- The correction of g requires gray-level interpolations. In many instances, trilinear interpolations were sufficient because of the fact that the quality of the reconstructed volumes is not optimal for measurement purposes [56].

- The evaluation of picture gradients has to be consistent with the gray-level interpolation. When a trilinear gray level interpolation is chosen, a finite difference scheme should be used.

To mention an order of magnitude of the computation time required by such an approach, a cubic ROI of 150 voxel lateral size is analyzed within 10 min on a regular PC.

5.5.2
Resolution Analysis

The results obtained in Section 5.2.5 are also valid in three dimensions. In particular, Eq. (5.29) still applies. If the correlation length of the texture is significantly larger than the voxel size – which is desirable if the noise sensitivity is to be limited, but not too large so that matrix **M** is still invertible – the standard displacement uncertainty σ_u reads

$$\sigma_u = \frac{\sqrt{6}Ap\sigma}{\sqrt{\langle(\nabla f)^2\rangle}\ell^{3/2}} \tag{5.63}$$

where ℓ^3 is the number of voxels in the considered FE, p is the physical size of one voxel, and A is a dimensionless constant dependent on the interpolation function [38, 45].

A $128 \times 128 \times 128$ voxel ROI of the volume to be studied in Section 5.5.2 (Figure 5.21a) is chosen to illustrate the result given by Eq. (5.63). This ROI defines f. A Gaussian white noise ($\sqrt{2}\sigma = 1$ gray level) is added to the latter to create a corrupted volume g as was performed in Section 5.2.5. To register these two

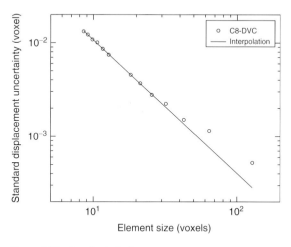

Figure 5.20 Standard displacement uncertainty σ_u as a function of the element size ℓ in C8-DVC when $\sqrt{2}\sigma = 1$ gray level. The interpolation (solid line) is described by a power law $A^{\alpha+1}/\ell^\alpha$, where $\alpha = 1.42$, in good agreement with Eq. (5.63) for which $\alpha = 1.5$.

volumes, C8-DVC is carried out by using different element sizes. The standard deviation of the displacement field σ_u is shown in Figure 5.20, and the compromise between resolution (σ_u) and spatial resolution (ℓ) is (again) observed. In particular, when the displacement resolution is extrapolated to very small element sizes, it reaches values that make any measurement unreliable or impossible (i.e., when matrix **M** is no longer invertible).

The results of Figure 5.20 are very well described by a power–law interpolation as long as a large number of degrees of freedom do not belong to any boundary of the ROI. When a degree of freedom belongs to a boundary, a fraction of the information needed to get a reliable estimate is missing, namely, 1/2 for a face, 3/4 for an edge, and 7/8 for a corner, so that the noise sensitivity is expected to be significantly higher. For elements whose size is less than 32 voxels, a power law $A^{2.42}\ell^{-1.42}$ is identified, which is close to a power -1.5 predicted by Eq. (5.63).

It should be mentioned that some caution should be exercised when analyzing reconstructed volumes. The reconstruction itself may degrade the performance of the correlation code [56]. For lab (divergent) X-ray sources, an additional bias may occur because of temperature variations of the X-ray tube [57].

Box 5.6: Fatigue Crack Characterization

A nodular graphite cast iron sample is imaged by computed microtomography during a fatigue experiment. Figure 5.21 shows the reference volume f and the deformed volume g when corrected by the measured displacement field. The residual map shows that, except in the areas traversed by the crack where a displacement discontinuity occurs and was not accounted for by C8-DVC, the registration was successful.

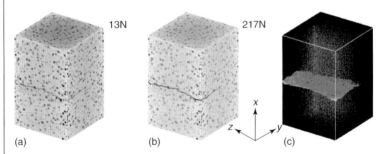

Figure 5.21 (a) Reference volume, (b) volume under load corrected by the estimated displacement field, and (c) the corresponding residual map obtained by subtracting the two volumes after displacement correction.

The three displacement components u_x, u_y, and u_z are shown in Figure 5.22. The displacement range is the highest for the component u_x along the loading direction, thereby indicating a dominant mode I situation. Figure 5.23 shows maps of the crack opening displacement Δu_x across the crack surface. This type of information gives access to the opening/closure process of *any* point of the crack surface.

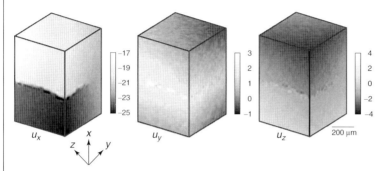

Figure 5.22 3-D rendition of the displacement field components inside the cracked sample expressed in voxels (1 voxel ↔ 3.5 µm). (Please find a color version of this figure on the color plates.)

Figure 5.23 shows maps of the crack opening displacement Δu_x across the crack surface. This type of information gives access to the opening/closure process of any point of the crack surface.

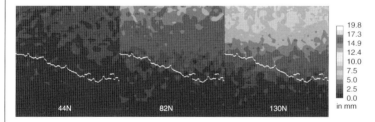

Figure 5.23 Maps of crack opening displacement Δu_x extracted from the jump in u_x displacement across the crack surface for different values of applied load. The white line indicates the position of the crack front [57]. (Please find a color version of this figure on the color plates.)

From a theoretical point of view, a crack is open in mode I as soon as $\Delta u_x > 0$. In practice however, the measurement resolution has to be considered. For the chosen element size, it was estimated to be of the order of 0.09 voxel (or 330 nm), and an opening greater than three times this value (0.27 voxel or 1 µm) is needed to state that the crack is open. Values below this threshold appear in black in Figure 5.23. For a load level of 44 N, numerous contact points are observed in the wake of the crack front. Conversely, only a narrow zone of the crack front seems to resist opening for an 82 N level. Last, the crack is fully open when the load level is equal to 130 N.

Another quantitative way of interpreting the displacement fields is to use them to extract the values of SIFs along the crack front [46]. These results are not presented here. The interested reader will find additional results dealing with 3-D analyses of cracked samples in Refs [31, 57, 58].

Tutorial Exercise 5.5: Spatiotemporal Analyses

Nowadays, it has become quite common to acquire a few hundreds, or even thousands, of images during a single experiment. It is possible, although cumbersome, to process each image as compared to the reference one. Moreover, little benefit is gained out of such long temporal series.

The use of DVC has shown that it is possible to measure 3-D displacement fields. Inspired by such a treatment, it is appealing to consider the global problem of determining the displacement field $u(x, y, t)$, where time t is a coordinate that can be treated in the same way as a spatial one: that is, a stack of images $f(\mathbf{x}, t)$ is a 3-D "volume" of data. Determining the displacement field $\mathbf{u}(\mathbf{x}, t)$, relating $f(\mathbf{x}, t)$ to a reference image $f(\mathbf{x}, t_0)$, relates the deformed "volume" to the reference "volume," $f_0(\mathbf{x}, t) \equiv f(\mathbf{x}, t_0)$. As compared to DVC, this spatiotemporal treatment appears to be similar. The only difference is that the displacement $\mathbf{u}$ has no component along t. This is "pure luck", since the temporal gradient of the reference image is equal to zero! Thus at the expense of removing one dimension to the displacement field, the same DVC code could be used to treat a long temporal series of images. Is this realistic?

Solution:

Let us assume that the reference image is changed to $f_1(\mathbf{x}, t) \equiv f(\mathbf{x}, t - 1)$; then the displacement field that would be measured, say $\mathbf{v}(\mathbf{x}, t)$, is the one that would relate $f(\mathbf{x}, t)$ to $f(\mathbf{x}, t - 1)$. In other words, field $\mathbf{v}$ is rather the *velocity* field (in the sense of a finite difference between consecutive images), or $\mathbf{v} = \dot{\mathbf{u}}\delta t$, if δt is the interframe period.

The above remark is much more than just an aesthetic argument. The discretization of the displacement field now is spread over $2\ell_t$ images, all of which contribute to a given degree of freedom. Hence, for the same uncertainty, the spatial size of the elements ℓ_x or ℓ_y can be reduced at the expense of the temporal discretization scale ℓ_t. Hence, a net gain is expected in the spatial resolution [21]. This result can be compared to the common procedure of averaging a series of images to reduce the noise level. Here, however, the proposed approach exploits the full benefit of DIC.

Another approach would be to consider a spatiotemporal formulation of the minimization of the square difference, as was performed spatially in Sections 5.2.3 and 5.3.3.1 by either considering a space--time decomposition of the displacement field [59] or the velocity field [21].

5.6
Summary

As shown in this chapter, DIC in its various versions provides 2-D displacements (2D-DIC), 3-D displacements of object surfaces (3D-DIC), and even 3-D displacements in the bulk (DVC). These displacement fields are measured with resolutions that are usually very small (i.e., in the range of 1/10 to 1/100 of a pixel, or even lower in some cases) even though the technique is reasonably easy to use in practical situations. With the development of new and more powerful imaging systems, DIC will find new fields of application, provided the artifacts and intrinsic noise associated with any acquisition device are understood, accounted for, and minimized whenever possible.

In some of its current developments, it can be shown that any a priori knowledge on the type of kinematics (e.g., given by mechanical considerations) can be incorporated at the measurement stage with global approaches (I-DIC). If this a priori information turns out to be inconsistent with the studied phenomenon, correlation residuals are very useful in terms of warning. They may be used to enrich the kinematic basis (e.g., when a continuous description is used to analyze cracked samples, see Section 5.5.2 and Ref. [31]).

When imaging surfaces or even volumes, the measured degrees of freedom become equal to or more numerous than those used in numerical simulations. This fact opens the way for identifying parameters of constitutive laws and fracture models, and for validating not only material models but also numerical simulations. Some simple examples were discussed, but they only give a flavor of the potential of such techniques that enable the gap between experiments and simulations to be bridged in a consistent (in terms of kinematics) and thorough way (in terms of full-field comparisons).

The next step of the developments is then to drive experiments by using mechanically meaningful parameters (e.g., SIFs [29]) extracted in real time from full-field measurements so that more complex and more representative tests might be performed to identify material parameters but also to validate numerical simulations at the design and certification stages of structures.

5.7
Problems

The answer to each unsolved problem can be found in one of the cited authors' papers.

5.7.1
Mean Strain Extractor

This first problem can be seen as a continuation of the Tutorial Exercise 5.3. Let us assume that one would like to measure at best a mean strain, with the least uncertainty, from a displacement field measured by Q4-DIC. The mean strain, $\bar{\epsilon}$, reads

$$\bar{\epsilon} = \frac{1}{|\mathcal{D}|} \int_{\mathcal{D}} \epsilon(\mathbf{x}) d\mathbf{x} \tag{5.64}$$

First, ignoring fluctuations in the uncertainty of the nodal displacements, and considering that the domain is a simple rectangle, show that the above quantity is a weighted sum of the displacements, and give the weight field.

Hint: Within each element, the strain components reduce to finite differences. For the average strain over the entire domain, observe that for all internal nodes, the total weight of the nodal displacement vanishes.

Estimate the global uncertainty on $\bar{\epsilon}$ assuming a white noise on the nodal displacement.

As an alternative approach, one may consider fitting (in the least-squares sense) the displacement field by a constant strain. Show that the latter can be written as

$$\tilde{\epsilon} = \frac{\langle u_i x_i \rangle - \langle u_i \rangle \langle x_i \rangle}{\langle x_i x_i \rangle - \langle x_i \rangle \langle x_i \rangle} \tag{5.65}$$

Estimate its uncertainty, and compare it to the first estimate. Which one would you recommend?

Hint: It is again a weighted sum, although the weights are now different. Thus the computation of the uncertainty is straightforward. Looking at the scaling of the uncertainties with the size of the analyzed domain should easily answer the question of your preferred measurement tool.

An "extractor field" is a field $\mathbf{e}$ such that $\tilde{\epsilon} = e_i u_i$. Show that the above expressions can be rewritten in this form, and give the expression of e for $\bar{\epsilon}$ and $\tilde{\epsilon}$.

The covariance matrix is no longer assumed to be proportional to identity, but assumes a general form $C_{ij} = \langle \delta u_i \delta u_j \rangle = 2\sigma^2 M_{ij}^{-1}$. How would the above estimate $\tilde{\epsilon}$ be written? Express the result in terms of an extractor field.

Hint: The previous expression involves a scalar product. Consider now the following scalar product: $\mathbf{a} \circ \mathbf{b} \equiv a_i C_{ij}^{-1} b_j$. If the image noise is Gaussian, then the latter extractor field is the best one that can be designed!

5.7.2
On the Use of a Global Approach When Analyzing Experiments on Beams

How do you measure displacement fields with a global approach when following an experiment on beams?

It is beam-DIC [60]. It consists in implementing a beam kinematics. For instance, when monitoring a cantilever, only six degrees of freedom are needed in the elastic regime of a long beam (i.e., Euler–Bernoulli hypothesis).

What would you do if local buckling occurred?

When local buckling occurs, the correlation residuals can be used to detect it [60], and a more detailed analysis is needed [61]. In that last case, 3-D displacement measurements are also a solution to consider.

5.7.3
Propagation of Uncertainties

Let us assume that Q4-DIC [38] is used to measure displacement fields of the surface of a cracked sample. The displacement field is subsequently postprocessed by resorting to a least-squares technique to determine mode I and II [62] SIFs. Determine the uncertainty of the identified SIF values as functions of the picture noise assumed to be white Gaussian of standard deviation σ. The analysis will be carried out by assuming that the crack tip position is known.

Hint: This is another application of the results obtained in Section 5.3.3.2. See Ref. [45] for the answer.

As a final comment, let us note that an SIF evaluation has the same algebraic structure as the mean strain estimate (Section 5.7.1) and optimal extractor can be designed following the same strategy [46].

5.7.4
Measuring Displacement Fields with a Global Approach in the Presence of Cracks

How would you develop a DIC code when cracks are monitored?

The first answer is to resort to the integrated approach presented in Section 5.3.3.2. To find an alternative route, it is worth having a look at the way this question is addressed in computational mechanics [63, 64]. The next step is to understand why the enrichment strategy is relevant in the framework of DIC. Finally, the implementation is straightforward [65], right? It is called *eXtended digital image correlation (X-DIC)*. Its integrated counterpart is called *eXtended and integrated digital image correlation (XI-DIC)* [66].

For those who are very courageous, the next question is whether DVC can be extended the same way.

Hint: The (positive) answer can be found in Ref. [31].

5.7.5
DIC Coupled with Finite-Element Analyses

In Section 5.3.3.2, an integrated approach to DIC was introduced to measure the displacement field in the vicinity of a crack tip. A closed-form solution was used. In many practical situations, the solutions to a mechanical problem are obtained numerically by resorting, say, to FE simulations. By still assuming linear elasticity, devise different strategies to identify elastic properties [47] or even SIFs [66] by combining DIC with FE analyses.

Hint: The main point is to make sure that the measured displacement field is statically admissible and satisfies the considered constitutive equation (here linear elasticity). This condition was automatically satisfied when the fields given in Eq. (E.46) were considered. In the present case, an FE analysis (associated with the mechanical problem) is coupled with global DIC (also FE based). The global residuals to minimize have two terms: the first one related to the gray level conservation, and the second one associated with static admissibility (i.e., it consists in minimizing the equilibrium gap). When done, you have developed a digital image correlation based on finite-element shape functions (FEI-DIC) procedure. Congratulations!

What can be done with such tools?

For instance, it allows the identification of elastic properties [47] or even the measurement of SIFs [66].

5.7.6
Application of Integrated DIC to a Brazilian Test

The Brazilian test is a diametral compression performed on a cylinder-shaped specimen. Although a compressive load is applied, in the elastic regime the diametral plane is subjected to a uniform extension, and thus it is a convenient way of performing a tensile test without the troublesome problem of grip alignment for brittle and quasi-brittle materials. Under plane strain or stress assumption, its analytical solution is known [35].
How would you measure the displacement field in such an experiment?

Hint: The knowledge of the expected displacement field allows you to formulate an integrated DIC approach. Do not forget the rigid body motions.

How would you make sure to locate properly the cylinder center and its radius, the compression axis?

Hint: The picture may be helpful, and the location of the axis may be one of the unknowns to be sought.

How would you estimate Poisson's ratio of the specimen in this experiment?

Hint: Note that the analytic displacement consists of the addition of two terms, one independent and one dependent on ν. Suppose both parts are treated as independent fields, what could be said about their relative amplitudes?

The stress field diverges at the contact points, and hence one may fear deviations from linear elasticity. How could one avoid this possible 'glitch' in the analysis?

Hint: Is it necessary to consider the entire section?

6
Rough Surface Interferometry

Kay Gastinger, Pierre Slangen, Pascal Picart, and Peter Somers

6.1
Introduction

In the earlier days, interferometry was only applied to mirror-like objects. Many objects in industrial applications have nonspecular surfaces with a roughness in the range of the illumination wavelength or above. With the introduction of holography by Gabor [1], interferometry could also cope with nonspecular objects.

Nowadays, interferometric techniques are used in experimental mechanics to evaluate structural displacements in mechanical testing and also to give high-resolution displacement maps during nondestructive testing (NDT). Since these techniques are noncontact and noninvasive, there is no perturbation of the object by the measurement tool. Moreover, displacement maps with high spatial resolution can be measured. Application of these optical techniques to mechanics is sometimes called *"photomechanics."*

The lateral resolution depends on the size of the imaged area and the resolution of the camera detector array used to image the object. Displacement resolution is mainly linked to the wavelength of the laser light. It is common practice to work in the visible range, but it is now possible to extend these techniques into the infrared and therefore to increase the measurement range to larger displacements.

All interferometric techniques applied on optically rough surfaces are based on speckles as information carriers. The basics of speckles are described in Chapter 1. In this chapter, speckle statistics for laser and low-coherent speckle patterns are presented. But the main goal of this chapter is to introduce four interferometric techniques that are based on a full-field approach and can be applied to rough surfaces.

Electronic speckle pattern interferometry (ESPI), also called *digital speckle pattern interferometry* (DSPI), is a laser-based technique used for the detection of small deformations and vibrations of the object surface with subwavelength accuracy. ESPI is implemented by producing interference between an optical wave front scattered from the object and a fixed reference wave, yielding the displacement for each point in the image of the object.

Low coherence speckle interferometry (LCSI) introduces depth-resolved functionality to ESPI. Thus deformations of interfaces inside an object can be measured with subwavelength accuracy.

Speckle pattern shearing interferometry (SPSI) or *shearography* measures the slope of the surface deformation with subwavelength accuracy by interferometrically comparing the out-of-plane displacement of pairs of points on the object. Shearography is implemented by producing interference between two sheared images of the object.

Digital holographic interferometry (DHI) is the expected evolution of classical holographic interferometry into numerical processing of holograms. The hologram is derived from digital images and the computed holographic image is then compared with the image of the object under deformation. The sensitivity range is the same as that for ESPI but digital Fresnel holography directly provides full information on the encoded object. Especially, the optical phase can be simply computed and is the key for full-field and contactless optical metrology. Furthermore, digital Fresnel holography is a versatile tool that is very well suited for studying the mechanical behavior of mechanical structures. Low-cost lasers and robust color image sensors are now available off the shelf and pave the way for full-field multidimensional deformation measurements using digital color holography.

Each of these techniques produces phase maps. These phase maps are then converted into displacement maps by taking into account the interferometer geometry.

From the displacement maps, it is then possible for the user to derive strain and stress, or to compute mechanical properties such as Poisson's ratio or Young's modulus. Local strain gauges give only one-dimensional strain information integrated over a small area. The techniques presented here enable full-field measurements of all three strain components at the object surface with high spatial resolution. Moreover, the displacement maps can also be implemented in finite element modeling (FEM) to verify the modeling results, to help refine the boundary conditions, or to enable an inverse approach. Furthermore, vibration measurements are introduced for three of the techniques. Vibration analysis is used for modal analysis, acoustic studies and NDT.

6.2
Speckle

When illuminating an optically rough surface coherently (e.g., with a laser), the light backscattered from the surface is nonuniform. A granular pattern occurs with a few bright and mainly dark spots – known as a *speckle pattern*. Speckle patterns observed in freespace propagation are called *objective speckles* and speckle patterns observed through an imaging system are called *subjective speckles*. The description of subjective speckle patterns must incorporate diffraction as well as interference because of the imaging aperture of the system.

6.2.1
Speckle in Monochromatic Light

In the following description, we assume monochromatic and perfectly linearly polarized light. Speckles occur because of the coherent superposition of elementary waves. These elementary waves are generated from a statistically distributed array of microscopic scatterers at the surface of the sample. The statistical properties of laser speckle patterns are described in more detail by Goodman [2].

For both subjective and objective speckles, the amplitude and phase of the interference pattern in space are randomly distributed. The speckle pattern at the observation point (x, y, z) consists of contributions from different scattering regions of the surface of the sample. The phasor amplitude $A(x, y, z)$ in the observation point is therefore given by the sum of a large number N of elementary phasor contributions a_k, that is,

$$A(x, y, z) = \sum_{k=1}^{N} \frac{1}{\sqrt{N}} \alpha_k(x, y, z), \quad k = 1, 2, \ldots, N \qquad (6.1)$$

If we can assume that

- the amplitude and phase of each elementary phasor are statistically independent;
- the amplitude and phase of each elementary phasor are statistically independent from all other elementary phasors;
- the phase is randomly distributed in the interval $\langle -\pi, \pi]$;
- the number N of scatterers is large; and
- the speckle pattern is fully resolved,

this complex addition of the elementary phasor contributions is identical to the classical statistical problem of the random walk in the complex plane. The statistical properties of the speckle pattern can now be calculated. The probability density functions of the intensity $p(I)$ and phase $p(\varphi)$ are then given by

$$p(I) = \frac{1}{\langle I \rangle} \exp\left(-\frac{I}{\langle I \rangle}\right), \quad (I \geq 0) \qquad (6.2)$$

$$p(\varphi) = \frac{1}{2\pi}, \quad (-\pi \leq \varphi \leq \pi) \qquad (6.3)$$

where $\langle I \rangle$ denotes the mean intensity in the speckle pattern. Its intensity distribution is given by a negative exponential distribution, while the speckle phases are uniformly distributed. Figure 6.1 shows an example of a coherent speckle pattern and the histogram of the intensity distribution.

The speckle contrast C_{sp} is given by

$$C_{sp} = \frac{\sigma_I}{\langle I \rangle} = 1 \qquad (6.4)$$

with σ_I being the root mean square (rms) intensity fluctuation, which is equal to the mean intensity. The spatial structure of a speckle pattern is described using an analogy to the van Cittert–Zernike theorem. It says that the spatial coherence can

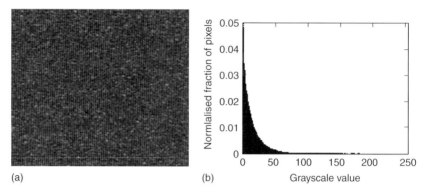

Figure 6.1 Example of a coherent speckle pattern. (a) Image of the speckle pattern. (b) Histogram of a coherent speckle pattern showing the negative exponential intensity distribution.

be calculated by solving an equivalent diffraction problem for focusing through the corresponding angular aperture. By solving this diffraction problem, the average speckle size at the image side can be calculated. For subjective speckles observed via a circular aperture, it is given by

$$l_{ct} = 1.22\lambda(1+M)\#F \quad (6.5)$$
$$l_{cl} = 8\lambda(1+M)^2\#F^2 \quad (6.6)$$

where

- l_{ct} is the transverse speckle size, also referred to as the *transverse spatial coherence length* or *speckle diameter*;
- l_{cl} is the longitudinal speckle size, also referred to as *longitudinal spatial coherence length* or *speckle length*;
- M is the magnification of the imaging system, given by $M = (s-f)/f$ with s as the distance from the imaging aperture to the imaging plane and f as the focal length of the lens; and
- $\#F$ is the F-number of the imaging system given by $\#F = f/D$ with f as the focal length and D the aperture diameter of the imaging system.

For a circular aperture, the transverse speckle size is the diameter of the first zero in an Airy function of the equivalent diffraction problem. The longitudinal speckle size is calculated from the first zero of a sinc function.

6.2.2
Speckle in Low-Coherent Light

A low-coherence speckle pattern occurs when a rough surface is illuminated by a light source with a short coherence length. Low-coherent illumination changes the statistical properties of the speckle pattern described above. Only the reflected elementary waves with an optical path length difference (OPD) within the coherence length of the source interfere. This reduces the contrast in the speckle pattern, similar to not-fully-resolved or not-fully-developed speckle patterns.

A low-coherence speckle pattern can be described as an incoherent sum of several speckle patterns, as shown by Goodman [2]. However, Goodman only considered the case of surface scattering objects. The following section extends Goodman's investigations for a volume scattering object.

Figure 6.2a shows a recording of a low-coherent speckle pattern coming from a volume scattering object. A scattering layer of about 300 μm thickness is applied on an aluminum substrate. The coherence length of the source is about 30 μm. Figure 6.2b illustrates the histogram of the intensity distribution for a medium intensity of 50 grayscale values. The negative exponential distribution for coherent speckle patterns as shown in Figure 6.1 is changed. It can be seen that the speckle contrast is reduced.

A speckle pattern coming from a volume scattering object is a depth integral of the interference of coherent light coming from randomly distributed scatterers. For increasing depth, the scatterers contribute with decreasing intensity as a result of the reduction of the amount of reflected light.

The intensity distribution of a low-coherent speckle pattern is shown in Figure 6.2b. The intensity distribution can be calculated using a simplified model, where the total thickness of the depth scattering object is divided into a number of sublayers. Each of these individual speckle patterns has a depth equal to the coherence length of the light source. From each sublayer, a speckle pattern is generated. The intensity of all these independent layers is incoherently added as described by Goodman [2].

The result of such modeling is significantly different from the negative exponential distribution known from surface scattering objects illuminated by coherent light.

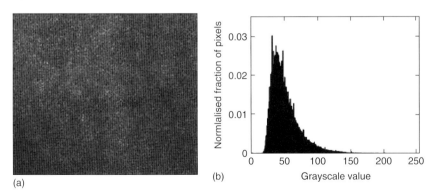

Figure 6.2 Example of a low-coherence speckle pattern. (a) Image of the speckle pattern. (b) Histogram of the intensity distribution.

The intensity distribution for the sum of several uncorrelated speckle patterns is given by Goodman [2] as

$$p(I) = \sum_{k=1}^{N} \frac{\kappa_k^{N-2}}{\prod_{\substack{p=1 \\ p \neq k}}^{N} (\kappa_k - \kappa_p)} \exp\left(-\frac{I}{\kappa_k}\right), \quad (I \geq 0) \tag{6.7}$$

For uncorrelated speckle patterns, κ_k denotes the mean intensity $\langle I_k \rangle$ of the individual speckle pattern [2, p. 26]. N denotes the number of independent speckle patterns to be added.

We need to consider that the speckle patterns coming from a larger depth have a lower intensity. Therefore, we introduce 10 speckle patterns with a realistic distribution of the mean intensity of the individual speckle patterns in depth (Figure 6.3a). We expect a strong reflection from the top (first layer). Inside the scattering medium, we expect an exponential reduction in the intensity due to attenuation. The back surface is assumed to give a stronger reflection, which is indicated by a stronger mean intensity than that of the last speckle pattern.

Figure 6.3b shows the calculated intensity distribution. The gray line shows the model and indicates only minor deviations. By including more layers, the model can be refined. To obtain an exact model, the sum in Eq. (6.7) should be replaced by an integral.

The speckle phases of the individual speckle patterns are uniformly distributed. The speckle patterns are added incoherently, making coherent phase investigations unnecessary. The theoretical development of a refined model is however beyond the scope of this book.

For the low-coherence sources discussed in this chapter, the mean transverse speckle diameter is only slightly influenced. Assuming a plane coherence area, a transverse displacement of the object within the spatial coherence region does not change the average distance to the source. A longitudinal displacement, on the other hand, changes this distance. In a speckle pattern generated by a coherent

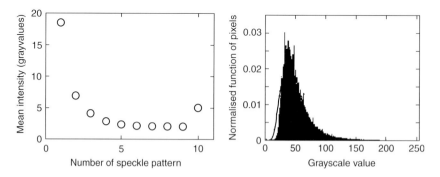

Figure 6.3 Model of the intensity distribution in a low-coherent speckle pattern. (a) Modeled mean intensity of the individual speckle patterns. (b) Adapted intensity distribution.

source, the imaging system limits the longitudinal speckle size. If the temporal coherence of the source is shorter than this longitudinal speckle size, the temporal coherence limits the speckle length. This limits the size of the maximum object deformation that can be measured.

6.3 Electronic Speckle Pattern Interferometry–ESPI

6.3.1 Introduction

ESPI was developed from holographic interferometry in the early 1970s. The first ESPI papers were almost simultaneously published by research groups in England by Butters and Leendertz in 1971 [3], in the United States by Macovski *et al.* in 1971 [4], and in Austria by Schwomma in 1972 [5]. Since then, ESPI has experienced extensive development through the last three decades. Today, ESPI is used in many industrial applications, from quality control in tyre production to characterization of micro-electromechanical systems (MEMSs). There is still quite some research work going on in the field by some dozens of groups around the world.

The theoretical background and applications of the technique are intensively documented in the literature. The interested reader should consult Jones and Wykes [6], Løkberg and Slettemoen [7], Doval [8], and Rastogi [9].

The following section introduces the principle of operation, describes the theoretical basics, and gives practical measurement examples. The focus is on the applications of the method in the field of solid mechanics.

6.3.2 General Principle

Traditionally, ESPI is used for NDT and small-amplitude vibration analysis. The comparison of the object surface before and after its excitation often reveals defects on or even below the surface. The following excitation methods are used:

1) vacuum (vacuum chamber) or pressure
2) thermal energy (i.e., heater, microwave, or flash)
3) mechanical stress (i.e., tension, compression, or torsion)
4) vibration

These excitation methods are described in more detail in Section 6.5.6, where we focus on NDT. ESPI is suited for NDT as well; however, in the following section we focus on its application for mechanical testing.

The principal setup of most ESPI instruments is a Mach–Zehnder type interferometer, shown in Figure 6.4. ESPI is based on the digital video recording of the interference between the reference and the object field – hereafter called the *primary interference pattern* or *specklegram*.

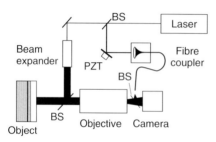

Figure 6.4 Principle setup of an ESPI system.

The laser beam is split by a beam splitter into a reference beam and an object beam. The beam expander spreads the object beam to illuminate the surface of the object. A speckle field is scattered back and imaged through the objective lens onto the camera. As shown previously, this object field has a random distribution of amplitude and phase. This results in a strong spatial variation of these variables, $A = f(x, y)$ and $\varphi = f(x, y)$. The reference beam travels via a piezo actuator and is coupled into an optical fiber. The divergent beam from the exit of the fiber is combined with the object beam. The resulting primary interference pattern becomes phase-sensitive and is recorded by the camera. The reference beam can be either a speckle field or a uniform wave; however, subsequently we assume a uniform reference wave. To obtain maximum fringe contrast, the reference beam must appear to diverge from the center of the viewing lens aperture [4], or more accurately the output pupil of the optical system. This is valid for all ESPI setups, except when spatial phase-shifting (SPS) is applied.

By using a piezoelectric crystal (PZT) in the reference arm (Figure 6.4), the phase of the primary interferograms can be modulated. This is useful in time average (TA) ESPI and for temporal phase-shifting (TPS). The video signal can be filtered in an analog filter, and a personal computer digitizes and processes the video images.

Some other systems are also commonly used to shift the phase of light with subwavelength phase shifts. The calibration algorithms are described in Chapter 3.

Finally, it should be stated that the principal ESPI setup sketched in Figure 6.4 measures pure out-of-plane displacement. Configurations for other sensitivity directions are considered in one of the following sections.

6.3.2.1 Interference Equation

The interference in each speckle can now be considered as the interference from a Michelson interferometer. The interference intensity $I(x, y)$ in the primary interferograms is given by

$$I(x, y) = I_r(x, y) + I_o(x, y) + 2\sqrt{I_r(x, y) I_o(x, y)} \cos(\varphi_r(x, y) - \varphi_o(x, y)) \quad (6.8)$$

where φ_r and φ_o denote the phase of the reference and the object wave, respectively. The resulting interference phase in each speckle $\varphi_{ro} = \varphi_r - \varphi_o$ is now changed by movements of the object or other changes, causing an OPD.

6.3.3
Evaluation

As the main methods phase shifting and phase stepping have already been described in Chapter 3, we will only discuss the other procedures used in speckle interferometry, that is, subtraction and time average methods.

6.3.3.1 Subtraction Mode

Different operational modes of ESPI are introduced, for example, subtraction mode, time-averaged ESPI [7], and double-pulsed ESPI [10]. In this section, we focus on subtraction-mode ESPI or, more specifically, phase-shifting ESPI. This technique is mainly used for static deformation measurements.

By combining the primary interference patterns before and after the deformation, phase changes between the recordings give rise to new secondary interference fringes (also called *correlation fringes*). Considering two states of the object (state 1 "initial" and state 2 "excited"), the two primary interference patterns I_1 and I_2 are given by

$$I_1 = I_r + I_{o,1} + 2\sqrt{I_r I_{o,1}} \cos(\varphi_s) \tag{6.9}$$

$$I_2 = I_r + I_{o,2} + 2\sqrt{I_r I_{o,2}} \cos(\varphi_s - \Delta\varphi) \tag{6.10}$$

The spatial dependency of the variables $I = I(x, y)$ and $\varphi = \varphi(x, y)$ is omitted for readability. φ_s denotes the start phase (speckle phase) at the initial state of the object given by $\varphi_s = \varphi_{ro,1}$. The phase change between the states 1 and 2 is denoted $\Delta\varphi$.

These speckle interferograms can be subtracted, giving the following equation for the secondary interference fringe pattern, assuming perfect spatial correlation (i.e., no rigid body motion) between the two primary speckle patterns, that is,

$$I_2 - I_1 = 2\sqrt{I_r I_o}(\cos(\varphi_s - \Delta\varphi) - \cos(\varphi_s)) \tag{6.11}$$

It can be seen that the self-interference terms vanish and only the modulation of the interference term is left. However, because of the random nature of the object intensity I_o and the phase φ_s, this interferogram contains speckle noise, decreasing the accuracy of the measurements.

Numerous studies in the literature [11, 12] try to optimize the fringe contrast and to reduce the speckled appearance of the secondary fringes. However, speckle noise limits the accuracy of intensity subtraction ESPI to about 1/10 of a fringe.

6.3.3.2 Time Averaged Method

When an object, illuminated by a continuous-wave (CW) laser, is vibrating harmonically, the acquired speckle intensity pattern is modulated with the vibration frequency. If the integration time of the camera system used for acquisition is much longer than the vibration period, the measured speckle intensity pattern follows a zero-order Bessel function of the first kind J_0. The information of interest is in the argument φ_d of the Bessel function, which is the vibration amplitude

$d(x, y)$ expressed as a phase change:

$$\varphi_d = \frac{4\pi}{\lambda} d(x, y)$$

Using φ for $\varphi_o - \varphi_r$, the randomly distributed primary optical or speckle phase, and substituting $I_B = I_o + I_r$ and $I_M = 2\sqrt{I_o I_r}$ in Eq. (6.8), the intensity in any point of the modulated speckle interference pattern is given by

$$I = I_B + I_M \cos \varphi \times J_0(\varphi_d) \tag{6.12}$$

The speckle phase φ can be considered as noise. Fringes showing the vibration pattern can be observed in real time when subtracting a speckle interference pattern taken under static conditions from the dynamic pattern. This removes the background intensity I_B, but since speckle phase φ and the modulation term I_M are still present in the difference equation, the fringes are very noisy (Figure 6.5).

Similar to static cases, phase stepping can be applied for time-averaged vibration measurements phase-shifted-time-averaging (PSTA); that is, using the well-known four-bucket phase-stepping algorithm. Here, four speckle interference patterns are acquired, using a phase step of $\pi/2$ between the acquisitions, that is,

$$I_1 = I_B + I_M \cos \varphi \times J_0(\varphi_d) \tag{6.13}$$
$$I_2 = I_B - I_M \sin \varphi \times J_0(\varphi_d) \tag{6.14}$$
$$I_3 = I_B - I_M \cos \varphi \times J_0(\varphi_d) \tag{6.15}$$
$$I_4 = I_B + I_M \sin \varphi \times J_0(\varphi_d) \tag{6.16}$$

Each of the patterns contains the same Bessel term containing the desired displacement information. From these four patterns, the phase φ and the background I_B can be eliminated, yielding the following expression for an intensity distribution I_{TAV} depicting the TA vibration pattern:

$$I_{TAV} = \sqrt{(I_4 - I_2)^2 + (I_3 - I_1)^2} = 2I_M |J_0(\varphi_d)| \tag{6.17}$$

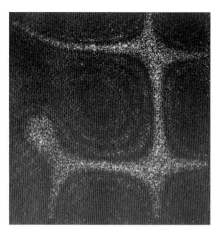

Figure 6.5 Time-averaged vibration fringes on a 200-μm Si solar cell wafer vibrating at 178 Hz. The wafer is supported by three metal pins and not clamped. (Courtesy REC Wafer Norway.)

The method described here is the one most commonly used. It is easy to implement with any phase-stepping system, both temporal and spatial (Figure 6.6).

Pulsed lasers or stroboscopic illumination can also be used to measure vibration maps. In this case, the fringe patterns are not relative to the J_0 Bessel function anymore.

Box 6.1: Speckle Averaging

ESPI results can be improved effectively by performing a series of measurements under the same conditions but with different speckle patterns, for instance, by changing the illumination. This approach has been successful, in particular, for improving the quality of speckle interferograms depicting vibration patterns [13]. When the illumination is changed slightly for each measurement in such a way that an independent speckle interference pattern is obtained, the resulting vibration patterns can be averaged. This method is called *speckle averaging*. This approach is also possible for static applications, in particular for NDT, where the probability of detection is more important than high accuracy.

6.3.4
Configuration

As in holography, it is possible to customize the ESPI setup to a desired sensitivity of displacement measurement.

In the setup described by Figure 6.4, the sensitivity for displacement measurements is purely out-of-plane and the relation between phase and displacement is

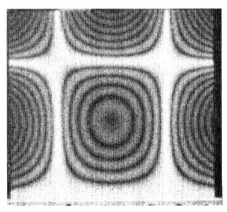

Figure 6.6 PSTA ESPI vibration pattern, showing Bessel fringes, depicting the out-of-plane vibration amplitude. The sample is a steel plate 150 × 150 × 1 mm, clamped at the bottom side, vibrating at 670 Hz. Speckle averaging (Box 6.1) has been applied by changing the direction of the illumination four times. (Courtesy Dan Borza, INSA Rouen – France.)

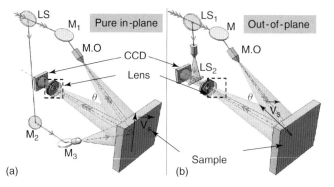

Figure 6.7 (a) In plane and (b) mixed in-plane and out-of-plane sensitivity setups.

given by

$$d = \frac{\lambda}{4\pi}\varphi \qquad (6.18)$$

where φ is the phase calculated by the phase-shifting algorithms described in Chapter 3.

6.3.4.1 Sensitivity Vectors

Generally, the illumination and observation beams do not follow the same optical path. The sensitivity vector is given by the bisector of the angle between the object illumination and observation directions. Figure 6.7 presents the main setups coping with pure in-plane (a) and mixed out-of-plane and in-plane (b) sensitivity. In these configurations, the relation between phase and displacement regarding the sensitivity vectors V_s can be written as

$$d = \frac{\lambda}{4\pi \sin\theta}\varphi_{x,y} \qquad (6.19)$$

and as

$$d = \frac{\lambda}{2\pi(1+\cos\theta)}\varphi_z \qquad (6.20)$$

respectively, where θ is the angle of incidence. Some interferometers are constructed to give all components of the displacement vector by a clever combination of illumination and observation beams [14–16].

6.3.5
Characteristics

6.3.5.1 Optimization of the Interference Signal

The fringe contrast in ESPI is more complex than described in Chapter 3 for classical interferometry. It depends on

- the electronic and optical noise in the system,
- the imaging aperture,

- the beam ratio,
- the degree of correlation of the speckle pattern between initial and excited state,
- the dynamic range of the video camera,
- the power of the light source,
- the intensity distribution of the object and reference waves, and
- the changes of the state of polarization and birefringence.

Basic theoretical work on this topic was done by Slettemoen [11, 17]. Lehmann [18] investigated the influence of a speckled reference wave, and Maack et al. [19] derived an expression for the spatial frequency spectrum of a speckle interferogram and its influence on the reference intensity. Burke [20] investigated the influence of the beam ratio for both TPS and SPS, experimentally. Owner-Petersen [21] discusses decorrelation effects and its influence on fringe visibility.

Practical Considerations The laser power should be high enough to illuminate the whole object and to give a uniform intensity distribution of the reference beam. The polarization of the object and reference beams must be kept parallel to produce a high-contrast interference signal. The competition is now between camera sensitivity, the frame rate linked to the shutter time, and the aperture size of the optical setup.

To test the interferometric sensitivity, a slight perturbation (i.e., mechanical vibration) is applied to the setup, resulting in a "boiling" of the speckles. If this can be obtained, the setup is ready to work and the aperture can be opened to reach a size of the speckle grains just resolved by the camera pixels. The beam ratio is adjusted at $r = I_r/I_o = 1$ using the mean intensity of the speckle pattern as object intensity.

If the laser power is not sufficient, the aperture must be opened, yielding unresolved speckles (many speckle grains on the same pixel). This is not advantageous for the interference contrast, as a large aperture reduces also the depth of field. For this case, the beam ratio can be adjusted at $r > 1$ since usually the attenuation in the reference arm is lower and we can "gain up" the interference contrast by increasing the reference intensity.

6.3.6
Applications

We present two different applications. The first one is dedicated to a study of a compact tension notch (CTN) sample in the mechanical lab by applying three-dimensional (3-D) displacement measurements (Figure 6.8). The other one is realized on the optical table to measure the in-plane displacement of a standard tensile test sample made of polymer (Figure 6.13).

Remark: As we are dealing with CW lasers, external perturbations (e.g., air turbulences, thermal effects, and building vibrations) have to be removed or strongly reduced by vibration isolation facilities (i.e., optical bench with compressed air supports, vibration insulated floor, etc.).

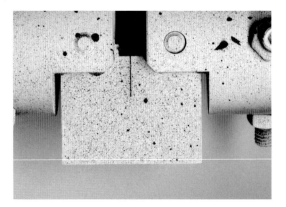

Figure 6.8 CTN sample.

6.3.6.1 Compact Tension Notch Sample

The sample, 30 mm × 30 mm, is realized in 16MND5 bainitic steel with a Young's modulus of about 210 GPa and a Poisson's ratio of 0.3. Its thickness B is 12.5 mm. It has been notched with a mean ratio a/W of 0.552. The last loading leading to notching has given a critical strain concentration K_{IC} of 14 MPa · m$^{1/2}$.

The surface to be studied is covered with white powder to give a perfectly diffusing surface and thus to avoid inhomogeneities of the object beam intensity.

The sample is illuminated with a collimated beam. The charge-coupled device (CCD) camera is equipped with an objective ($f = 50$ mm) to image the object in the center zone at a distance of about 50 cm. The aperture is set to have speckle grains around the size of the sensor pixels (7 μm). A smooth reference beam is delivered via a relay plate by an optical fiber mounted on the objective F-C[1]

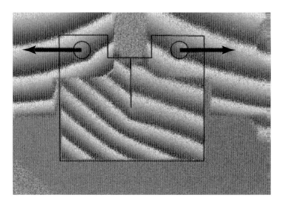

Figure 6.9 ESPI mod 2π phase map.

---

1) F-C mount is a conversion ring to mount optics with Nikon's F-mount on scientific cameras with C-mount.

6.3 Electronic Speckle Pattern Interferometry–ESPI

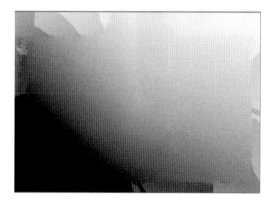

Figure 6.10 Unwrapped fringes on the CTN sample.

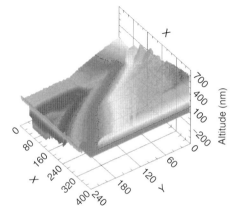

Figure 6.11 Displacement map. (Please find a color version of this figure on the color plates.)

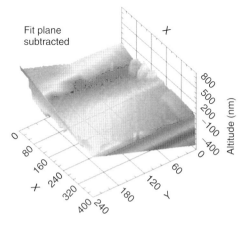

Figure 6.12 Tilt correction. (Please find a color version of this figure on the color plates.)

mount, ensuring a beam ratio I_r/I_o of about 1–2 over the whole detector area. Figures 6.9–6.12 present the subsequent results after the different processing steps. Four-bucket phase-shifting has been applied to produce quantitative results. The wrapped phase shown in Figure 6.9 is the result of the subtraction of two primary phase maps obtained at two different phase load steps. After fringe unwrapping, the displacement map is obtained from the optical setup geometry (apply Eq. (6.20) with $\theta = 25°$). Finally, as can be seen from Figure 6.9, a global tilt of the whole apparatus occurs during the test. This is clearly shown by parallel phase transitions over the whole sample surface. The operator can pick three points on the displacement map to compute and remove the tilt contribution. Finally, the real displacement map at the sample surface is obtained, which is presented in Figure 6.12.

6.3.6.2 Tensile Test Sample

This tensile test fully describes the basics of ESPI application to solid mechanics. The setup is simple to develop and uses two beams of laser light illuminating the polymer sample under study at the same angle from each side of the camera. Collimated beams are used to avoid errors on the side of the objects, as the sensitivity vector is on the bisector of the illumination and viewing beams (Figure 6.7a).

As the sample undergoes the tensile test, images of the phase-shifted specklegrams are recorded by the CCD camera. The initial phase map is then subtracted from the deformed phase map, live, or by postprocessing.

Figure 6.13 presents wrapped phase maps obtained at different load steps.

After unwrapping the fringe pattern and considering the illumination angle, it is possible to compute the deformation map of the sample.

Figure 6.14 presents the evolution of the x-deformation along the sample obtained from the in plane fringes for 90 N load.

Tutorial Exercise 6.1

The object under study is an aluminum sample of 2 cm × 1 cm × 1 mm undergoing a compression test. The aim of the study is to measure the out-of-plane displacement before buckling. The maximum displacement before buckling is about 30 μm. The sample is stuck between jaws and cannot be illuminated from the top or bottom. The available laser is just delivering the power to get a good image onto the CCD, but with a relatively short coherence length of about 30 mm at 532 nm.

Problems

1) Calculate the sensitivity of the measurement to cope with the displacement and explain the setup and the test procedure.
2) As the object is small, is it better to be closer or not? Explain your lens choice.
3) What is the expected fringe pattern?

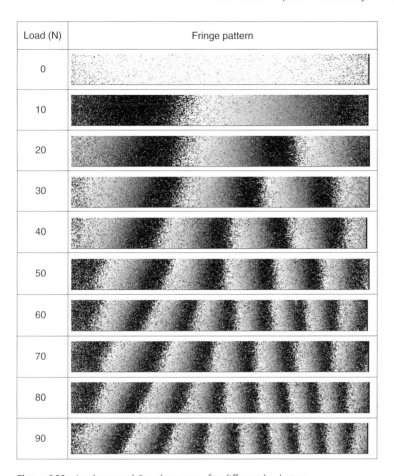

Figure 6.13 In-plane mod 2π phase map for different load steps.

Solution:

1) The sensitivity must be out-of-plane, and so Eq. (6.20) must be applied. To respect the short coherence length, the setup must balance the reference and object beam paths to stay within the coherence length of the laser. The reference beam can travel along a common path with the object beam, or must follow a delay line (done with mirrors, for example) to propagate over the same path length. As the displacement is about 30 μm, and the displacement sensitivity is about $532/2 = 266$ nm per fringe, there will be too many fringes after subtraction to make the measurement in one step. The test procedure is then quasistatic or step-by-step.

2) To avoid errors on the angles and thus a change of the sensitivity vectors along the object, you have to work with a telecentric lens or position the camera as far as possible from the object in order to obtain an approximately parallel beam observation.

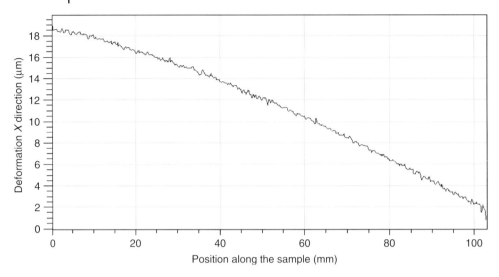

Figure 6.14 Evolution of the x-deformation along the sample.

3) The fringes are parallel to the longer side of the jaws and should be centered if the compression is well applied. Some inhomogeneities can appear at the jaw contacts due to jaw/sample surface defects or misalignments of the jaws during the test. The horizontal center line of the sample must then go inward or outward, depending of the initiated buckling. Some oval fringes can appear on the sides of the sample. An example for a small load is given in the figure that follows.

6.4
Low-Coherence Speckle Interferometry–LCSI

6.4.1
Introduction

LCSI adds depth resolution to ESPI and thus makes it possible to move speckle interferometry from the measurement of surface displacements to the measurement of displacements of interfaces inside an object.

This can be done by introducing a low-coherent light source into an ESPI configuration. The use of low-coherent sources implies that an interference signal occurs only if the optical path length of the reference and the object beams coincides within a few coherence lengths, also called the *coherence gate*. The corresponding volume is called the *coherence layer*.

If the coherence layer is sufficiently thin, interfaces inside the object can be resolved and changes there can be detected. LCSI thus provides depth-resolved, high-accuracy deformation measurements. The basic idea is to measure the deformation of interfaces or other structures inside an object with an accuracy below the wavelength of the light. These measurements are necessary to follow displacements inside an object while changing the ambient conditions (e.g., temperature, humidity, and pressure) or while introducing strain. Another topic can be the investigation of functional interfaces such as membranes in turbid media.

What are the challenges if we move the deformation measurements from the surface of the sample to an interface inside the sample? The first important task is the consideration of optical versus geometrical dimensions of the sample. Secondly, only a part of the reflected light comes from the depth of interest. This reduces the interference signal. Therefore, the optimization of the interference signal is an important task. A higher interference contrast increases the probing depth and the maximum size of the deformation as a result of minimizing the decorrelation effects. The last and maybe largest challenge is the interpretation of the measurement results. All phase changes experienced by the coherent part of the object wave along the optical path will influence the measurement. This can lead to misinterpretations.

The idea to combine ESPI with low-coherent sources was already suggested in 1981 by Neiswander and Slettemoen [22] to separate the signals from various structures in the cochlea in the inner ear. In 2001, the first LCSI setup was developed by Gülker *et al.* [23].[2] The progress of the following years is summarized by Gastinger [24, 25].

LCSI has many similarities to optical coherence tomography (OCT), in particular to full-field OCT. The review articles of Fujimoto *et al.* [26] and Fercher *et al.* [27] and a comprehensive collection of different papers on OCT given by Bouma and Tearney [28] are recommended for the interested reader.

In the following section, the setup and measurement principle of LCSI are introduced. LCSI is classified, and the differences with OCT, full-field OCT, and ESPI are described. We establish some theoretical approaches such as the interference equation. Furthermore, some rules of thumb for the optimization of the interference signal are given. Finally, two applications of LCSI are presented.

6.4.2
General Principle

LSCI adopts depth resolution from OCT and the measurement algorithms from phase-shifting ESPI. Besides being used to measure depth-resolved deformation

---

2) Introduced under the term *low-coherence TV holography* or *low-coherence ESPI*.

of features inside a material, an LCSI instrument can be used as a full-field OCT instrument as well as a standard ESPI instrument.

LCSI and full-field OCT are comparable in the idea to combine coherence gating and full-field measurements on the internal structures of objects. In contrast to OCT, LCSI aims at the measurement of dynamic phenomena, whereas the former is mainly applied for the measurement of the static structure of the object.

In contrast to ESPI, LCSI measures deformations inside a material. Coherence gating ensures the separation of the interference signal from the measurement signal coming from areas outside the coherence layer.

6.4.2.1 Measurement Principle

LCSI is used for out-of-plane deformation measurements inside semitransparent homogenous or multilayered objects. The random nature of the light reflected from rough surfaces, interfaces, and volume scatterers introduces speckled interference signals. The algorithms from ESPI and full-field OCT are therefore combined.

The measurement algorithm can be divided into three different parts:

1) Structural imaging
2) Sample alignment
3) Depth-resolved deformation measurement.

The first step is used to obtain the necessary a priori information about the sample and position of the interfaces under investigation. In the second step, the coherence layer is adjusted at the interface of interest.

In the last step, the actual LCSI measurements are carried out. The following sections will shortly describe the parts of the measurement algorithm. The instrument can be used in two modes. The full-field OCT mode is used in the first two steps of the measurement algorithm, whereas the LCSI-mode is used for the deformation measurement. The instrument modes are described in more detail in Section 6.4.4.

Structural Imaging In a semitransparent material, the scatterers are positioned at all depths and contribute to the interference term depending on the coherence function.

An important requirement for the deformation measurement on interfaces is that the largest part of the backscattered coherent light comes from the interface itself.

The measurement of deformations of an interface inside an object presumes a priori information about the sample itself: that is,

- optical position of the interface
- orientation and shape of the interface
- reflection coefficient of the interface
- obstacles along the optical path causing refractive index changes
- attenuation of the signal/probing depth.

All this information can be given qualitatively by operating the LCSI instrument as a full-field OCT interferometer. We can detect the position and orientation of the interfaces by scanning the coherence layer through the sample and detecting the interference amplitude.

To illustrate the measurement algorithm, we will introduce a test sample. Figure 6.15 shows its configuration, which consists of a highly reflective substrate (e.g., a metal) and a step-formed transparent layer (e.g., a polymer). The thickness of the polymer layer is in the range of ∼100 μm and the difference between thickness $d1$ and $d2$ is larger than the coherence length, typically larger than 30 μm.

The coherence layer is "translated" through the object starting at the top of the sample. The interfaces are probed by the coherence layer in the following order:

1) Top of layer 1 (thick section $d1$)
2) Top of layer 1 (thin section $d2$)
3) Interface between layers 1 and 2 below the thin polymer layer
4) Interface between layers 1 and 2 below the thick polymer layer.

The interface between layers 1 and 2 is detected at different depths for the left or right side. This is due to the longer optical path length in the thicker polymer layer.

Sample Alignment After the interface of interest has been identified, the sample (or the coherence layer) has to be adjusted. Also, the sample alignment is carried out by operating the LCSI instrument as a full-field OCT interferometer. To enable measurements on interfaces in a scattering sample, the interface under investigation has to be positioned inside and parallel to the coherence layer. Furthermore, to utilize maximum contrast in the interference signals, the interface should be positioned close to the maximum of the coherence function.

The coherence layer has to cover the interface over the whole measured area. This is done by moving the coherence layer to the interface under investigation. Often, this interface is not parallel to the coherence layer, so only a part of the interface is covered.

Here the scanning is stopped. Now the sample is iteratively tilted until the whole area is positioned parallel within the coherence layer. Figure 6.16 shows the ideal positioning of the coherence layer. A second scan in structural imaging mode now allows the positioning of the maximum of the coherence layer at the interface.

Figure 6.17 shows the position of the coherence layer in the test sample introduced in Figure 6.15 after sample alignment. It can be seen that the interface between layers 1 and 2 below the thinner polymer layer is perfectly aligned, while the same interface below the thicker polymer layer cannot be aligned at the same time.

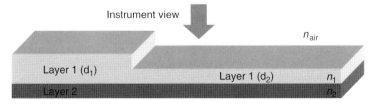

Figure 6.15 Configuration of the test sample.

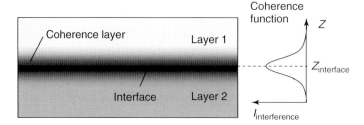

Figure 6.16 Ideal positioning of the coherence layer on the interface under investigation.

Figure 6.17 Position of the coherence layer after sample alignment.

Depth-Resolved Deformation Measurement Finally, after adjusting the sample the deformation measurements at the interface under investigation can be carried out. These are the actual LCSI measurements. Standard measurement algorithms of ESPI are applied while holding the coherence layer adjusted to a constant depth. To illustrate this part of the measurement algorithm, the sample configuration shown in Figure 6.17 is used. The interface between layers 1 and 2 below the thinner polymer layer is adjusted within and parallel to the coherence layer. Now the ESPI algorithms for deformation measurement are applied. The reference phase map is recorded. Then the sample is slightly rotated and the phase map after deformation is recorded. Now the phasemaps are subtracted. Figure 6.18 presents the expected measurement result of the tilt. Only the part of the sample with an interface positioned inside the coherence layer (right part of the image) shows the mod 2π

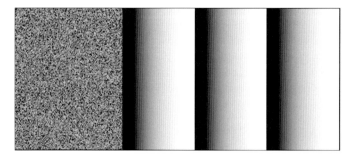

Figure 6.18 Demonstration of the measurement principle on a transparent step sample; expected mod 2π phase map for a horizontal tilt of the sample with the coherence layer adjusted at the interface between layers 1 and 2 under the thinner polymer layer. Areas outside the coherence layer produce only noise (left side of the object).

phase map for a horizontal tilt. Areas outside the coherence layer cannot interfere and contain only noise, as seen on the left part of Figure 6.18.

The measured deformations in LCSI are in the range from subwavelength up to several wavelengths. With OCT, these deformations are barely detectable. OCT detects the position of the maximum of the coherence envelope. Displacements in the order of the wavelength are hardly detectable. However, applying phase measurement algorithms from ESPI enables the LCSI instrument to detect these small deformations.

6.4.3
Interferometer Setup

The combination of ESPI and OCT is noticeable in the LCSI setup. It is adopted from the Mach–Zehnder interferometer configuration used in ESPI. But in contrast to ESPI, LCSI uses a low-coherent light source. The coherence layer can be positioned at the area of interest by scanning either the corner cube or the object. The basic configuration of an LCSI instrument is shown in Figure 6.19.

A super luminescence diode (SLD) is used as the low-coherent light source. The spectrum of the light source in LCSI is rather narrow, because the requirements on the depth resolution are lower than in OCT. On the contrary, it can be a disadvantage to have a coherence layer that is too thin, because this limits the size of the deformation that can be measured.

The typical coherence length is in the range of 30–50 µm, and the bandwidth of the source is therefore rather narrow, that is, between 10 and 30 nm, depending on the wavelength of the source. These specifications have to be optimized for each application depending on the distance between the interfaces to be separated and the size of the expected deformation. As a consequence of the intentionally used low-coherent source, particular care must be taken regarding dispersion, coherence, and resolution.

The object arm is identical in both LCSI and ESPI. In the reference arm, a corner cube placed on a translation stage is included to control the position of

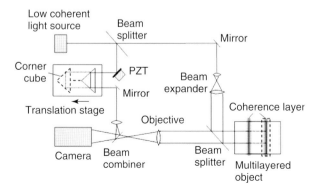

Figure 6.19 Principle setup of an LCS interferometer.

the coherence layer. In LCSI we assume a uniform reference wave. This setup can be used in two different ways by slightly varying the position of the source point of the reference beam. If the expanded reference wave is "virtually" coming from the center of the imaging aperture plane, the setup can be applied to TPS. A piezoelectric transducer (PZT) in the reference arm is included for this purpose. By simply displacing the source point laterally in the aperture plane, the setup can be utilized for SPS in the virtual aperture plane.

6.4.4
Evaluation

The three parts of the measurement algorithms require two different instrument modes.

For the first two steps of the measurement algorithm, the full-field OCT mode of the instrument is used. When the object is adjusted in the right position, the LCSI mode of the instrument is used to carry out the actual deformation measurement.

Both modes can be utilized with two different types of data processing depending on the type of phase shifting applied. SPS is applied if the deformation takes place at high speed or the environmental conditions are changing quickly. Phase information can thus be obtained from a single image. A fast Fourier transform (FFT) algorithm is then applied for the full-field OCT mode. TPS is applied if higher spatial resolution is required. In this case, a high-pass filtering algorithm can be used.

This section derives the interference equation for different object positions and introduces the two instrument modes.

6.4.4.1 Full-Field OCT Mode

In the full-field OCT mode, the interference equation for OCT is used. The mathematical treatment is based on Mandel and Wolf [29] and the review article by Fercher et al. [27]. We assume stationary ergodic signals and perpendicular illumination.

From Eq. (6.8) the interference equation is given by

$$I = I_r + I_o + 2 \operatorname{Re} \Gamma_{ro} \tag{6.21}$$

where Γ_{ro} is the complex coherence function between the reference field $V_r(t)$ and the object field $V_o(t)$ at the exit of the interferometer. In OCT, generally temporal coherence is used to separate signals coming from different depths of the object. We introduce thus

$$G_{ro}(\tau) = 2Re\Gamma_{ro}(\tau) = 2Re\langle V_r * (t) V_o(t+\tau)\rangle \tag{6.22}$$

as the cross-interference term. By using mirrors in both interferometer arms, the complex coherence function $\Gamma_{ro}(\tau)$ of the interferometer equals the complex coherence function of the source $\Gamma_{source}(\tau)$, up to a constant depending on beam splitter and mirror reflectivity.

However, when one of the mirrors is replaced by a depth scattering object, the object influences the coherence function of the interferometer. We choose to represent the local amplitude reflectivity of the object by a normalized response function denoted $h(t)$ [30]. The object field $V_o(t)$ is then given by

$$V_o(t) = V_{0o}(t) \otimes h(t) \tag{6.23}$$

where $V_{0o}(t)$ denotes the field incident on the object and $\otimes$ is the convolution operator.

The cross-interference term $G_{ro}(\tau)$ is then given by Fuji et al. [30].

$$G_{ro}(\tau) = 2\,\mathrm{Re}\,\Gamma_{ro}(\tau) = 2\,\mathrm{Re}\,\Gamma_{source}(\tau) \otimes h(\tau) \tag{6.24}$$

Using the convolution theorem, the spectral interferogram in the frequency domain can be expressed as

$$W_{ro}(\nu) - S(\nu)H(\nu) \tag{6.25}$$

where $W_{ro}(\nu)$ is the cross-spectral density function of the two interfering waves. The spectral functions of this equation are analogous to the Wiener–Khintchine theorem given by the Fourier transform of the corresponding functions in the time domain.

$$W_{ro}(\nu) = \mathrm{FT}[\Gamma_{ro}(\tau)] \tag{6.26}$$
$$S(\nu) = \mathrm{FT}[\Gamma_{source}(\tau)] \tag{6.27}$$
$$H(\nu) = \mathrm{FT}[h(\tau)] \tag{6.28}$$

$S(\nu)$ denotes the power spectrum of the source, and $H(\nu)$ is the sample transfer function or the spectral response of the object at frequency ν. The cross-interference term is then given by the real part of the inverse Fourier transform of $W_{ro}(\nu)$ and the interference equation is given by

$$I(\tau) = I_r + I_o + 2\,\mathrm{Re}\{\mathrm{FT}^{-1}[S(\nu)H(\nu)]\} \tag{6.29}$$

Full-field OCT is based on the same interference equation, but it has to be extended for three dimensions $I(x, y, z)$. As shown previously, the cross-interference term depends on the reflectivity given by the object function and the position of the coherence layer.

The spatial dimension of the coherence layer in the z-direction is calculated by the full-width at half-maximum (FWHM) of the coherence function of the source l_{FWHM}. In the given interferometer configuration in Figure 6.19, the light is double-passing the interferometer arms. Therefore, the so-called round trip coherence length l_c is introduced. For a narrow Gaussian spectrum and in vacuum, l_c can be calculated as in Fercher et al. [27].

$$l_c = \frac{l_{FWHM}}{2} = \frac{2\ln 2}{\pi}\frac{\lambda_0^2}{\Delta\lambda} \tag{6.30}$$

The maximum position of the coherence function is denoted as the probing depth z_d. The probing depth is given by the depth of zero OPD corresponding to the position of the reference mirror. When measuring inside a medium, the coherence length in vacuum l_c is divided by the refractive index of the medium n_m.

In practice, the OCT mode is carried out using either an FFT algorithm [31] or a high-pass filtering algorithm [24, 25].

6.4.4.2 LCSI Mode

In the LCSI mode, measurement algorithms from ESPI are utilized. As in ESPI, the object wave is speckled. However, the interference equation for ESPI, given in Section 6.3.2, has to be extended because of the use of low-coherent light.

The main application of LCSI is the deformation measurement of the internal interfaces of objects. Once the coherence layer is aligned with the interface under investigation, the reference mirror is not moved. Therefore, for deformations in the range of a few wavelengths, the probing depth can be assumed to be constant, that is, $z_d = $ constant, and the coherence layer is not displaced inside the object.

Therefore, the derivation of the interference equation for LCSI can be simplified by introducing the normalized degree of coherence $|\gamma_o|$ between the reference and the object field. Assuming a uniform reference wave, we get

$$I(x, y, z_d) = I_r + I_o(x, y) + 2\sqrt{I_r I_o(x, y)} |\gamma_{ro}(x, y, z_d)| \cos(\Delta\varphi(x, y, z_d)) \quad (6.31)$$

As in OCT, in LCSI the object light I_o consists of a coherent part I_{oc} and an incoherent part I_{oi}.

$$I_o(x, y) = I_{oc}(x, y) + I_{oi}(x, y) \quad (6.32)$$

where I_{oc} denotes the intensity of the object light being coherent with the reference light. The incoherent part I_{oi} is often called the *incoherent background*. This light is reflected from other interfaces or scatterers both above and below the coherence layer or is generated by multiple scattering.

We will now derive the interference equation for the sample configuration introduced in Figure 6.15. The first layer is transparent (e.g., polymer), a part of the incoming field is reflected at interface 1, and the transmitted field travels towards interface 2. The second layer is opaque (e.g., metal) and all light is reflected at interface 2. Thus no light reaches interface 3.

As shown in the previous section, the coherence layer is divided in two. We will analyze the interference equation in the two different cases shown in Figure 6.20.

The coherence functions for the corresponding sample configurations are shown on the left and right part of the object.

Case 1 On the left side of the sample, the coherence layer is adjusted in the middle of the transparent layer. All light is reflected from interfaces 1 and 2. These contributions are denoted I_{11} and I_{12} and contribute to I_{oi}. Furthermore, some light is multiply reflected between the interfaces of layer 1, namely, $I_{multiple}$, which also contributes to I_{oi}.

However, no light is reflected within the coherence layer and the coherent object intensity is $I_{oc} = 0$. Thus $|\gamma_o|(z) = 0$ and the cross-interference term vanishes.

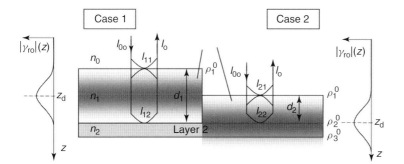

Figure 6.20 Interference equation for the sample object consisting of a transparent layer on top and a nontransparent layer below. Left: Coherence function $|\gamma_0|(z)$ for case 1 (thicker transparent layer). Center: Sample configuration with position of the coherence layer (gray shadowed). Right: Coherence function $|\gamma_0|(z)$ for case 2 (thinner transparent layer).

The interference equation is reduced to the sum of the self-interference terms:

$$I(x, y) = I_r + I_{oi}(x, y) = I_r + I_{11}(x, y) + I_{12}(x, y) + I_{\text{multiple}}(x, y) \qquad (6.33)$$

Case 2 On the right side of the image, the coherence layer is adjusted at the second interface. The maximum of the coherence function is adjusted at the probing depth z_d.

The light reflected at interface 1 is not coherent. It contributes to the incoherent background intensity I_{oi}.

Interfaces 2 and 3 are within the coherence layer. However, no light is reflected from interface 3 and it contributes neither to I_{oi} nor to I_{oc}.

A part of the light reflected from interface 2 travels directly through interface 1 towards the beam splitter and is coherent with the reference light. This part of the light contributes to I_{oc} and the cross-interference term. Furthermore, some light is multiply reflected between the interfaces. This light is not coherent with the reference light and contributes to the incoherent background.

The interference equation is then given by

$$I(x, y, z_d) = I_r + I_o(x, y) + 2\sqrt{I_r I_{22}(x, y)} \cos(\Delta\varphi(x, y, z_d)) \qquad (6.34)$$

where the object light is given by $I_o(x, y) = I_{21}(x, y) + I_{22}(x, y) + I_{\text{multiple}}(x, y)$.

If we introduce a scattering object, the equations are more complicated and the interference intensity is given by Eq. (6.29). The phase information is given by the object function. In particular, multiple reflections have to be considered.

This instrument mode is used for the actual LCSI measurements to measure the deformation of an interface or structure inside an object. We know that this can be done using the ESPI algorithm introducing phase shifting (Chapter 3). Because of the narrow bandwidth of the sources used in LCSI, both SPS and TPS are similar to the algorithms introduced for ESPI in Chapter 3.

6.4.5
Characteristics

We have to consider two significant differences between LCSI and ESPI. First, because of the use of low-coherent light, the statistics of the speckle pattern is changed. The intensity distribution of these speckle patterns is given in Section 6.2. The second difference is more crucial. The coherent part of the object light is significantly smaller than in ESPI. Only a fraction of the object light contributes to the interference signal, while the camera detects all light coming through the imaging system. The dynamic range of the interference signal is dramatically decreased.

Besides the parameters described for ESPI in Section 6.3.3, the following parameters have to be considered in addition when optimizing LCSI measurements:

- the degree of coherence between the object and reference light (determines the amount of incoherent background intensity);
- the shape and positioning of the coherence layer;
- the attenuation in the sample;
- dispersion effects.

The most efficient way to increase the interference contrast in LCSI is to increase the coherent object light intensity. Thus the adaptation of the optical system in an LCSI instrument for the utilization of the maximum SLD power is crucial.

The optimization parameters are specified below:

1) Select the most suited low-coherent light source:
   a. The center wavelength of the SLD should be adapted to the attenuation properties of the object.
   b. The FWHM of the SLD spectrum should fulfill two criteria:
      i. The coherence length should be large enough to cover the whole shape of the interface and the expected deformation range;
      ii. The coherence length should be small enough to separate interfaces and features inside the object.
2) Reduce the incoherent background using polarizers and spatial filters (SFs)
3) Utilize the whole dynamic range of the camera. The total intensity can be higher than in ESPI, depending on the degree of coherence of the object light reflected in the probing depth.
4) Optimize the beam ratio $r = I_r/I_o$:
   a. If the constraints on the light intensity are due to the limited power from the SLD, an increase of the beam ratio, by guiding a larger part of the available light into the reference arm, will improve the measurement results.
   b. If the constraints on the light intensity are due to the limited dynamic range of the camera, the beam ratio should be $r \geq 1$. That is analogous to our results for ESPI. However, depending on the degree of coherence in the probing depth, the object light intensity should be as high as possible, without reaching the saturation level of the camera and taking care that $r \geq 1$.

5) Optimize the position of the interface at the maximum of the coherence layer during excitation.
6) Compensate for dispersion effects. This is particularly important for fiber-based interferometers.
7) Adapt the shape of the coherence layer to the interface.

The optimization of the interference signal is described in detail by Gastinger and Winther [32].

The quantification of the phase measurement results obtained in LCSI can be challenging. Three different phase contributions make up the measured phase change:

1) Phase changes caused by the displacement of the interface under investigation;
2) Phase changes caused by refractive index changes in the optical path length along the total propagation path of the coherent reflected light;
3) Phase changes caused by the displacement of the scatterers inside the coherence layer.

Phase changes caused by the reflected light from the interface inside the coherence layer are the required measurement result. Light reflected from the surface or other interfaces positioned outside the coherence layer will not contribute to the coherence layer due to coherence gating. The other unwanted phase distributions have to be sorted out. This requires modeling of the interference signal. An approach for this is suggested by Gastinger *et al.* [33] using a one-dimensional transmission line model.

6.4.6
Applications

LCSI is a relatively novel technique. Therefore the range of applications introduced in the literature is limited. In this section, we concentrate on two different applications, namely, the nondestructive characterization (NDT) of adhesive bonded joints, and the measurement of the membrane deformation in a MEMS pressure sensor.

6.4.6.1 NDT of Interfacial Instabilities of Adhesive Bonded Joints

Defects at the interfaces of the adhesive layer and the substrate reduce the adhesive properties of bonded joints. These defects can be caused by imperfect pretreatment, surface topography, or other surface phenomena (e.g., corrosion or intermetallic particles). NDT of these interfacial instabilities is required to make it possible to test the sample over longer time periods and under different environmental conditions.

To investigate adhesion, a glass plate is bonded onto the substrate using a commercially available adhesive (semitransparent Araldite). By measuring through the glass plate, the optical signal reflected from the adhesive as well as from the substrate/adhesive interface is investigated. The sample configuration and the test procedure are illustrated in Figure 6.21.

6 Rough Surface Interferometry

A test procedure has been developed in order to detect the surface-related deformation pattern. A controlled mechanical force is introduced on the glass plate, which applies a stress to the adhesive layer and the aluminum/adhesive interface. The right image in Figure 6.21 shows how the glass plate is pulled with a defined mechanical force perpendicular to the aluminum surface under investigation. The microscopic delamination areas open up, and the interferometer can detect the resulting cavities. By relating the measured deformation pattern to the corresponding surface phenomena, mapped before the bonding, the adhesion properties can be evaluated.

Figure 6.22 shows a typical measurement result. The phase map in the left image shows a defect area in the upper center region. While the rest of the jointed

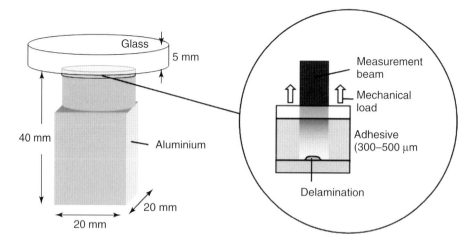

Figure 6.21 Demonstration of the measurement concept. Instabilities at the aluminum/adhesive interface cause deformation patterns that can be detected with high sensitivity.

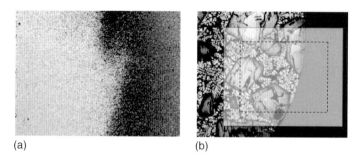

Figure 6.22 Defect at the adhesive/aluminum interface.
(a) Phase map after introducing mechanical load.
(b) Microscope image recorded before gluing. The solid frame shows the approximate measurement area, and the dashed line the most likely field of view.

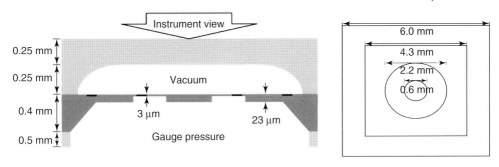

Figure 6.23 Pressure sensor. (a) Cross section. (b) Top view [34].

area appears bright, the defect area is dark. Comparing the measurement with a microscope image indicates where the defect occurred on the sample surface. The detected interfacial instability might correspond to an area where the pretreatment shows a defect on top of an intermetallic particle.

6.4.6.2 Membrane Deformation of a MEMS Pressure Sensor

The measurements are carried out on a MEMS pressure sensor [34]. A sketch of the sensor cross section and a top view can be seen in Figure 6.23. The dimensions of the sensor are $6.0 \times 6.0 \times 1.4 \, \text{mm}^3$.

A silicon membrane is covered by a glass wafer forming a 250-μm thick vacuum cavity. At atmospheric pressure, the membrane is bent; when reducing the gauge pressure, the membrane is displaced towards a planar position; at vacuum pressure, the membrane is flat.

The varying thickness of the Si membrane causes a circular piston at the center to be excited at low pressures. The thicker outer membrane is used to measure higher pressures up to 1 bar.

The flat membrane is illuminated with speckled illumination using a ground glass in the illumination path. This increases the angular spectrum of the reflected light and thus enables the measurement of larger deformations than with plane-wave illumination. All measurements are carried out through the glass window.

Figure 6.24 shows the mod 2π deformation phase maps for two different excitations measured with LCSI and ESPI. Figure 6.24a,b shows excitations at a low pressure, while (c,d) shows excitations at a high pressure.

In Figure 6.24a,c, LCSI measurements for the given pressure differences can be seen. Figure 6.24b,d shows ESPI measurements. It can be seen that the use of LCSI reduces the noise in the measurements significantly, in particular because influences caused by multiple reflections and distortions of the light path on the lower glass interface are removed.

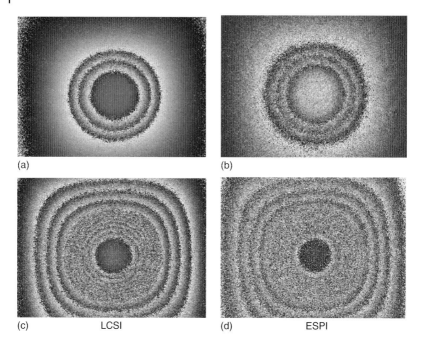

(a)   (b)

(c)   LCSI   (d)   ESPI

Figure 6.24 Measurement results of the deformation of the pressure membrane measured through the glass plate. (a,b) Pressure changed from 270 to 530 Pa. (c,d) Pressure changed from 1300 to 2000 Pa [34].

Tutorial Exercise 6.2

A square sample ($l = 10$ mm) consists of three layers. A glass plate of 100 μm thickness is glued onto a spherical surface by a semitransparent adhesive with $n_{adh} = 1.7$. Three different SLDs are available:

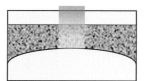

- SLD 1: $\lambda_c = 650$ nm, $\Delta\lambda = 70$ nm;
- SLD 2: $\lambda_c = 750$ nm, $\Delta\lambda = 35$ nm;
- SLD 3: $\lambda_c = 850$ nm, $\Delta\lambda = 18$ nm.

The field of view of the instrument is 2×2 mm² at the central area of the sample. The adhesive layer is 20 μm at the central position. The height difference of the spherical surface is 5 μm within the field of view. The range of the deformation of the interface is not known. The absorption spectrum of the adhesive has a minimum around a wavelength of 850 nm and about the same moderate absorption at the other two wavelengths.

Problems

1) Calculate the round-trip coherence length of the light sources inside the medium.
2) What requirements do you have on the FWHM of the spectrum of the source?
3) Which center wavelength would you select?
4) What is the best solution out of the three available SLDs?
5) How can you optimize the interferometer to give maximum contrast?
6) What is the required accuracy of the translation stage?
7) What do you need to consider when quantifying the measurement results?

Solution:

1) Use Eq. (6.30) and divide the resulting coherence length in vacuum by the refractive index of the adhesive n_{adh}.

   SLD1:$lc,m = 1.57\,\mu m$, SLD2:$lc,m = 4.17\,\mu m$, SLD3:$lc,m = 10.42\,\mu m$

2) The FWHM of the light source decides the thickness of the coherence layer inside the material. The thickness of the coherence layer should be so large as to cover the whole surface profile of the interface under investigation and the expected deformation range. On the other hand, it should be so small as to clearly distinguish the interfaces inside the sample. The profile of the interface has a height variation of 5 µm within the field of view of the instrument and the distance to the next interface is 20 µm. Therefore, SLD 2 would be most suited from this point of view.

3) The center wavelength should be adapted to the absorption properties of the sample. The given sample has a minimum absorption at 850 nm. It needs to be considered that the interference signal can be detected from an area about twice as long as the calculated coherence length. From this point of view, SLD 3 should be selected.

4) When selecting the most suited SLD, both criteria need to be considered. The separation of the interfaces has higher priority as long as we still detect the interference signal from the interface, since the quantification of the measurement results is more sophisticated. Thus, the best compromise is the selection of SLD 2.

5) It is important to optimize the shape of the coherence layer to the spherical shape of the object. This can be done by using a phase plate in the illumination arm or changing the wave front of the illumination wave correspondingly. A beam ratio slightly larger than 1 is recommended, depending on the absorption properties of the adhesive.

6) The requirements on the accuracy of the translation stage in LCSI are not as high as in OCT. The translation stage is used to map the object's internal structure and to set the maximum position of the coherence layer

at the interface of interest. Thus a positioning error of about one-fifth of the coherence length in vacuum (Eq. (6.30)) is acceptable.
7) All phase changes along the optical path influence the measurement. Such influences can be the refractive index changes in air, but also changes of the refractive index of the adhesive caused by the elasto-optic effect. Furthermore, an independent deformation (i.e., a tilt) of the first interface will change the optical path length, even though it does not contribute to the interference signal.

6.5
Speckle Pattern Shearing Interferometry – Shearography

6.5.1
Introduction

Measurements in industrial environments pose a special challenge for interferometric systems which are sensitive by nature to changes in optical path length between the system and the object of interest. Air turbulence, thermal perturbations, acoustic noise, and vibrations are often encountered in industrial environments. Even under laboratory conditions, out-of-plane rigid body displacements cannot always be avoided during structural testing. Shearography is a full-field speckle-based method that is better suited to cope with these problems than other interferometric methods.

6.5.2
General Principle

Shearography was invented by Leendertz and Butters [35] and developed further by Hung [36]. The most important feature that determines the favorable properties of this method is the concept that both the object and the reference beam originate from the object under test. The object is imaged twice onto a sensor, usually a CCD, by an imaging system that contains a device implementing a relative shift between the two images. As a result, any point in the overlap area of the two images receives light from two points on the object, as shown in Figure 6.25: points P and Q are imaged as P′ and Q′ by the lens only, and shifted as P″ and Q″ by the lens plus the shearing element. Points P′ and Q″ coincide because of the shearing element in the imaging system. The relative shift between the two images is called the *shearing distance P–Q*, referred to the object; the method is called *speckle pattern shearing interferometry* but is usually referred to as *shearography*.

As the light source that illuminates the object under test is assumed to produce coherent light, interference occurs between the two contributing beams. The resulting intensity at any point of the sensor depends on the local optical path difference between the two beams. All points at the CCD form a speckle interference pattern,

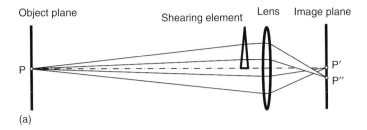

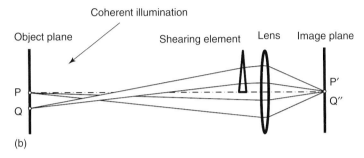

Figure 6.25 Principle of shearography. (a) Point P in the object plane is imaged as P' and P'' in the image plane. (b) Points P and Q in the object plane are imaged twice in the image plane as P', Q', and shifted as P'', Q''. P' and Q'' coincide with each other in the image plane. P'' and Q' are not shown for clarity.

which is the result of two identical speckle patterns that are laterally shifted with respect to each other. Since speckle phase is random, the speckle interference pattern, which is a function of the optical path difference, or phase, is also random [37].

As with other full-field speckle interferometry methods, shearography is sensitive to displacements along the sensitivity vector, which is the bisector between the illumination direction and the viewing direction. However, if both points of the object move in the direction of the sensitivity vector, the optical paths of object and reference beams change by the same amount, resulting in the same intensity at the sensor: within certain limits, the system is not sensitive to the out-of-plane displacement caused by rigid body movement of the object. Only if there is a relative change between the two optical paths, the interference result does change: the intensity at any point on the sensor is a function of the displacement gradient of the corresponding two points on the object. When the shearing distance is small, the displacement gradient approximates the displacement derivative in shearing direction of the out-of-plane displacement. An in-depth description of shearography and its applications can be found in [38].

The intensity $I(x, y)$ resulting from interference between an object and a reference wave at a point x,y can be expressed by

$$I(x, y) = I_o(x, y) + I_r(x, y) + 2\sqrt{I_o(x, y) I_r(x, y)} \cos(\varphi_r(x, y) - \varphi_o(x, y)), \quad (6.35)$$

where I_o and I_r are the intensities of object and reference beams and φ_o and φ_r are their phase values. Shearography is self-referencing: the object beam interferes with a shifted version of itself. Assuming a shear in only the x-direction, Eq. (6.35) can be written as

$$I(x,y) = I_o(x,y) + I_o(x+\delta x, y) + 2\sqrt{I_o(x,y)I_o(x+\delta x, y)}\cos(\varphi_o(x,y) - \varphi_o(x+\delta x, y)) \quad (6.36)$$

where δx is the shearing distance in the x-direction. For reasons of simplicity, the coordinates are often omitted, and phase difference between the object and the reference is denoted as φ. Subscript r is used again for the reference field, keeping in mind that it is a shifted version of the object field. By substituting $I_B = I_o + I_r$ and $I_M = 2\sqrt{I_o I_r}$ a simplified version of Eq. (6.36) is obtained:

$$I = I_B + I_M \cos\varphi, \quad (6.37)$$

where I_B and I_M denote background and modulation intensity, respectively. All terms I_B, I_M, and φ are random for a speckle interference pattern.

6.5.3
Evaluation

Basically, there are two ways of evaluating shearography measurements. The first one is to observe the so-called correlation fringes obtained by subtraction of two speckle interference patterns, one acquired before and the other after loading the object. Another approach is to calculate the phase at each pixel, using phase-stepping, followed by subtraction of phase results obtained before and after loading the object (Chapter 3).

A result typical for the subtraction method is shown in Figure 6.26. Two sets of correlation fringes are visible on a relatively undisturbed background, indicating a

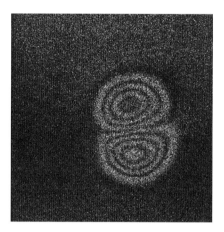

Figure 6.26 Two sets of correlation fringes, one for negative displacement gradients and the other for positive gradients.

defect area with reduced stiffness. One set depicts positive displacement gradients, and the other indicates negative gradients. It is not possible to determine the sign of the gradients from this single interferogram.

In order to obtain quantitative results, phase-stepping (Chapter 3) can be applied for shearography, similar as for ESPI. Quantitative results are mod 2π. All phase retrieval methods that are applicable for ESPI can be applied for shearography too, even the so-called spatial carrier frequency method [39], also referred to as *spatial phase stepping*. A shearing implementation proposed by Pedrini et al. [40] is based on a Mach–Zehnder type interferometer, where the reference speckle field can be shifted and tilted independently. The method requires speckles that are at least six times the pixel size. Sequential or temporal phase-stepping requires the conditions to remain stable between phase steps, which is not necessary for spatial phase stepping where the phase can be derived from a single speckle interference pattern. However, the requirement to use large speckle implies less efficient use of available light.

A result obtained with a phase-stepped shearography system is shown in Figure 6.27. The highest positive displacement gradient value is depicted by white, and the lowest value by black. Gray represents a zero change. The result is wrapped, that is, phase jumps of 2π are present in the phase distribution. There is no phase ambiguity: the result, which is typical for shearography, shows a defect area in a relatively undisturbed background consisting of two areas with opposite sign of the displacement gradient in shearing direction.

Phase jumps can be removed by spatial phase-unwrapping, or by applying temporal phase unwrapping [41] (Chapter 3). If care is taken that the phase change during any measurement interval is within the range $\langle -\pi, +\pi]$, phase difference distributions can be accumulated in real time, resulting in phase-unwrapped

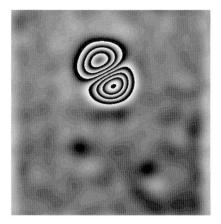

Figure 6.27 Wrapped phase distribution representing the deformation gradient of a loaded object containing a defect. Two unambiguous sets of fringes due to wrapping, one for negative gradients and the other for positive ones.

intermediate and final results. In addition to the advantage of being able to present phase-unwrapped intermediate results in real time, this method can also handle much larger phase changes, even beyond values in the order of the longitudinal speckle size where speckle decorrelation would occur. Speckle decorrelation poses a fundamental limit to the measurement range when using conventional processing based on comparison between the last phase distribution and the reference one, taken for the initial state of the object.

As for ESPI, shearography results can also be improved by speckle averaging. An implementation possible only for shearography is to manipulate the shearing mirror in a Michelson or Mach–Zehnder (Section 6.5.4) interferometer in order to obtain relatively small variations of the shearing distance around its average value, referred to as *micro-shearing*. For each orientation of the shearing mirror, an independent speckle interference pattern can be obtained and the results can be averaged or combined in a more intricate manner by postprocessing. The shift necessary to obtain a useful new speckle interference pattern depends on the speckle size, and can be easily determined by auto-correlating the micro-sheared reference patterns. In order to keep the sensitivity change caused by the variation of the shearing distance as small as possible, small speckles are preferable.

When temporal phase-unwrapping is used, the results of a series of similar test runs, each performed with an independent speckle pattern, can be accumulated. Since temporal phase-unwrapping produces unwrapped results during each of the tests, results can be filtered and accumulated conveniently.

6.5.4
Configurations

A speckle pattern shearing interferometer can be implemented in several ways. The most common configurations are realized by either by incorporating a wedge (Figure 6.25) or biprism element in the imaging system to obtain two laterally shifted images or by using a configuration based on a Michelson (Figure 6.28) or Mach–Zehnder (Figure 6.29) type of interferometer, where one of the mirrors can be tilted to obtain a variable lateral shift. Systems where the shift is based on a wedge, by partly covering the imaging lens, are relatively simple; they contain very few components and can be made compact and rugged. Other simple systems can be based on biprisms, glass plates with different tilts in front of the imaging lens, or a birefringent element that splits an incoming beam into two output beams. All shearing methods discussed here to produce a shifted version of the image on the sensor, superimposed on the original one, yield the same type of mechanical information: the displacement gradient in shearing direction of the out-of-plane displacement in the direction of the sensitivity vector.

Michelson-based configurations often use a piezo-controlled mirror to implement phase-stepping, thereby mechanically changing the length of the optical path (Figure 6.28). In the Mach–Zehnder configuration shown in Figure 6.29, the optical path can be changed by stepping one of the mirrors oriented at $45°$. Stepping the mirror introduces a small shift of the beam entering the lens, without

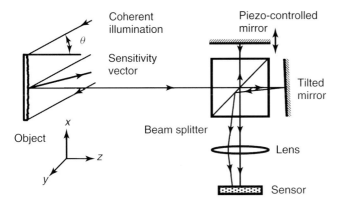

Figure 6.28 Shearography system based on the Michelson configuration.

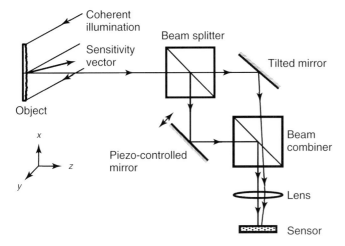

Figure 6.29 Shearography system based on the Mach–Zehnder configuration.

changing its direction; so there is no shift of the image at the sensor caused by phase-stepping. Piezoelectric devices are not perfectly linear and therefore require calibration. They also exhibit hysteresis and drift, which have to be accounted for.

Alternative ways to avoid these drawbacks, at the expense of higher complexity, are multichannel systems using one or more cameras. All phase-stepped speckle interference patterns may be projected onto one camera, or each pattern may have its own camera. These systems use polarization-based phase-stepping implemented with wave plates or other polarization-sensitive devices [42]. The major problem for these systems is alignment of the specklegrams involved [43], whereas the major drawback of temporally phase-stepped systems is phase error due to unwanted phase changes between phase steps. The latter are therefore more appropriate for use with applications with slowly varying displacements and for measuring harmonic vibrations, while the former can also be used for single-event dynamic applications. The same holds for spatially phase-stepped systems that acquire all information required to calculate a phase distribution in a single image [39, 40].

6.5.5
Characteristics

A major advantage of the common path nature of shearography is its relatively low sensitivity for common mode disturbances such as air turbulence in the optical path. For shearography systems, this common path consists of the optical path between the object and the shearography system as well as the optical path between the illumination source and the object. For ESPI, the reference beam follows a different path than the object beam, meeting different conditions, which may lead to unwanted changes in the optical path difference between the object and reference beams.

Because of the almost common path of the object and reference beams, the requirements for the illumination source can be less stringent than for other interferometric methods. In general, the difference between the optical path lengths of the beams is in the order of the shearing distance. For low values of the illumination and observation angles and relatively flat objects, the difference is even much smaller. As a consequence, relatively inexpensive illumination sources can be used, such as diode lasers that are not temperature-stabilized.

An attractive property of many shearography systems is that the sensitivity can be influenced by changing the shearing distance: the smaller the distance, the lower the sensitivity, but the more closely the measured gradient approximates the displacement derivative. This property should be considered in connection with the size of the phenomena that are measured. If the shearing distance is equal to the lateral size of an area at which a tilt has occurred, a further increase will not lead to increased sensitivity for that phenomenon: there is a phenomenon-related upper limit for the sensitivity. If the shearing distance is much larger than the phenomenon of interest two ESPI-like out-of-plane indications will be visible, one positive and the other negative. In that case, the sensitivity is the same as for ESPI but the indications no longer represent the local displacement gradient.

For shearing distances that are small compared to the size of the phenomenon of interest, a shearography system can be considered as a 2-D array of sensors with finite size, each of them independently measuring the average tilt over the shearing distance of a point on the object. As such, the finite size of the sensor limits the spatial resolution of the system since average tilt is measured over the shearing distance. This is in contrast with ESPI where the result at each point of the sensor represents the displacement of a corresponding point on the object. However, small details such as indications caused by surface cracks can be made clearly visible, as they are imaged twice when the shearing distance is much larger than the crack width and the shearing direction is perpendicular to the crack.

Because the out-of-plane displacement gradient is measured instead of the displacement itself and the sensitivity can be controlled, the measurement range of shearography can be larger than for ESPI: since ESPI generates more fringes, the resolution limits for the fringe pattern are reached earlier.

Shearography is based on interference between two speckle patterns, whereas most other speckle interferometric methods use a smooth reference, either a spherical wave or a plane wave. This is a drawback; since in a speckle field most speckles are dark, the probability that a bright speckle in the object pattern interferes with a bright one in the reference pattern is low, whereas for ESPI all speckles interfere with a uniform reference. However, even though this is a disadvantage, the modulation, serving as a quality indicator, is still sufficient to produce acceptable results in many points at the sensor. A comparison between the two approaches, made for phase-stepped systems, can be found in [18]. Efficiency is defined in this study as the percentage of pixels with a modulation above a certain threshold and for which the intensity levels stay below the saturation level. Using a threshold value of one-eighth for the normalized modulation intensity (modulation level/saturation level) the efficiency for phase-stepped systems with a smooth reference wave is reported to be 0.779 against 0.560 for systems based on a speckle reference wave.

Another fundamental drawback relates to the shearing distance in connection with the period of spatially periodic phenomena in the shearing direction. It is not possible to measure such phenomena when the shearing distance is a multiple of the spatial period of the phenomenon. In those applications, a shearing distance of $(n + 0.5)$ times the spatial period of the phenomenon should be chosen, where n is a positive integer or zero. For instance, when performing measurements on periodic structures such as the honeycomb panel shown in Figure 6.32c,d, a shearing distance equal to a multiple of the honeycomb cell size should be avoided.

6.5.6
Applications

Shearography systems are used for a wide range of applications, in laboratories, in industrial environments, as well as in the field. The most important applications at present are undoubtedly in the field of NDT, but strain measurements [38, 44] and vibration analysis are also possible. Applications can be divided into two classes: static and dynamic, both applicable to the fields mentioned above.

6.5.6.1 Excitation or Loading of the Object

Depending on the type, namely static or dynamic, different loading and processing methods are used. For static applications, mechanical loading or loading by applying heat or pressure differences is used; for dynamic applications, loading can be implemented by mechanical means such as electromagnetic devices or piezo-driven devices. Alternatively, also sound is used for loading. When developing an application for shearography, it is important to pay proper attention to the loading mechanism. Obviously, only a loading system that produces a measurable out-of-plane displacement for a given object configuration can lead to success. In the field of NDT, application of heat, either in the form of a powerful short pulse, or by a few seconds of exposure to a static heat source is often the preferred loading method because these methods are relatively simple. Depending on exposure time and on the material type used in the structure, the presence of defects at some depth can lead to measurable out-of-plane displacements at the surface.

Another very effective method of mechanical loading can be implemented by applying small pressure differences using a vacuum chamber attached to the surface of the object under test, causing defect-related out-of-plane displacements. Larger vacuum loading facilities are used as well, for instance for the inspection of helicopter blades [45].

Vibration loading can also be used to detect defects or damage in metallic and composite structures. For instance, defects in metallic bonded structures can be found by sweeping the excitation frequency through a certain range while observing the vibration pattern. Resonance vibrations have a higher amplitude, which will cause a defect to become visible during the sweep, depending on the resonant frequency related to the defect area. Composite structures may degrade during environmental or fatigue testing, affecting the stiffness of the structure, which in turn will influence resonance frequency and vibration patterns.

Shearography as an NDT tool can be used throughout the life cycle of products: during the phase of materials research and product development where the properties of materials and structures are investigated; during manufacturing to ensure product quality; and in service to inspect for damage and defects.

6.5.6.2 Static Applications

Static applications are mostly implemented by comparing the phase distribution before and after loading the object. In many cases, the measured phase difference related to the out-of-plane displacement gradient is larger than 2π, which implies wrapping in the resulting phase difference distribution. An experienced operator may very well be capable to evaluate more complex wrapped phase distributions, but unwrapped versions are easier to interpret and can also be filtered more easily because they are generally continuous. An example is given in Figure 6.30, where a composite structure containing some artificial defects is shown. The structure consists of a fiber-reinforced plastic skin bonded to a honeycomb core. The structure was tested twice using simple thermal loading with a heat gun, producing equivalent but not exactly the same wrapped results. However, after phase-unwrapping, the results are almost identical.

A result of a test using temporal phase unwrapping is shown in Figure 6.31, which is the last of a series of intermediate phase-unwrapped results that appear in real time during the test. All intermediate phase difference distributions represent phase changes within the range $\langle-\pi,+\pi]$ during a measurement interval and are accumulated to obtain the unwrapped phase difference distribution between the initial and final state. The total phase change is several wavelengths. The sample is an aeronautical structure consisting of a metal skin of 0.5 mm bonded onto a honeycomb core. The structure contained artificial defects by local removal of the

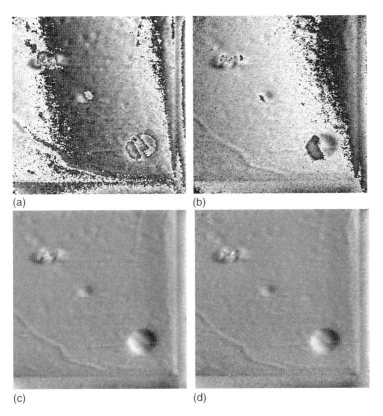

Figure 6.30 Artificial defects in a composite structure, tested twice under similar conditions using thermal loading. (a,b) Wrapped results. (c,d) Unwrapped versions of (a) and (b) that are easier to interpret.

honeycomb, simulating a debonded area. The size of the defect area shown in Figure 6.31 is Ø 25 mm.

Figure 6.32 shows some additional results obtained for the same sample. Figure 6.32a shows an overview of the thermally loaded panel, taken at a distance of 2 m, which is quite large considering the negative effects caused by air turbulence that occurred during the tests due to heating with a heat gun. Despite these effects, it was possible to achieve acceptable results at this distance, which were enhanced by speckle averaging over four runs, implemented by laterally moving a diffuser in the illumination beam between the four tests. The results show several indications caused by artificial defects simulating debonded areas up to Ø 25 mm. Figure 6.32b shows a closeup taken of one of the defect areas at the same distance using an imaging lens with a longer focal length, also averaged over four runs. Shearing distance for both images is approximately 10 mm, in an almost horizontal direction.

Figure 6.31 Artificial defect in a metallic structure, size ⌀ 25 mm. Aluminum skin 0.5 mm is bonded onto aluminum honeycomb core. Data was processed by temporal phase-unwrapping.

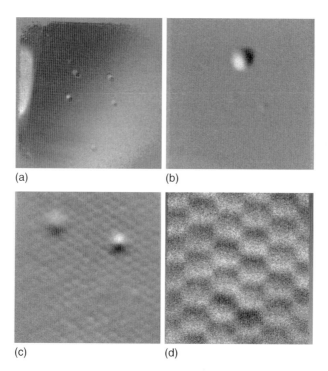

Figure 6.32 Metallic aeronautical structure with artificial defects inspected at distances of 2 and 1 m. (a) Overview taken at 2 m, field of view $0.64 \times 0.64\,m^2$. (b) Detail showing one of the defects, taken at 2 m, field of view $160 \times 160\,mm^2$. (c) Closeup taken at 1 m, showing two minor defects, field of view $80 \times 80\,mm$. (d) Closeup taken at 1 m, showing underlying honeycomb structure, field of view $35 \times 35\,mm$. All measurements are speckle-averaged over four runs with different speckle illumination.

When using speckle averaging, it is possible to visualize details that would otherwise be hardly visible and thus increase the probability of detection. Figure 6.32c shows two minor defects extending over only a few honeycomb cells and also reveals the internal structure of the panel. Figure 6.32d shows the honeycomb structure covering a field of only 35×35 mm^2. Both images were taken at a distance of 1 m. The shear applied was vertical. To visualize these details, it was necessary to increase thermal loading. Both results were obtained by averaging the phase distributions of four independent runs.

Another example of an application of shearography in the field of NDT is the detection and characterization of impact damage in carbon-fiber-reinforced plastic (CFRP) structures. CFRP is a high-strength composite material that is increasingly applied in aerospace structures. The material is sensitive to impact damage which may extend below the surface of the structure and may hardly be visible from outside. An optical method such as shearography can be applied to detect and characterize such damage under field conditions.

Examples are given in Figure 6.33, showing damage in a flat CFRP plate and a T-blade stiffened panel damaged by an impact at the center of the blade. The T-blade is adhesively bonded to the skin. Figure 6.33a clearly shows the presence of the T-stiffener and the location of the blade. An area of reduced stiffness indicates a debonding between the skin and the stiffener. Note the effect of applying shear in horizontal direction: all stiffness discontinuities due to the presence of the T-stiffener are clearly visible, which would not have been the case when shear would have been applied in vertical direction.

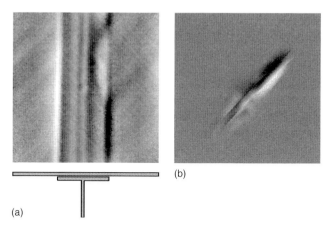

Figure 6.33 Impact damage in two CFRP panels. (a) T-blade stiffened CFRP panel (damage at center top right), horizontal shearing. The T-blade providing locally high out-of-plane stiffness is clearly visible through the skin. The debonded area between skin and stiffener causes reduced stiffness in the defect area, leading to changed out-of-plane displacement gradients. (b) Flat CFRP panel. Shearing distance was 10 mm vertical. Two types of damage are visible: internal damage oriented at 45°, apparent by gradually changing gradient indications; and a distinct couple of crack indications at shearing distance, oriented at 45°, with opposite sign.

The damage in the flat CFRP plate shown in Figure 6.33b consists of internal damage, reducing the local stiffness of the sample, and a surface crack. The shearing distance was set at 10 mm vertically, allowing measurement of displacement gradients over 10 mm. Some of the indications show much greater detail, due to the presence of the crack, which leads to two sharp indications with opposite signs, located at a distance equal to the shearing distance and oriented at 45°, which is the fiber direction at the surface of the plate. Both samples were loaded thermally with a heat gun in order to obtain out-of-plane displacements.

Additional examples of the application of shearography in the NDT field can be found in application-oriented references [46–48].

6.5.6.3 Dynamic Applications

Time Averaged Method Another field in which shearography can be applied is the analysis of vibration patterns, either for modal analysis, acoustic studies, or in NDT.

An example of a vibration pattern measured with a phase-stepped shearography system is given in Figure 6.34. The sample is identical to the one used for the ESPI experiment, Figure 6.6, and the same vibration mode was measured. The sample was clamped at one side and was excited acoustically by a loudspeaker with a frequency of 670 Hz. The result was obtained by acquiring four phase-stepped speckle interference patterns and by processing in the standard way, based on Eq. (6.17). The shearing distance is 10 mm, in vertical direction. Zero displacement gradients are indicated by a zero-order fringe, shown in white. Figure 6.34b shows the average vibration pattern resulting from a series of nine independent measurements, using micro-shearing. Figure 6.34c,d shows the results of further processing for these series. For each of the pixels in the nine vibration patterns, the results are sorted according to their modulation, resulting in nine new compound vibration patterns, ranging from the lowest (c) to the highest (d) value of the modulation per pixel.

In contrast to ESPI, the modal fringe pattern obtained with shearography depicts the displacement gradient instead of the displacement, so the special locations in a modal pattern should be interpreted differently for the two methods. At the nodal lines, the displacement is zero, so at these locations a zero-order fringe can be observed in the fringe pattern obtained with ESPI. Positions of maximum displacement, that is, the antinodes, are at the centers of concentric fringe systems for ESPI.

The displacement gradient is zero in all directions for antinodes. This property can be used to identify antinodes with shearography: a zero-order fringe indicating zero gradient in the chosen shearing direction runs through the antinode, irrespective of the shearing direction. So antinodes are at the crossings of zero-order fringes obtained for all shearing directions.

At a nodal line, the displacement gradient is maximum perpendicular to that line. The maximum gradient measured with a shearography system can be found

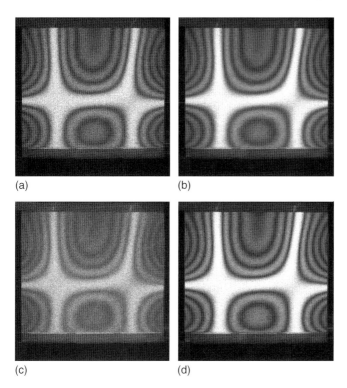

Figure 6.34 Speckle-averaged vibration patterns showing Bessel fringes, depicting the y-gradient of the out-of-plane vibration amplitude. The sample is identical to the one used for the ESPI experiment (shown in Figure 6.6) measured at the same mode and processed using Eq. (6.17). (a) Single run. (b) Speckle-averaged over nine runs using micro-shearing. (c) Pixels sorted for modulation: compound image based on pixels with lowest modulation out of nine. (d) Compound image based on pixels with highest modulation out of nine.

at the center of a concentric fringe system, so a nodal line runs through this center, and its direction is perpendicular to the shearing direction.

Using shearography for dynamic applications has a similar advantage over methods such as ESPI as for static applications. Especially in those cases when relatively freely suspended test objects are measured or other unfavorable test conditions exist, it can be hard to obtain good-quality vibration results with ESPI, while shearography can still produce acceptable results.

Pulsed Illumination When a harmonically vibrating object is stroboscopically illuminated by a CW laser, phase-stepped interference patterns for a selected state in the vibration cycle can be acquired. Each of the phase-stepped speckle interference patterns is acquired during a different vibration cycle. If illumination and image acquisition are synchronized and the relative phase between vibration and illumination is varied, phase differences between states can be visualized, and even full temporal characterization can be realized. Similarly, pulsed lasers can be used for freezing fast phenomena, either for simple comparison between two states or for full temporal analysis. Multichannel systems or systems using spatial phase-stepping can be used to reduce synchronization requirements. Some examples are presented in Figures 6.35 and 6.36.

The experiments have been carried out on an aluminum plate of 240 × 240 × 0.5 mm, elastically supported by four springs attached to the corners of the plate. The specimen was excited at the center by a shaker that was driven by a waveform generator with a sine wave of 900 Hz, which was one of the resonance frequencies of the plate. The specimen was illuminated by a divergent beam from a pulsed Nd : YAG laser, delivering approximately 30 mJ at a repetition rate of 10 Hz. The pulse duration was approximately 15 ns.

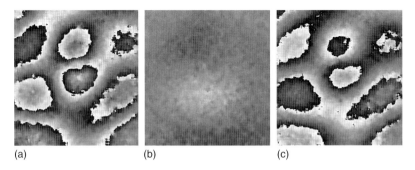

Figure 6.35 Vibration patterns obtained with pulsed laser illumination. Wrapped intermediate vibration states of an aluminum plate, size 240 × 240 × 0.5 mm, vibrating at 900 Hz. (a) During the first half of the vibration period. (b) Halfway down the vibration period. (c) During the second half of the vibration period.

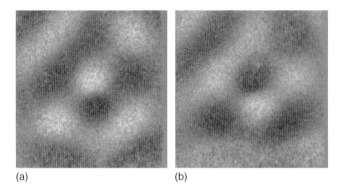

Figure 6.36 Accumulated phase difference distributions producing unwrapped results. (a) During the first half of the vibration period. (b) During the second half of the vibration cycle.

Optical phase at random positions in the vibration cycle was measured by a CCD camera in a two-channel shearography system. The camera was synchronized to the pulsed illumination. Simultaneously, mechanical vibration phase was measured by a general-purpose counter circuit, triggered by synchronization pulses from the laser and the waveform generator that was used for excitation of the sample. After sorting the vibration patterns, using the mechanical vibration phase as the sorting key, the interference patterns were processed in sorted order using temporal phase-unwrapping [41]. This procedure allowed accumulation of phase differences to obtain unwrapped phase data at any instance of the vibration phase [49].

Figure 6.35 shows phase differences between three intermediate vibration states and the reference state, which was close to the static state where vibration amplitude was zero. Figure 6.35a shows the results for an intermediate vibration state during the first half of the vibration period; Figure 6.35b,c shows the results for vibration states halfway down the vibration period and during the second half of the vibration cycle, respectively. These results are part of a large series allowing full spatial and temporal characterization of the vibration pattern. For most of the measured intermediate states, the displacement-related phase differences were outside the interval $\langle -\pi, +\pi]$, resulting in phase wrapping. Shear has been applied in vertical direction. Figure 6.35 clearly shows the reversal of phase for first and second halves of the vibration cycle.

When a sufficient number of samples per period are taken, phase differences between samples can be within the range of $\langle -\pi, +\pi]$, avoiding wrapping and allowing phase differences to be accumulated over all or a selection of measurement intervals. Following this procedure, the results for a selection of measurement intervals during the positive and the negative part of the vibration period were accumulated. Figure 6.36a, taken from another test run, shows the accumulated phase difference distribution for the positive portion of the vibration cycle; Figure 6.36b shows the results for the negative part. Again, the reversal of phase for first and second halves of the vibration cycle is clearly shown.

Tutorial Exercise 6.3

Figure 6.37 depicts a simulated ESPI out-of-plane indication for a structure with a defect extending in the x-direction, sized 100×10 mm. Derive the indication that would have been obtained with a shearography system using (i) a vertical shearing distance of 5 mm and (ii) a horizontal shearing distance of 5 mm.

Solution:

See Figure 6.38a for vertical shear: the defect area is visible over its full length. See Figure 6.38b for horizontal shear: there are only indications at the far ends of the defect zone. In all other locations, the displacement gradient in the horizontal direction is zero.

278 | 6 Rough Surface Interferometry

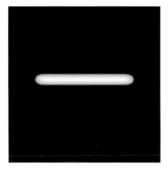

Figure 6.37 Simulated ESPI out-of-plane indication for a defect of 100 × 10 mm.

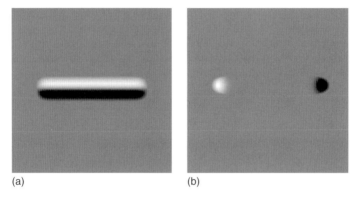

Figure 6.38 Shearography indications for the defect depicted in Figure 6.37. (a) When using a vertical shear of 5 mm. (b) When using a horizontal shear of 5 mm.

6.6
Digital Holography

6.6.1
Introduction

Holography was discovered by the Hungarian physicist Dennis Gabor [1] when he was working on electron microscopy. Gabor received the Nobel Prize for Physics in 1971 for this discovery. However, holography took off in 1962 with the arrival of the first lasers, with the demonstration by Leith and Upatnieks [50] and Denisyuk [51]. Holography is a fruitful mix between interference and diffraction of light. Interferences encode the amplitude and the optical phase of any 3-D object, and diffraction acts as a key to decode the encoded object since it builds a wave that seems to originate from the original object. The holographic recording method is a key for contactless metrology because it provides high sensitivity and high spatial resolution to study physical phenomena in their four dimensions (three spatial, and one temporal). It is a method that turns our habits in metrological matters upside down, which researchers, engineers, and technicians cannot ignore nowadays.

After several years of maturation, it is now possible to record digital holograms with electronic devices that have good spatial resolutions (CCD or complementary metal–oxide–semiconductor (CMOS) sensors). Thus the digital hologram is a matrix of discrete values encoded in bits. Work realized in 1972 by Kronrod constituted the first attempt to reconstruct numerically an object encoded in a matrix hologram [52]. However, digital holography was realized in 1994 by Schnars and Jüptner [53]. Indeed, advances in microtechnologies resulted in building matrix sensors with small pixels to fulfill the Shannon theorem for holographic recording. Furthermore, digital image processing is now so efficient that digital holograms can be obtained through rapid computation. Digital holography exhibits high potential for noninvasive and contactless optical metrology. In this context, investigating structures submitted to thermal, mechanical, or pneumatic loadings is of utmost interest. The analysis is simultaneously qualitative, by the visualization of mod 2π phase map, and quantitative by digital processing of these phase maps.

Over the past few years, digital Fresnel holography has stimulated many researchers and has been applied in several domains. For a few years, many spectacular applications were demonstrated such as microscopic imaging and phase-contrast digital holographic microscopy [54]; 3-D object recognition and information securing [55]; polarization imaging [56]; surface shape measurement and contouring; material property investigations [57, 58]; vibration analysis with pulsed lasers and time averaging [59, 60]; and, finally, multidimensional dynamic investigations [61]. The theoretical background of all these applications was presented in previous works, which exposed the theory and reconstruction algorithms for digital holography, according to the different possible recording schemes: that is, in-line holography and off-axis Fresnel holography [62], digital Fourier holography [63], and reconstruction with Fresnelets in off-axis holography [64].

This section presents the methods of digital Fresnel holography. We first describe the theoretical basics of digital holography. Then we focus on interferometric methods for the investigation of the mechanical behavior of structures using digital holography. We present especially methods to measure simultaneously multidimensional deformations.

6.6.2
General Principle

The basic principle to record digital Fresnel holograms is given in Figure 6.39. The digital hologram is characterized by interferences between the object wave and the reference wave. In a lensless optical configuration, which means that there is no lens to image the object onto the sensor, digital holography is a general recording/reconstruction process. This includes the particular case of speckle interferometry for which the imaging lens allows an image to be formed on the recording sensor. The reconstruction process of the object follows the principle of

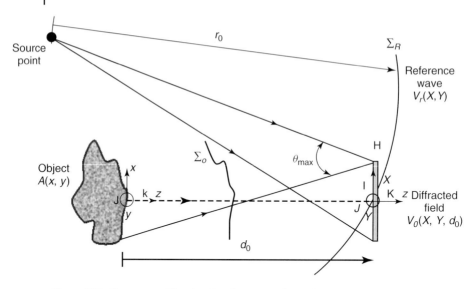

Figure 6.39 Free space diffraction, interference, and notations.

light diffraction. Thanks to digital recovery, the amplitude and phase reconstruction in the object plane can be computed.

Let us consider an extended object, sized $\Delta A_x \times \Delta A_y$, illuminated by a coherent monochromatic wave and a set of reference coordinates attached to the object (x, y, z) and to the recording plane (X, Y, z). This object diffracts a wave onto the recording plane located at a distance $d_0 = |z_0|$. The object surface generates a wave front that will be noted according to Eq. (6.38):

$$A(x, y) = A_0(x, y) \exp(i\psi_0(x, y)) \tag{6.38}$$

Amplitude A_0 describes the object reflectivity, and phase ψ_0 describes its surface or shape. Owing to the natural roughness of the object, phase ψ_0 is random and uniformly distributed over the range $\langle -\pi, +\pi]$. When taking into account the diffraction theory under the Fresnel approximations [65], the diffracted field at distance d_0 is expressed by the following relation:

$$V_o(X, Y, d_0) = -\frac{i \exp(2i\pi d_0/\lambda)}{\lambda d_0} \exp\left(\frac{i\pi}{\lambda d_0}(X^2 + Y^2)\right)$$
$$\times \int\int A(x, y) \exp\left(\frac{i\pi}{\lambda d_0}(x^2 + y^2)\right) \exp\left(-\frac{2i\pi}{\lambda d_0}(xX + yY)\right) dxdy \tag{6.39}$$

The diffracted field is equal to the Fresnel transform of the complex amplitude of the object surface. In the recording plane, this wave can be written according to Eq. (6.40):

$$V_o(X, Y, d_0) = a_o(X, Y) \exp(i\varphi_o(X, Y)) = \sqrt{I_o(X, Y)} \exp(i\varphi_o(X, Y)) \tag{6.40}$$

with a_o being the modulus of the complex wave and φ_o its phase. Note that, since the object is not smooth, the diffracted field at distance d_0 is a *speckle* field.

This means that the amplitude and phase repartition in the recording plane has a random nature. Thus, φ_o is again random and uniformly distributed over the range $\langle -\pi, +\pi \rangle$.

Let V_r be the complex amplitude of the reference wave front at the recording plane. We have then

$$V_r(X, Y) = a_r(X, Y) \exp(i\varphi_r(X, Y)) = \sqrt{I_r(X, Y)} \exp(i\varphi_r(X, Y)) \tag{6.41}$$

with a_r denoting its modulus and φ_r its optical phase. The reference wave front is generally issued from a source point: it is a spherical diverging wave with oblique incidence on the recording sensor. Let (x_s, y_s, z_s) be the coordinates of the source point in the reference coordinate system of the hologram plane ($z_s < 0$). The reference phase can be written in the Fresnel approximation as follows [65]:

$$\varphi_r(X, Y) \cong \frac{\pi}{\lambda z_s} \left[(X - x_s)^2 + (Y - y_s)^2 \right]. \tag{6.42}$$

After expanding Eq. (6.42), the optical phase can be written in the following form:

$$\varphi_r(X, Y) = -2\pi(u_0 X + v_0 Y) - \frac{\pi}{\lambda r_0}(X^2 + Y^2) + \varphi_s \tag{6.43}$$

where $(u_0, v_0) = (x_s/\lambda z_s, y_s/\lambda z_s)$ are the carrier spatial frequencies, $r_0 \approx |z|$ is the curvature radius of the wave, and φ_s is a constant. In the case where $(u_0, v_0) \neq (0, 0)$, we get "off-axis digital holography," and when $(u_0, v_0) = (0, 0)$, we get "in-line digital holography." Usually, one adjusts the optical setup so that the reference wave may be uniform, which means $a_r(X, Y) = \text{const}$. The total intensity received by the recording sensor is then written in the form of interferences:

$$H = I = |V_r + V_o|^2 = |V_r|^2 + |V_o|^2 + V_r^* V_o + V_r V_o^*$$
$$= I_r + I_o + 2\sqrt{I_r I_o} \cos(\varphi) \tag{6.44}$$

with $\varphi = \varphi_o - \varphi_r$. The analytical form of Eq. (6.44) is similar to that obtained in a previous section of this chapter concerning speckle interferometry. However, the recorded hologram does not contain any observable image of the object. But one can recover the original object because of the coherent mixing of the two waves. The hologram is the sum of three orders: the first two terms in the first line of Eq. (6.44) represent the zero order, the third term is the $+1$ order, and the fourth term is the -1 order.

The Shannon theorem applied to off-axis digital holography, resulting in the spatial separation of the three diffraction orders, leads to the optimal recording distance [66]. It is given for a square object sized $\Delta A_x = \Delta A_y = a$ by

$$d_0 = \frac{4 \max(p_x, p_y)}{\lambda} a, \tag{6.45}$$

where p_x, p_y are the pixel pitches of the recording sensor, including $N \times M$ pixels. For a circular object shape with radius a

$$d_0 = \frac{\left(2 + 3\sqrt{2}\right) \max(p_x, p_y)}{\lambda} a \tag{6.46}$$

Ideally, the spatial frequencies of the reference wave must be adjusted to $(u_0, v_0) = (\pm 3/8p_x, \pm 3/8p_y)$ for a square object and $(u_0, v_0) = (\pm(1/2 - 1/(2+3\sqrt{2}))/p_x, \pm(1/2 - 1/(2+3\sqrt{2}))/p_y)$ for a circular object. For a square object, one might also choose $u_0 = 0$ or $v_0 = 0$ [66].

6.6.3
Evaluation

6.6.3.1 Discrete Fresnel Transform

The reconstruction of the amplitude and phase of the encoded object is based on the numerical simulation of light diffraction on the numerical aperture included in the digital hologram. One must take into account the sampling of the recording space, $(X, Y) = (np_x, mp_y)$ where $(m; n) \in (-M/2, +M/2-1; -N/2, +N/2-1)$; then one must take into account the sampling of the reconstructed plane. For a reconstruction distance equal to $-d_0$, the reconstructed field A_R is given by the discrete Fresnel transform [53, 62]:

$$A_R(x, y, -d_0) = \frac{i \exp(-2i\pi d_0/\lambda)}{\lambda d_0} \exp\left[-\frac{i\pi}{\lambda d_0}(x^2 + y^2)\right]$$
$$\times \sum_{k=-K/2}^{k=K/2-1} \sum_{l=-L/2}^{l=L/2-1} H(lp_x, kp_y) \exp\left[-\frac{i\pi}{\lambda d_0}(l^2 p_x^2 + k^2 p_y^2)\right] \quad (6.47)$$
$$\times \exp\left[+\frac{2i\pi}{\lambda d_0}(lp_x x + kp_y y)\right]$$

It can be shown that the diffracted field in the +1 order is given by [60, 66]:

$$A_R^{+1}(x, y, -d_0) = \lambda^2 d_0^2 \exp\left[-i\pi \lambda d_0 (u_0^2 + v_0^2)\right]$$
$$\times V_r^*(x, y) A(x - \lambda u_0 d_0, y - \lambda u_0 d_0). \quad (6.48)$$

The +1 order is then localized at spatial coordinates $(\lambda d_0 u_0, \lambda d_0 v_0)$. The computation of Eq. (6.47) leads to complex-valued results, from which the amplitude image (modulus) and the phase image (argument) can be extracted. The spatial resolution in the reconstructed field is found to be $\rho_x = \lambda d_0/Np_x$ and $\rho_y = \lambda d_0/Mp_y$ [66]. If the reconstructed plane is computed with $(K, L) \geq (M, N)$ data points, then the sampling pitches in the reconstructed plane are equal to $\Delta\eta = \lambda d_0/Lp_x$ and $\Delta\xi = \lambda d_0/Kp_y$ [53, 62]. Note that it depends on the wavelength of the light. This case corresponds to zero-padding of the hologram, meaning that we add zeros around the hologram to extend its size to $K \times L$ pixels. The sampling of the object plane is now simply equal to $x = l\Delta\eta$ and $y = k\Delta\xi$, with l and k varying from $-L/2$ to $L/2 - 1$ and $-K/2$ to $K/2 - 1$, respectively. The algorithm for the discrete Fresnel transform is given in Figure 6.40, for a reconstruction distance $d_r = -d_0$.

6.6.3.2 Convolution Algorithm

The discrete Fresnel transform is adapted to a large range of object sizes and shapes. In such computation, the sampling pitch in the reconstructed plane depends on physical parameters such as the wavelength. The second possibility

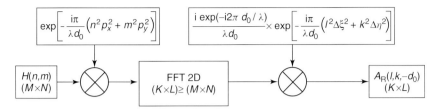

Figure 6.40 Algorithm for the discrete Fresnel transform.

to reconstruct the object is based on the convolution formulae of diffraction. An exhaustive description was provided by Kreis et al. in 1997 [62]. This means that the reconstructed field is obtained by this convolution equation (⊗ means convolution) at any distance d_r:

$$A_R(x, y, d_r) = H(x, y) \otimes h(x, y, d_r), \tag{6.49}$$

where $h(x, y, d_r)$ is the kernel associated to the diffraction along distance d_r. The convolution kernel can be the impulse response of the free-space propagation. Such a kernel leads to a transfer function which is the Fourier transform of the impulse response. The mathematical expression of the kernel is given by Goodman [65] as

$$h(x, y, d_r) = \frac{id_r}{\lambda} \frac{\exp\left[2i\pi/\lambda\sqrt{d_r^2 + x^2 + y^2}\right]}{d_r^2 + x^2 + y^2} \tag{6.50}$$

The angular spectrum transfer function can also be used as the transfer function of the reconstruction process. In this case, the mathematical expression is given by Eq. (6.51) [62]:

$$G(u, v, d_r) = \exp\left[2i\pi d_r/\lambda\sqrt{1 - \lambda^2 u^2 - \lambda^2 v^2}\right] \tag{6.51}$$

The practical computation of Eq. (6.49) can be performed according to the properties of the Fourier transform, thus leading to a double Fourier transform algorithm:

$$A_R = FT^{-1}[FT[H] \times FT[h]], \tag{6.52}$$

which includes three FFTs when using the impulse response, or only two when using the angular spectrum transfer function:

$$A_R = FT^{-1}[FT[H] \times G]. \tag{6.53}$$

The computation with double FFT algorithms leads to a sampling pitch in the reconstructed plane which is equal to that of the sensor; thus it is independent of the wavelength. The synoptic of the convolution algorithm with the double Fourier transform method using the angular spectrum transfer function is given in Figure 6.41.

The next section presents experimental setup and results for the use of digital holography for multidimensional deformation measurements in the field of solid mechanics and contactless inspection.

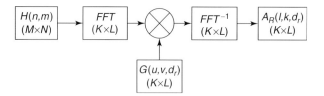

Figure 6.41 Double FFT algorithm using the angular spectrum transfer function.

6.6.4
Applications

Over the past two decades, digital holography has proven to be a very useful industrial tool for the measurement of full-field surface deformations of naturally rough objects [57–59, 61]. Digital holography is highly suitable for the investigation of mechanical properties of some mechanical components or assemblies.

In order to do this, the phases of two reconstructed wave fields recorded under different loading conditions of the object are computed from the complex amplitude. Thus, in digital holography the phase is available without the comprehensive phase shifting methods used in the other techniques. The displacement field is obtained from the difference of the two optical phases.

Mechanical technicians particularly need a multiple-component measurement of the displacement field of the studied structure. Multidimensional mechanical deformation measurements are topics that have stimulated many researchers. Methods for the simultaneous measurement of the 3-D deformation were first described by Kriens [67] and Linet [68]. In [68], double exposure holograms were recorded and spatially multiplexed on a photographic plate and then physically reconstructed by light diffraction for the quantitative analysis. Schedin proposed a 3-D dynamic deformation evaluation [69] using a single laser source and three illumination beams. The three reference beams included a delay line to take into account the coherence of the laser. The setup was complex since it was based on spatial multiplexing of image-plane holograms and needed a 3-D spatial carrier. The use of digital color holography for multidimensional measurements was proposed recently using a two-color and a three-color digital Fresnel hologram [70–72]. This section describes two different methods for the simultaneous measurement of two-dimensional (2-D) and 3-D deformations. The setups are based on digital two-color or three-color holography using, respectively, a spatial multiplexing scheme and a stack of photodiodes detection scheme.

6.6.4.1 2D Deformation Measurement Using a Two-Color Digital Holographic Interferometer

Figure 6.42 shows the optical setup that includes two CW lasers (green, G, $\lambda_G = 532$ nm and red, R, $\lambda_R = 632.8$ nm). The sensor is a monochrome CCD, 1024×1360 pixels with 4.65 µm pitches. The two object beams are collimated and illuminate the object with two symmetrical angles ($\theta_R = -\theta_G = \theta$) thus giving 2-D

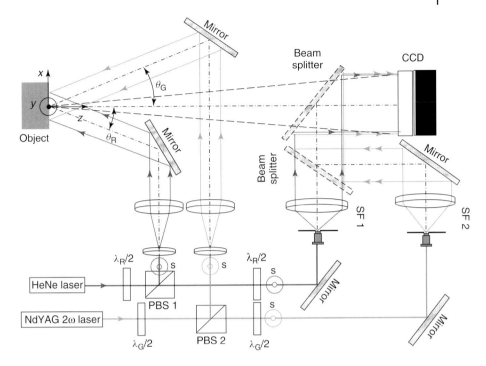

Figure 6.42 Experimental setup for two-color digital holography.

sensitivity. The setup is then sensitive to the in-plane and out-of-plane displacements of the object. A Mach–Zehnder interferometer produces mixing between the two doublets of colors. Each laser beam is split into an illuminating beam and a reference beam which are copolarized for each wavelength. The smooth and plane reference waves are produced through two spatial filters (SF1 and SF2). Thus each reference wave is the spatial carrier of each hologram. Since the monochrome sensor is not able to record the two colors simultaneously at each pixel, the spatial frequencies of the reference waves (R and G) are adjusted so that the two-color holograms may be spatially multiplexed in the field of view. The off-line holographic recording is carried out using the two SFs in which each collimating lens is displaced out of the afocal axis by means of two micrometric transducers (not shown in Figure 6.42).

An application of the method is demonstrated through an investigation of mechanical causes of cracks inside the capacitance of an industrial printed circuit board (PCB) component. The component (capacitor) is cracked during the clamping of the PCB inside its electronic box. The investigation of the causes of this anomaly is made possible since digital holography is well adapted for contactless metrology and microdeformation measurements. The mechanical simulation of the clamping is carried out by progressive loading of the rear panel of the PCB through the clamping region. This reproduces quite faithfully the real industrial situation. Figure 6.43 shows the component of interest and the region that is inspected by

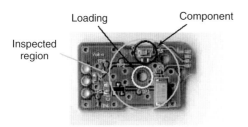

Figure 6.43 PCB component under test.

the two-color digital holographic setup. This PCB is placed at $d_0 = 908$ mm from the sensor, and the illuminated useful zone is $\Delta A_x \times \Delta A_y = 30$ mm $\times$ 30 mm.

Let us consider that the object produces a 3-D displacement vector according to

$$U = u_x \mathbf{i} + u_y \mathbf{j} + u_z \mathbf{k}, \tag{6.54}$$

where $\{u_x, u_y, u_z\}$ are the displacement fields along the x, y, and z directions, respectively. The relation between the displacement vector $U = u_x \mathbf{i} + u_y \mathbf{j} + u_z \mathbf{k}$ and the illuminating geometry is $\Delta \varphi_\lambda = 2\pi U \times (\mathbf{K}_e^\lambda - \mathbf{K}_o)/\lambda$ for each wavelength, where $\mathbf{K}_e$ is the illuminating vector (depending on θ_R, θ_G) and $\mathbf{K}_o$ the observation vector. When taking into account the illumination geometry, the phase changes measured for each color are given by

$$\Delta \varphi_R = \frac{2\pi}{\lambda_R} \left[+\sin\theta u_x - (1 + \cos\theta) u_z \right] \tag{6.55}$$

and

$$\Delta \varphi_G = \frac{2\pi}{\lambda_G} \left[-\sin\theta u_x - (1 + \cos\theta) u_z \right] \tag{6.56}$$

Then the weighted sum and difference of the phase changes lead to the extraction of the z and x components, respectively:

$$u_z = -\frac{\lambda_R \Delta \varphi_R + \lambda_G \Delta \varphi_G}{4\pi (1 + \cos\theta)} \tag{6.57}$$

and

$$u_x = \frac{\lambda_R \Delta \varphi_R - \lambda_G \Delta \varphi_G}{4\pi \sin\theta} \tag{6.58}$$

Figure 6.44 shows the reconstructed full-field images obtained from the double FFT algorithm with $d_r = -d_0 = -908$ mm (algorithm of Figure 6.41). Figure 6.44a presents the R hologram, and Figure 6.44b presents the G hologram. Figure 6.44c shows the two-color image obtained by combining these two results with a weighting of 2 : 1.

The optical phase change between the loading steps of the PCB can be computed, and, since the physical object scale is preserved, differences between the red and green phase changes can be computed in order to get the 2-D measurement. Figure 6.45 shows numerical fringes obtained mod 2π after mechanical loading of the PCB component, for the R and the G channels. Figure 6.46 shows the in-plane

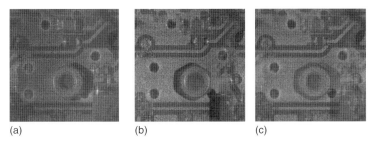

Figure 6.44 (a) *R* image. (b) *G* image. (c) Two-color image. (Please find a color version of this figure on the color plates.)

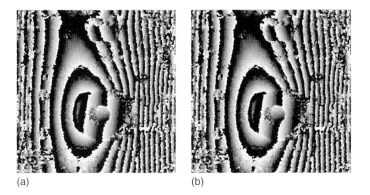

Figure 6.45 Wrapped phase change along (a) *R* and (b) *G* channels.

and out-of-plane displacement fields obtained after processing the holograms. The capacitor is localized in the region where the flexion is maximum.

After numerically processing these results, Figure 6.47 shows the region of the component under test, which shows highly nonuniform deformations. These deformations induce cracking of the component.

6.6.4.2 Real-Time Three-Sensitivity Measurement

For some mechanical applications, it is often necessary to get simultaneously 3-D deformation measurements: for example, when studying the mechanical behavior of heterogeneous structures [58]. In order to get a 3-D deformation measurement with digital color holography, one needs three wavelengths. In the method presented in the previous section, each reference wave of each color has suitable spatial frequencies, but in the case of three wavelengths, this leads to a more complicated experimental setup because each reference wave must be adjusted separately.

Furthermore, owing to the spatial multiplexing, the occupation of the useful bandwidth cannot be optimal, which thus degrades the spatial resolution, whereas it is not the case if one uses a color sensor based on a stack of photodiodes (*http://www.foveon.com*). The advantage of such a sensor is that the simultaneous

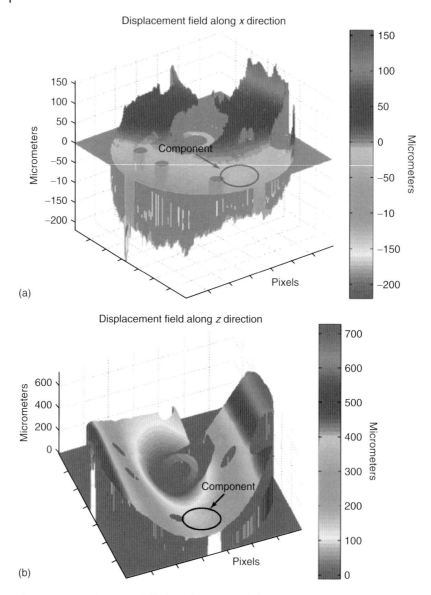

Figure 6.46 Displacement field along the (a) x- and (b) z-directions computed from a set of 90 recorded holograms.

recording of colors at each pixel location is possible with high resolution. In this kind of sensor, the spectral selectivity is related to the penetration depth of photons in silicon. The average penetration depth of blue photons at 425 nm is about 0.2 µm, that of green photons at 532 nm about 2 µm and that of red photons at 630 nm about 3 µm. Thus, building junctions at depths around 0.2, 0.8, and 3.0 µm

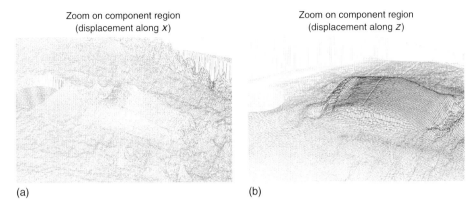

Figure 6.47 (a) In-plane and (b) out-of-plane displacement fields in the region of the capacitance after postprocessing.

provides workable spectral separation for true color imaging. Nevertheless, the spectral selectivity is not perfect because red and green photons can be detected in the blue band, blue and red photons in the green band, and green photons in the red band, the probability of finding blue photons in the red band being small.

So, in order to overcome this limitation, the three wavelengths must be well separated. For example, using a blue light source at $\lambda_B = 457$ nm, a green light source at $\lambda_G = 532$ nm, and a red source light at $\lambda_R = 671$ nm ensures that there is no color mixing in the three layers [62]. Furthermore, such an architecture guarantees maximum spatial resolution since the number of effective pixels for each wavelength corresponds to that of the full sensor.

Figure 6.48 shows the experimental setup for the real-time three-sensitivity deformation measurement using three-color digital holography. It uses three continuous diode-pumped-solid-state (DPSS) lasers. The color sensor in a color camera made up of three stacked layers of photodiodes with a single 8-bit per channel digital output, including $(M, N) = (1060, 1420)$ pixels with pitches $p_x = p_y = 5\,\mu m$. Each laser beam is split into an illuminating beam and a reference beam, which are copolarized for each wavelength. Each collimated laser beam illuminates the object under test with $\theta_R, \theta_G,$ and θ_B angles, respectively, for the red, green, and blue lines, leading to the three sensitivities. The R and G beams are included in the $\{x, z\}$ plane, whereas the B beam is included in the $\{y, z\}$ plane.

The R, G, and B reference beams are combined into a unique beam thanks to the use of two dichroic plates. The first one reflects the B beam and transmits the G beam; the second one reflects the G and B beam, but transmits the R beam (Figure 6.48). The smooth plane reference waves are produced through a single SF, which includes an achromatic lens. Thus, the three reference beams impact the sensing area with the same incidence angle. Off-line holographic recording is carried out by translating the object perpendicularly to the optical axis to produce suitable spatial frequencies [66, 72]. The mechanical object under test, sized $\Delta A_x \times \Delta A_y = 25 \times 35\,mm^2$, is placed at a distance $d_0 = 1630$ mm from the color sensor.

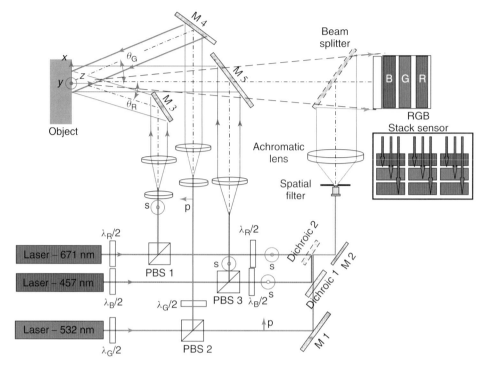

Figure 6.48 Experimental setup (M1–M5: flat mirrors, object–sensor distance is 1630 mm). (Please find a color version of this figure on the color plates.)

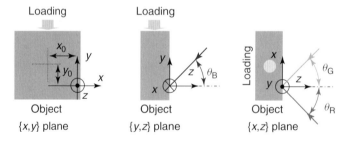

Figure 6.49 Details on the illumination geometry.

The illumination geometry is described in Figure 6.49.

In the setup, we have $K_o \cong k$, $K_e^R = \sin\theta_R \mathbf{i} - \cos\theta_R \mathbf{k}$, $K_e^G = -\sin\theta_G \mathbf{i} - \cos\theta_G \mathbf{k}$, and $K_e^B = -\sin\theta_B \mathbf{j} - \cos\theta_B \mathbf{k}$. Inverting the sensitivity matrix given by the scalar product between U and $(K_e^\lambda - K_o)$ leads to each individual displacement component. Consequently, the 3-D displacement field $\{u_x, u_y, u_z\}$ of the object submitted to some mechanical loading can be extracted from the previous relation and from the set of digital color holograms according to Eq. (6.59):

$$\begin{bmatrix} u_x \\ u_y \\ u_z \end{bmatrix} = \frac{1}{2\pi\alpha}$$

$$\begin{bmatrix} 1+\cos(\theta_G) & -1-\cos(\theta_R) & 0 \\ \sin(\theta_G)(1+\cos(\theta_B))/\sin(\theta_B) & \sin(\theta_R)(1+\cos(\theta_B))/\sin(\theta_B) & -\alpha/\sin(\theta_B) \\ -\sin(\theta_G) & -\sin(\theta_R) & 0 \end{bmatrix}$$

$$\times \begin{bmatrix} \lambda_R \Delta\varphi_R \\ \lambda_G \Delta\varphi_G \\ \lambda_B \Delta\varphi_B \end{bmatrix} \quad (6.59)$$

where $\alpha = \sin(\theta_R)(1+\cos(\theta_G)) + \sin(\theta_G)(1+\cos(\theta_R))$ and $\Delta\varphi_R, \Delta\varphi_G, \Delta\varphi_B$ are the optical phase changes between two deformation states, which are obtained from the reconstructed holograms for the R, G, and B beams, respectively. In the setup, the angles are adjusted to $\theta_R = 31°, \theta_G = 31°,$ and $\theta_B = 45°$. Figure 6.50a–c shows respectively the R, G, B reconstructed objects using the double FFT algorithm.

Figures 6.51a–c show respectively the R, G, and B wrapped phase changes computed between two reconstructed RGB color holograms at two different loading steps.

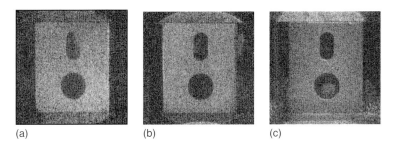

(a)   (b)   (c)

Figure 6.50 (a) R, (b) G, and (c) B – reconstructed objects using the double FFT algorithm.

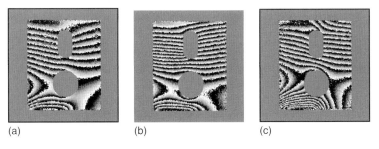

(a)   (b)   (c)

Figure 6.51 (a) R, (b) G, and (c) B – wrapped phase changes computed using three-color digital holograms.

According to Eq. (6.59), the three components of the displacement between the two loading steps can be computed with the full simultaneous RGB holograms. Figures 6.52a–c show respectively the $\{u_x, u_y, u_z\}$ components of the displacement field obtained from the set of data.

Such results show the ability of digital color holography to provide simultaneous multidimensional displacement field measurements. This gives opportunities for studying 3-D vibrations and the 3-D mechanical behavior of unconsolidated materials.

6.6.4.3 Vibration Analysis with Digital Fresnel Holography

As digital holography is quite suitable to measure static deformations of any structure submitted to some mechanical loading, it is also suitable for analyzing moving deformations, that is, vibrations of structures. There are mainly two modes that can be experimentally implemented for vibration analysis. The first one is called the *time-averaging mode* and the second one is related to a stroboscopic regime.

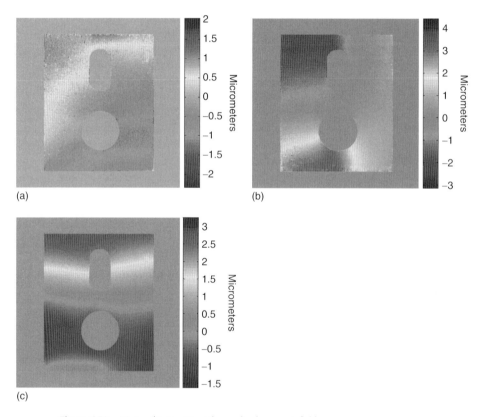

Figure 6.52 (a) u_x, (b) u_y, (c) and u_z – displacement field obtained from the set of data. (Please find a color version of this figure on the color plates.)

6.6.4.4 Time-Averaging Mode

The pure time-averaging situation corresponds to an exposure time much larger than the vibration period. In this case, the reconstructed object is modulated by the Bessel function J_0 [60]. This method is very simple and quite easy to implement. It has very high potential for NDT applications. As an example, Figure 6.53 shows the identification of defects in a dome loudspeaker at high frequencies.

It can be seen that, near the resonance frequency (14 520 Hz), the mode has good symmetry, whereas at the resonance frequency (14 530 Hz) it shows an unwanted dissymmetry.

The method can also be used to analyze vibrations of a structure to study its behavior rather than its defects. An example can be found in the clarinet reed under forced oscillation. This means that the object is not in free oscillation as would be the case when playing music with a clarinet. Figure 6.54 shows the behavior of the clarinet reed at frequencies 1880, 3260, and 4280 Hz. Both flexion and torsion modes can be observed.

When increasing the excitation frequency, the behavior of the reed appears to be quite different. This is shown in Figure 6.55 for frequencies 4420, 4500, and 4940 Hz.

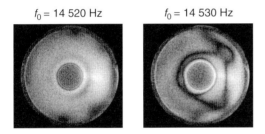

Figure 6.53 Identification of defects in a dome loudspeaker near 14 530 Hz.

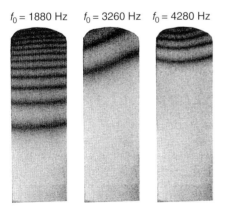

Figure 6.54 Behavior of the clarinet reed at 1880, 3260, and 4280 Hz.

294 | 6 Rough Surface Interferometry

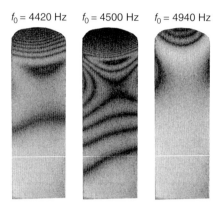

Figure 6.55 Behavior of the clarinet reed at 4420, 4500, and 4940 Hz.

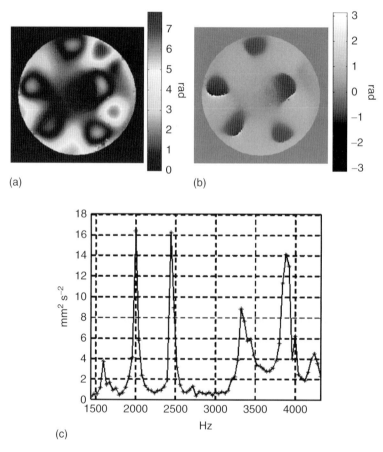

Figure 6.56 (a) Amplitude, (b) phase at 3880 Hz, and (c) mean quadratic velocity of the loudspeaker.

6.6.4.5 **Stroboscopic Regime**

In order to retrieve the vibration amplitude and phase of a pure sinusoidal vibration, in contrast to the time-averaging situation, the exposure time has now to be much smaller than the vibration period. For this, one has to use a pulsed laser with a few nanoseconds pulse width, a stroboscopic illumination with a CW laser with acoustooptic or mechanical shuttering, or direct laser diode modulation. Note that mechanical shuttering is useful only for low-frequency vibrations ($<50\,\text{kHz}$). In order to avoid distortion on the phase of the reconstructed object, which is due to the finite pulse width of the light, it is necessary that the condition $T/T_0 < 1/\Delta\varphi_m$ be fulfilled [61]. Here T is the exposure time, T_0 is the vibration period, and $\Delta\varphi_m$ is the maximum phase change due to the vibration.

As an example, Figure 6.56 shows the measured vibration amplitude and phase (in radians) of a loudspeaker, 60 mm in diameter and excited at 3880 Hz. When tuning the frequency, it is also possible to evaluate the mean quadratic velocity of the membrane. This is illustrated in Figure 6.56, where the peaks due to the resonance frequencies can be clearly seen.

When the object under consideration is vibrating along three directions, the setup of Figure 6.42 can be used to get a simultaneous twin-sensitivity measurement [61]. Then the two spatially multiplexed holograms include the amplitude and phase of the vibration along the x- and z-directions. A practical example can be found in the characterization of an automotive car joint constituted with elastomer material. Figure 6.57a shows the measured in-plane and out-of-plane vibration amplitudes and phases for a frequency of 680 Hz.

Figure 6.57b shows the mean quadratic velocities extracted from the set of data for an excitation frequency varying from 200 to 1000 Hz. Compared to the classical vibrometers which use a point optical probe, the digital holographic setup provides direct 2-D full-field information on the vibration behavior [61].

Tutorial Exercise 6.4

Explain to a nonspecialist, and without any technical terms, the physical process underling the generation and the computation of a digital hologram.

Solution:

The propagation of light from the object to the recording sensor is a 2-D light distribution that encodes the surface shape of the object. The propagation of the light is known as *diffraction*. The encoding of the object is done by recording a mix between this light and another light which we call *reference light*. Such mixing is known as *interference*. This light is necessary for the recording. If it does not exist, we are in the case of photography and there is no possibility to measure any displacement of the object. The recovery of the object uses the image recorded by the sensor. We simulate the backpropagation of the light from the sensor to the object plane. The computation gives a numerical result with a real and an imaginary part. The displacement is encoded in these numerical values.

296 | 6 Rough Surface Interferometry

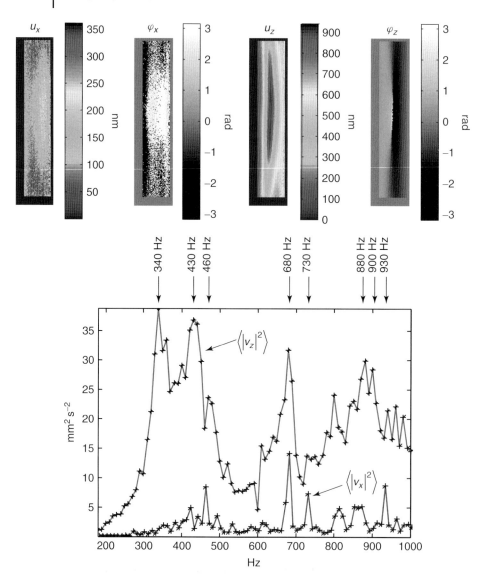

Figure 6.57 2-D vibration amplitude and phase at a frequency of 680 Hz and mean quadratic velocities.

▪ Tutorial Exercise 6.5

Explain to a nonspecialist, and without any technical terms, how to measure a mechanical deformation using digital Fresnel holography.

Solution:

Record a digital hologram in the current state of the object; then apply the mechanical loading and record a digital hologram in the new state of the object. Then compute the complex image using the discrete Fresnel transform and its phase. The difference between the two phases shows mod 2π fringes; they are the signature of the displacement field between the two states. The displacement field will be computed after unwrapping the phase difference and by taking into account the illumination and observation geometry through the sensitivity vector.

Tutorial Exercise 6.6

The recorded hologram is a speckle pattern in which the local fringe pitch is related to the local angle between the reference wave and the object wave, so that (Figure 6.39) $i = \lambda/2 \sin(\theta/2)$. Find the maximum angle to fulfill the Shannon theorem. Compute its value for the wavelength $\lambda = 0.6328\,\mu m$ (HeNe laser) and a pixel pitch $p_x = p_y = 4.65\,\mu m$.

Solution:

The pixel pitch is the sampling pitch in the recording plane and i is the local interfringe. The Shannon theorem leads to $\max(p_x, p_y) < i/2$. It follows that $\theta \leq 2\sin^{-1}(\lambda/4\max(p_x, p_y))$. With the given values, the maximum angle is equal to $\theta = 3.9°$. This very low angular tolerance needs an optical setup using precision mechanics.

Tutorial Exercise 6.7

We record a digital hologram with a sensor having $(N, M) = (1360, 1024)$ pixels with pitches $p_x = p_y = 4.65\,\mu m$. We compute this hologram with $L = K = 2048$ pixels. Find the spatial resolution if the recording distance is $d_0 = 1000\,mm$ and the wavelength is $\lambda = 0.6328\,\mu m$. Find the sampling pitches in the reconstruction plane and qualitatively discuss the result. What happens if we compute the hologram with $L = K = 512$ pixels?

Solution:

The spatial resolution is given along the x- and y-directions by $\rho_x = \lambda d_0/Np_x$ and $\rho_y = \lambda d_0/Mp_y$. The numerical evaluation with the parameters leads to $\rho_x = 100.1\,\mu m$ and $\rho_y = 132.9\,\mu m$; the spatial resolution is better in the x-direction since the sensor contains more pixels in this direction. With $L = K = 2048$, the sampling pitches are $\Delta\eta = \Delta\xi = \lambda d_0/Lp_x = 66.4\,\mu m$. The sampling pitches are smaller than the spatial resolution and this is due to the zero-padding of the hologram to reach $K \times L$ data points in the reconstructed plane. The spatial resolution is fixed by the recording conditions, whereas the sampling pitches of the reconstructed object can be adjusted using zero-padding. Using $L = K = 512$ pixels means that we restrict the hologram to only 512×512 pixels. Thus the spatial resolution is now modified and equal to $\rho_x = \lambda d_0/512 p_x = \rho_y = 265.8\,\mu m$ and the sampling pitches have the same value: $\Delta\eta = \Delta\xi = 265.8\,\mu m$. This means that we have degraded the image quality in the reconstructed plane.

6.7
Summary

Techniques applying interferometry to rough surfaces are widely used in mechanical testing and NDT. The main applications are in the measurement of surface deformation, vibration, and strain.

The optical techniques presented in this chapter provide full-field measurement results with high lateral resolution. Because of the noncontact approach of optical techniques, the measured object and the testing conditions are not disturbed. Furthermore, extensive object preparation, as necessary with conventional strain gages, is avoided.

The four techniques presented have different application areas. The following overview should help select the most suitable one for a specific application:

Electronic speckle pattern interferometry (ESPI) measures 3-D deformation fields of an object surface. Both out-of-plane and in-plane configurations are available. The advantages of the technique are the high accuracy (subwavelength) of the deformation measurement and the high quality of the measurement results. The main limitations are the complexity of the measurement setup and the high sensitivity to environmental disturbances.

Low coherence speckle interferometry (LCSI) is applied to measure the out-of-plane component of the deformation field of the object's internal interfaces. Both the spatial resolution and the accuracy of the deformation measurements are in the same range as for ESPI. The ability to measure the object's internal deformations opens the study of the internal defects in the object and allows the measurement of deformations of its internal interfaces. The main limitation is the necessity to have good a priori information about the object. Furthermore, the disturbances of the light along the whole optical path need to be modeled to obtain quantitative results.

Speckle pattern shearing interferometry (SPSI) is a full-field speckle-based method that measures the slope of an out-of-plane deformation field in the shearing direction, at the surface of a loaded object. All points in the overlap area of two images, sheared with respect to each other, are measured with similar subwavelength sensitivity as for ESPI. The lateral resolution of the measured displacement gradients is determined by the shearing distance. Since both interfering speckle fields are scattered from the object, shearography is relatively insensitive for common-mode environmental disturbances such as air turbulence in the almost common optical path laser–object–sensor. As a consequence, shearography can be used for NDT applications under laboratory conditions as well as in the field. The main limitation of shearography is its relatively high noise level compared to ESPI, due to a lower number of reliable information carriers.

Digital holographic interferometry (DHI) is a universal method that can be applied in many domains such as microscopy, fluid mechanics, object recognition, and solid-state mechanics. The main characteristic of this tool is that it encodes both the amplitude and the phase of the object under interest. This gives opportunity to measure deformation and vibrations of any object submitted to a mechanical loading. The use of digital color holography is a direct way to real-time multidimensional deformation measurements, and especially to the simultaneous measurement of the 3-D displacement field. The setup can be quite simple and accessible to nonspecialists. The high density of data points permits spatial derivatives to be computed, which is an essential point in many mechanical analyses.

It is essential to control the environmental conditions for highly accurate measurements with the techniques presented here. As a rule of thumb, it is recommended to mechanically attach the instrument as stable as possible to the object to be measured. Arrangements for tensile test machines are commercially available on the market. Furthermore, image correlation routines can be used to improve the measurement range by compensating rigid-body motions if necessary.

The results of the presented techniques are the displacement and the displacement gradient maps. To obtain strain maps, the initial shape of the object must be known or measured. The material parameters can then be used to calculate the stress field.

The nature of the intensity distribution of a speckle field leads to local lack of information disturbing the measurement results. This can be critical for the detection of localized strain fields, covering an area of less than 10×10 pixels. However, since the field of view of all techniques can be adapted to the feature size, we can apply image processing (i.e., median filtering). The speckle size limits the spatial resolution. However, the lateral resolution will always be better than conventional photographic image correlation where several speckles are necessary to obtain one measurement point.

All techniques are based on phase shifting to increase the accuracy of the deformation measurements. The resulting phase information is calculated as wrapped (mod 2π) phase maps. In NDT, wrapped phase maps are suited to detect the defect location. For deformation measurements, unwrapping and filtering of the results is recommended. Algorithms are introduced in Chapter 3 in this book. The smooth unwrapped phase information then gives access to the deformation field of a loaded object and can easily be used even by nonexperts in interferometric techniques.

In many cases, the results of the presented techniques are too sensitive or the spatial resolution is too high for today's requirements. The lateral resolution can be decreased by increasing the speckle size. This can be done by decreasing the aperture of the imaging system or by using a light source with a longer wavelength. The latter reduces also the accuracy of the phase measurements. This increases the stability of the setup to external perturbations. Data routines can also be used to resample the measured data and adapt them, for instance, to the mesh of an FEM model.

In contrast to other methods described in this book, all interferometric techniques presented in this chapter require coherent illumination. An intrinsic favorable property of the interferometric approach is high measurement sensitivity, at the expense of also relatively high sensitivity to environmental conditions, compared to techniques that are not based on interference.

Problems

ESPI

6.1 What are the effects on the fringes while reducing the aperture size of the optical system?

6.2 Describe the setup to cope with deformation measurement occurring both inplane and out-of-plane.

6.3 If the deformation is really small, and therefore generates only one fringe, explain why it is not necessary to unwrap the phase map.

6.4 In the out-of-plane setup, if everything is well balanced from the optical point of view, what can explain the fact that there is no fringe after subtraction process?

LCSI

6.5 What type of measurements can be carried out by LCSI?

6.6 What is the difference between an ESPI and an LCSI setup?

6.7 How can you optimize an LCSI instrument?

6.8 Why do we see only noise on the left side of the phase map in Figure 6.18?

Shearography

6.9 Why does air turbulence in the optical path laser–object–camera have a negative effect on interferometry measurements?

6.10 Why is shearography more robust than other speckle-based interferometry methods?

6.11 An out-of-plane displacement in micrometers is given by the following function:
$$f(x) = 0 \text{ for } x = 0 \text{ to } 60\,\text{mm}$$
$$f(x) = \sin^2(6(x-60)2\pi/360) \text{ for } x = 60 \text{ to } 90\,\text{mm}$$
$$f(x) = 0 \text{ for } x = 90 \text{ to } 150\,\text{mm}$$
Cross-section of simulated out-of-plane displacements.

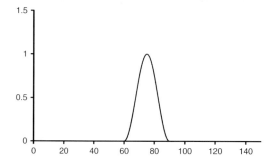

Determine the 1-D response in the shearing direction of a shearography system for this displacement. Use shearing distances of 2, 10, and 50 mm.

Reconstruct the out-of-plane displacement from the respective displacement gradients. What is your comment on the accuracy of the reconstruction of the displacement in relation to the shearing distance?

6.12 Speckle averaging for static applications involves addition of a series of phase distributions acquired with different independent speckle patterns. Would it be possible to implement speckle averaging by accumulating phase distributions that are wrapped? Explain why, or why not.

6.13 Figure 6.34 shows a modal pattern obtained for a vibrating plate, 155×155 mm, clamped over 20 mm at the bottom, leaving a vibration area of 155×135 mm. The shearography experiment was carried out with the time averaged method using a vertical shear of 10 mm. Assume that the three lower fringe patterns extend in the dark area below them. Explain the meaning of the two white fringes in the vertical direction and the white fringe in the horizontal direction. Can we conclude from the fringe pattern that we have a symmetrical vibration pattern referred to the vertical central line?

6.14 What are the out-of-plane displacements at the top of both white vertical fringes? Draw a graph of the out-of-plane displacements over the white vertical fringes. Consider the white vertical fringes to be perfectly vertical, and the horizontal one to be perfectly horizontal. Assume zero displacement at the clamp.

6.15 Draw a graph of the out-of-plane displacements over the central vertical cross-section and over the left and right vertical borders. Ignore cutoff of the fringe systems near the clamp: tilt and displacement are zero along the clamp.

Digital Holography

We are interested in studying the mechanical behavior of an aluminum wafer $\Delta A = 40$ mm in diameter with digital Fresnel holography. For this we use a green laser ($\lambda = 532$ nm) and a CMOS sensor with $(M, N) = (1024, 1414)$ pixels with pitch 5 µm. We use the discrete Fresnel transform with $(K, L) = (2048, 2048)$ to reconstruct the object and to compute the optical phase. The setup is similar to that of Figure 6.42 by considering only the green interferometer with $\theta_G = \theta = 45°$. The object is clamped along its x-direction and a mechanical loading is applied on its rear surface along its z-direction.

6.16 Determine the distance and spatial frequencies for optimal recording.

6.17 How could you experimentally adjust the spatial frequencies to the optimal value?

6.18 Compute the spatial resolution and the pixel pitches in the reconstruction plane; how many data points will occupy the reconstructed object?

6.19 Consider notations of Figure 6.42 and give the expression of the illuminating vector and the observation vector, and deduce the sensitivity vector of the setup.

6.20 Give the relation giving bright fringes versus parameters of illumination, laser wavelength, and displacement vector in the set of reference coordinates (x, y, z). Deduce the physical distance between two consecutives bright fringes.

7
Fringe Projection Profilometry
Jan Buytaert and Joris Dirckx

7.1
General Introduction

Profilometry, or the measurement of the three-dimensional (3-D) shape of objects, has many applications in science and engineering. For instance, profilometry or topography (*Greek*: *topo* = place or surface, *graphy* = to record or to describe) is used to monitor the manufacturing quality of optomechanical components, to detect obstacles for robot and vehicle guidance, and to measure the stress and strain on the surface of deformable objects [1, 2]. First, a short summary is given of the wide variety of techniques at hand to position the technique of fringe projection profilometry, which is mainly dealt with in this chapter.

7.1.1
Non-Optical Topography

One of the oldest techniques for shape measurement is the use of *tactile probes*. In recent decades, such probes have been perfected to the level of submicrometer measurement accuracy, and *hooked* probes make it possible to measure strong concave shapes or even inner bores of hollow structures with a small access opening. These *coordinate measuring machines* (CMMs) suffer from major disadvantages. The technique is inherently slow and physical contact with the surface is necessary; so the technique cannot be used on soft or scratch-sensitive surfaces.

Since the 1980s, tomographic techniques such as *micro X-ray computed tomography* (μCT) have emerged. This high-resolution technique has the enormous advantage that it cannot only measure the shape of a surface but can actually measure an entire 3-D volume, including inner surface areas that cannot be seen from the outside. In tomography, there is, however, a trade-off between resolution and the object size, so micrometer resolution is obtained only on relatively small objects (of a few centimeters) [3]. In addition, the equipment is very expensive; obviously, there are a number of safety issues connected with using X-rays; and most importantly, the generation of a 3-D scan can take several hours. Finally, the object needs to have suitable material properties: it needs to be transparent for the radiation used, and large differences in absorption can lead to artifacts.

7.1.2
Optical Full-Field Profilometry

A main advantage of optical profilometric techniques is that they are noncontacting. The general basic disadvantage is that the surface needs to be visibly accessible and must have suitable optical properties. For many techniques, the surface needs to be diffusely reflective and sometimes the use of a suitable coating is necessary to use the techniques [4].

Some techniques, such as time-of-flight measurements or heterodyne interferometry, can be used to measure distances at a single point, and an entire surface profile can be measured as well by point-scanning. The major application of optical techniques for profilometry lies however in the full-field methods.

7.1.2.1 Coherence-Based Techniques
Within the class of *coherence-based techniques*, the important ones are *digital holography* and *electronic speckle pattern interferometry* (ESPI) (Chapter 6). These coherence-based techniques can provide extremely high measurement accuracy, down to the nanometer range. As the wavelengths of lasers are accurately known, the techniques also have the important advantage of being self-calibrating. By using synthetic wavelengths (a mixture of two slightly different laser wavelengths), the techniques can also be desensitized to obtain artificial wavelengths up to millimeters or even centimeters in order to adapt the measurement range to large objects. A fundamental disadvantage of the techniques, however, is that they place strong requirements on the object and setup stability. Immobility down to a fraction of the wavelength of light is needed, even if far less measurement precision is required. Even when a synthetic wavelength of a millimeter is used, the information is still generated by the interference of submicrometer light waves. Although high-power lasers have become widely available so that large objects can be measured, this fundamental drawback puts a limit on the application of coherence-based profilometry.

7.1.2.2 Triangulation-Based Techniques
Mainly because of the stability problem of coherence-based techniques, optical profilometry based on simple geometrical optics has grown into a very important field and in recent years has become the most popular means to measure the shape of 3-D surfaces. All these techniques are based on triangulation. Either the surface points of the object are imaged from two different directions, or a structured light pattern is projected from one direction and observed from another, and the angles between the two directions are used to calculate the 3-D coordinates of the surface points.

In the first category of triangulation techniques, feature information on the object surface is used. Such identifiable features can be obtained in different ways.

A limited number of discrete markers can be adhered to the surface: paint with some clearly visible structure (speckle paint) can be sprayed onto the object; or the image texture of the surface itself can be used. The position of the feature points can be measured individually from two directions to calculate the point coordinates. Recent advances in *digital image correlation* have made it possible to automatically find and calculate all surface coordinates of the object from an image pair recorded from different directions (Chapter 5). These methods are used in techniques such as *stereoscopy* and *photogrammetry* [5, 6]. In-plane deformation can even be measured from one image of the undeformed and one of the deformed state of an object, using just one camera in one observation direction [1]. An advantage of the approach lies in the fact that measurement data are obtained in the object coordinate space, not in the measurement coordinate space. In this way, absolute coordinates are obtained without problems such as *phase jumps*. A major disadvantage is, however, that these techniques are only useful for a limited class of objects: if no clear markers or image texture are present on the surface, and if the object does not allow the use of speckle paint, the techniques cannot be applied. Moreover, the measurement precision depends on the local optical structure and texture of the surface, which may differ between zones on the same object. Moreover, applying speckle paint is regarded as a rather difficult *art* to get good results.

The other main category of geometric profilometry techniques does not require specific structures to be present on the object surface. The optical structure of interest now consists of a *structured light pattern* (such as a line or line grid) which is projected onto the object surface [6]. The surface geometry of the object under study will deform the projected structured light pattern when viewed from an angle. The surface shape information is encoded in the (image of the) deformation of this pattern. The main advantages of these techniques are that in principle they can work on any object, they are simple, and they can be very fast. The measurement resolution scales with the object size, but resolutions down to the micrometer range can be obtained, even on large objects. The required setup stability needs not be higher than the desired measurement resolution. So, if one measures with micrometer resolution, only micrometer stability is needed. For a certain resolution, large surfaces exceeding the field of view need to be patched and stitched together from several measurements. As seen in Chapter 3, a disadvantage can be that some ambiguity (called *phase jumps*) can occur in the measurement data, and to get best results it may be necessary to have a diffusely reflective surface. The structured light projection method is often also denoted as *active triangulation*. It covers projected coded light methods, line-scanning methods, and fringe projection methods.

The simplest way to implement the line-scanning technique is to project a single line of light and scan this over the object surface. When this line is observed at an angle with the projection direction, it will be seen as curved and deformed as a result of the object shape. Hence, the curvature of the line contains the information about the object shape. To obtain the 3-D object surface, a two-dimensional (2-D) imaging camera is needed that records all line shapes when the line is moved/scanned across the object surface by translation or rotation. The next step consists of analyzing the deformation of each line shape from a straight line to derive the underlying shape [7].

To avoid the time-consuming process of scanning, a whole grid of lines can be projected at the same time. Again, these lines get deformed and so their respective phases become modulated by the surface shape when the projected image on the surface is recorded from an angle. The surface height is thus coded in the phase of a periodic intensity profile [8]. The scheme presented in Figure 7.1 gives an overview of the different methodologies that are available to obtain the object height profile from a deformed, projected grid intensity profile seen on a surface.

The simplest optical setup for grid projection profilometry consists of the direct recording of an image of the deformed projected grid lines with a camera (Figure 7.1(1)). In this case, one only needs a projection system for the generation of the grid lines on the object and a camera that is placed at a fixed angle with the projection direction. A basic disadvantage of direct grid image recording techniques is the fact that the camera needs to resolve the individual grid lines without aliasing effects between the grid lines and the camera pixel columns. This means that the camera resolution needs to be at least double the observed projected grid pitch. Since the observed spacing between the projected grid lines also depends on the object surface shape, the camera resolution needs to be even higher when objects with steeply inclined zones on its surfaces are to be measured. The gray scale information of the camera is not used to obtain height information, just the X–Y position of the grid lines. To make it even more basic, binary grid projection suffices in some cases.

Once the image of the deformed grid lines is recorded, it can be processed in several ways. One approach requires just one recording (one shot) of the modulated grid lines on the object to be able to extract the height by means of Fourier transformation and filtering (Figure 7.1(3)) [9, 10]. This Fourier transform profilometric technique is simple in concept, and on modern personal computers it can be very fast (Section 7.3). Its main advantage is its measurement speed; its main disadvantage is the loss of spatial resolution due to the filtering involved.

Another approach to get surface information directly from the grid line image is to record multiple images with different grid positions to obtain phase-shifted grid lines. The object height can be retrieved directly from the position of the grid lines using phase-shifting (Figure 7.1(4)) [11], or *moiré* interference fringes can be generated digitally in the computer (Figure 7.1(5)) [12]. Two different mathematical techniques are used to provide the height result. The main disadvantage of both techniques is, again, that the camera needs to resolve the individual grid lines. In the case of the moiré fringe images, one needs to remove *grid line noise* from the images to be able to calculate the height at every pixel. Therefore, a larger number of images with different grid phases needs to be recorded (Figure 7.1(6)) [13, 14], or low-pass spatial filtering is necessary, thereby reducing the X–Y resolution (Figure 7.1(7)) [15, 16]. The terms *moiré fringes* and *grid noise* are explained further in this section and chapter.

Using a slightly more complicated optical setup, most of the disadvantages of the previous techniques can be overcome, and high spatial resolution can be

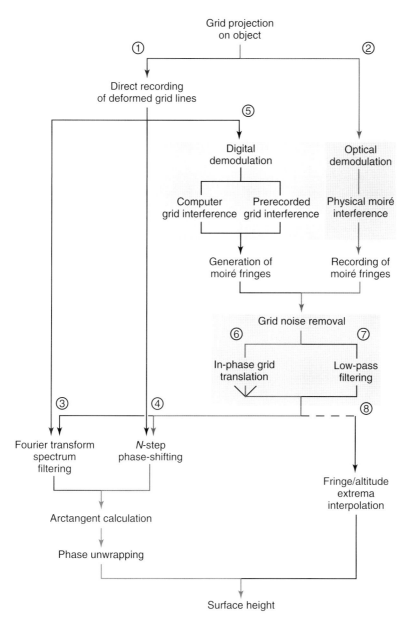

Figure 7.1 Schematic overview of the different grid projection techniques for profilometry. Path along (1) and (3): Fourier transform profilometry. Path along (2), (6) and (4): two-grid projection moiré profilometry.

combined with high height measuring resolution. This technique is generally known as *phase-shifting moiré interferometry*, and, because of its advantages in terms of precision, this technique is discussed in full in Section 7.4.

In moiré profilometry, the image of a grid is once again projected onto the surface of an object, but now the deformed grid lines are observed through a second (identical) straight grid. The process can be implemented using two separate identical grids and projection lenses – a process called *projection moiré* (Figure 7.2) [17, 18] – or by just casting a shadow of a grid using a point light source and viewing the shadow from an angle through the same grid – a process called *shadow moiré* (Figure 7.3) [19–22]. Because of the angle between the observation and the projection direction, the bright and dark lines of the projected grid will coincide or interfere with the opaque lines of the observation grid in some zones, whereas they will coincide with the transparent lines of the observation grid in others. Simplified – taking only the grid line centers into account – one can say that in zones where a dark line of the projected grid coincides with a transparent line of the observation grid, the intersecting object surface point will be seen as dark. A zone where the object turns dark due to this interference is called a *dark moiré fringe*. In zones where a bright projected grid line coincides with a transparent observation line, one will see light on the intersecting object surface, and such a

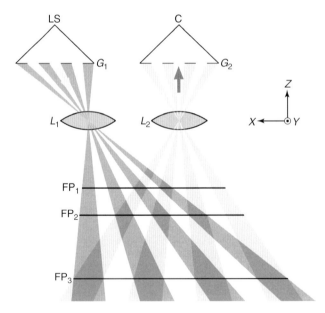

Figure 7.2 Projection moiré setup. Two identical grids G_i and lenses L_i are used – one to project grid lines on the object with an extended light source LS and one to image the deformed projected lines through the second grid. Geometrical interference fringes are thus seen by the camera C on the object surface through the second grid. In a symmetrical setup, these fringes represent positions of equal height, as the dark fringe planes FP_x are parallel.

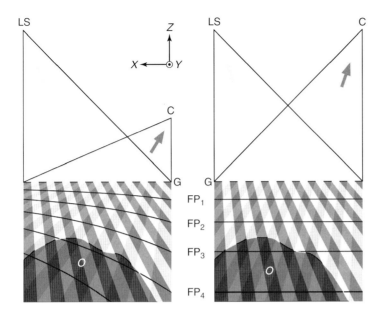

Figure 7.3 Shadow moiré setup. A point light source LS casts a shadow of the grid G onto the object surface O, and the deformed shadow lines are observed through the same grid by a camera C. On the object side of the grid, dark and bright zones virtually intersect the object. These zones are the so-called fringe planes. Left: asymmetric setup, creating curved dark fringe planes FP_x. Right: symmetric setup, generating parallel straight (dark) fringe planes FP_x representing positions of equal height.

zone is called a *bright moiré fringe*. In reality, in between the fully dark and bright moiré fringes, an intermediate state with gray scales will exist. It should be noted that the interference fringes are still *contaminated* with the original structure of the observation grid (Figure 7.5a). This is the so-called grid noise, which is always present in a *moiré interferogram* recording. It needs to be removed before further processing of the information is possible. As explained further on, the fringes can be understood as the intersection of the object with a set of dark and bright fringe planes, which are parallel to the XY plane (Figures 7.2 and 7.3). The inter-fringe plane distance determines the object height difference between two subsequent fringes, and will therefore play a fundamental role in the measurement accuracy of the technique. As will be demonstrated in the mathematical derivations, the fringe plane distance depends on the grid period as well as on the angle between projection and observation direction. Once the grid noise has been removed, clear moiré fringes appear in an image of the surface as alternating light and dark bands or regions. Such a recording is called a *moiré topogram* (Figure 7.5c). The intensity centers of the fringes on the object represent lines of equal height on the surface, like in a topographic map. The gray scale values of the fringes contain, however, much more detailed height information than just their extrema positions. To extract the height from these gray scales, recording of several clear fringe images with different fringe phases is needed. The different ways of performing phase modulation or encoding the height information in the fringes have led to a great variety of *phase-evaluation methods* (PEMs). Many PEMs use phase-stepping algorithms (PSAs) as a tool for phase calculation and height extraction. PSAs obtain the phase from a series – three or more – of phase-shifted intensity recordings, which are combined in the argument of trigonometric functions (Chapter 3).

The first main advantage of the moiré technique is that the individual grid lines do *not* need to be resolved by the camera. The camera only needs to record the much larger resulting interference fringes. The second main advantage of this technique is that the gray scale values in the fringes can be used to calculate object height without any spatial averaging or filtering. So, in contrast to the previously discussed techniques, all three dimensions of the camera are now used to measure the object height profile: the X–Y resolution of the camera is fully used to obtain a measurement on every pixel, and the gray scale range of the recording device is used for the height measurement resolution [18]. With modern cameras with dynamic ranges of 16 bits and more, this means that the height-measuring resolution can be equal to the in-plane spatial resolution, so a high-resolution measurement of the Z-coordinate is obtained on every pixel coordinate (x, y).

Because the Fourier method is the best in terms of measurement speed and the optical moiré fringe generation method is the best in terms of measurement accuracy, the rest of the chapter will concentrate on these two methods. The methods are, respectively, indicated in gray and light gray in Figure 7.1.

7.2
Grid Projection Profilometry: the Basics

In the following, the coordinate system defined in Figure 7.6 is used. The surface height of an object can be described by a function $O(x, y)$, which relates the object height z with coordinate points (x, y). It is the aim of profilometry to recover this shape function $O(x, y)$.

To measure the shape of the object, some structured light pattern is projected onto its surface. A simple and commonly used pattern is a grid of straight lines. Other patterns – such as a dotted motif – can also be used but this only complicates the explanation. This text limits itself to straight vertical line grids [23]. In practice, most grids consist of equally spaced black (opaque) and white (transparent) lines, so the grid has a *square* or *box car* wave transmission profile. Vertical line grids with a horizontal *sinusoidal* transmission profile are somewhat less common in practice. The reason is only of a technical nature: square-wave grids (also called *Ronchi rulings*) can be manufactured using etching processes, while sinusoidal grids are made using photographic techniques. Consequently, such sinusoidal grids have less transparency and a smaller modulation factor. Recently – through the introduction of liquid-crystal display (LCD) projectors as grid generators – the use of sinusoidal grids has gained interest [10, 24–26]. In the following theory, we mainly limit ourselves to sinusoidal line grids. This will keep the equations simpler, as square grids have to be described by an infinite sum of sine terms.

In practice, there is only a small difference between the results obtained for both grid types [18].

In the following, line grids are considered parallel to the XY plane, with the grid lines themselves parallel to the Y-axis and the sinusoidal profile along the X-axis. Now suppose that such a grid structure is projected along the Z-axis onto the surface of the object. If one views this projection also along the Z-axis, one will simply see straight lines with a fixed (magnified) grid period. If, however, the grid is projected obliquely along a direction that makes an angle Ω with the Z-observation axis, one will see deformed lines: the straight lines being distorted by the shape and geometry of the object. Mathematically speaking, the frequency of the grid has been modulated by the object function $O(x, y)$. The intensity distribution of the frequency-modulated grid structure can be represented by a function $f(q(\Omega) O(x, y))$, where q represents the combination of some known geometrical factors depending on the angle between projection and observation direction. Simply taking the inverse function f^{-1} of the observed frequency-modulated grid would suffice to retrieve the object surface function $O(x, y)$.

In practice, however, the recording becomes more complicated. When the grid is projected onto the object, a light source is used and the intensity of that source may vary over the object surface. Even more importantly, the local reflectivity of the object surface will probably vary from point to point. Hence, an amplitude modulation function $a(x, y)$ has to be introduced, which is multiplied with the intensity distribution f of the deformed grid. Next, background light also varying from point to point will be present. Therefore, a function $b(x, y)$ is added, resulting in the observed intensity distribution I:

$$I(x, y) = a(x, y) f(q(\Omega) O(x, y)) + b(x, y) \qquad (7.1)$$

Thus, the problem is one of dealing with three unknown functions $O(x, y)$, $a(x, y)$, and $b(x, y)$. Consequently, at least three independent equations are needed to solve the problem for the function $O(x, y)$. This is achieved by introducing *phase shifts* in and recording several images of the projected grid lines, cf. *phase-shifting algorithms* in Section 7.4 and Chapter 3. If one has some a priori knowledge about the three unknown functions, it is possible to solve the problem using just one recording. This is done in *Fourier transform profilometry* (FTP), which is discussed in Section 7.3.

7.3
Fourier Transform Profilometry

7.3.1
Theory

A very fast and powerful full-field structured light projection technique is FTP, which was first proposed by Takeda [8, 9]. It requires only one recording of the deformed grid lines to extract the height information of a surface through Fourier analysis (Figure 7.1(3)), so its major advantage over other profilometric techniques is the acquisition speed. It is important to emphasize that, in FTP, the deformed projected grid lines are called *fringes*. The phase modulation of these deformed lines or fringes is directly related to the height $O(x, y)$.

If the functions $O(x, y)$, $a(x, y)$, and $b(x, y)$ vary slowly in comparison to the period of the grid structure that is projected onto the object surface, it is possible to separate the functions in the frequency domain. The grid function will have a specific frequency, slightly modulated by the object function, while the intensity variations due to $a(x, y)$ and $b(x, y)$ have very different low frequencies. The grid function can then be separated from the other components using a correctly chosen filter window. It is important to notice, however, that this technique can only be successful if one has the a priori certainty that the frequency of the grid function differs enough from the frequency components in the other functions. If some overlap occurs, this will inevitably lead to errors in the calculation of the shape function. Suppose that one is measuring the shape of a striped shirt, and that these stripes have the same frequency as the grid that is projected onto them, then one will see a frequency-modulated grid together with the unmodulated stripes, and it will be impossible to separate both in the frequency domain. Only if the gray scale variations in the object have very different frequencies from the projected grid structure, the Fourier method can deliver very good results.

Suppose a sinusoidal grid with period p is projected as a structured light pattern. A sine pattern is the preferred grid shape to allow easy calculation. The transmission function t of such a grid is given by

$$t(x) = \frac{1}{2} + \frac{1}{2} \sin\left(\frac{2\pi}{p} x + \phi\right) \tag{7.2}$$

with ϕ an initial arbitrary phase. When this grid is projected using a light intensity I_0 onto the object surface, the frequency of the grid lines becomes modulated by the object surface shape when it is viewed at an angle Ω. Remember that the object surface is described by a function $O(x, y)$, which represents the object height $O(x, y) = z$ at every point (x, y). It is the aim of the method to determine this function. The recorded intensity distribution of the distorted sinusoidal grid can be described as

$$I(x, y) = a(x, y) \sin \varphi(x, y) + b(x, y) \tag{7.3}$$

with $a(x, y)$ again the position-dependent amplitude modulation and $b(x, y)$ the intensity offset, both caused by variation in the object reflectivity or inhomogeneities in the projection illumination intensity I_0. $\varphi(x, y)$ is the frequency-modulated phase of the projected distorted grid lines or fringes. Hence, this phase can be written as the original grid phase with constant period p, to which a position-dependent phase $\Delta\varphi(x, y)$ is added:

$$\varphi(x, y) = \frac{2\pi}{p} x + \Delta\varphi(x, y) \tag{7.4}$$

This phase modulation $\Delta\varphi$ is caused by surface depth variations and is thus related to $O(x, y)$. Rewriting expression (7.3) to a sum of complex exponential functions, one gets

$$I(x, y) = \tilde{a}(x, y) \exp\left(i\frac{2\pi}{p}x\right) - \tilde{a}^*(x, y) \exp\left(-i\frac{2\pi}{p}x\right) + b(x, y) \quad (7.5)$$

with $\tilde{a}(x, y) = \frac{1}{2i} a(x, y) \exp\left(i\Delta\varphi(x, y)\right) \quad (7.6)$

where $i^2 = -1$ and the complex conjugate is denoted by *. Taking the one-dimensional (1-D) Fourier transform of Eq. (7.5) gives

$$F_I(f_x, y) = \text{FT}\left(I(x, y)\right) = \int_{-\infty}^{+\infty} I(x, y) \exp\left(-i2\pi f_x x\right) dx \quad (7.7)$$

$$= \tilde{A}\left(f_x - \frac{1}{p}, y\right) - \tilde{A}^*\left(f_x + \frac{1}{p}, y\right) + B(f_x, y) \quad (7.8)$$

where $\tilde{A}$, $\tilde{A}^*$, and B are complex Fourier amplitudes. Note that only the frequencies of the deformed grid lines (around the grid frequency $1/p$) are of interest. Also, note that frequencies related to the object structure and the noise in the image will be present. Using half-band frequency filtering combined with zero-term suppression, one can select only the term $\tilde{A}(f_x - (1/p), y)$ from expression (7.8).

Calculating the inverse Fourier transform of the selected term, one obtains

$$J(x, y) = \text{FT}^{-1}\left[\tilde{A}\left(f_x - \frac{1}{p}, y\right)\right] \quad (7.9)$$

$$= \frac{1}{2i} a(x, y) \exp\left(i\Delta\varphi(x, y)\right) \exp\left(i\frac{2\pi}{p}x\right) \quad (7.10)$$

$$= \frac{1}{2i} a(x, y) \exp\left(i\varphi(x, y)\right) \quad (7.11)$$

Finally, the desired phase distribution is obtained by taking the arctangent:

$$\varphi(x, y) = \tan^{-1}\left(\frac{\text{Im}\left(J(x, y)\right)}{\text{Re}\left(J(x, y)\right)}\right) = \arctan\left(\frac{\text{Im}\left(J(x, y)\right)}{\text{Re}\left(J(x, y)\right)}\right) \quad (7.12)$$

which, after phase unwrapping (Chapter 3) and subtraction of the linearly increasing phase $(2\pi/p)x$ (Eq. (7.4)), gives the *phase map* $\Delta\varphi(x, y)$. $\Delta\varphi$ relates directly to the object surface height:

$$O(x, y) = k.\left(\varphi(x, y) - \frac{2\pi}{p}x\right) \quad (7.13)$$

$$= k.\Delta\varphi(x, y) \quad (7.14)$$

The scaling factor k can be derived from the geometry of the setup [27–30], and depends on the grid period p, the magnification factor M used in the projection, and the angle Ω between the projection and observation direction. What complicates matters is that k thus depends on the local height and phase $\Delta\varphi$. The formula derived by Takeda (for both parallel projection and observation axes as for *crossed optical axes*) is given by [27]:

$$k\left(\Delta\varphi(x, y)\right) = \frac{\ell}{\Delta\varphi(x, y) - \frac{D\cos\Omega}{pM}} \quad (7.15)$$

where ℓ is the grid to object plane distance along the Z-axis and D is the intergrid distance along the X-axis. However, several authors have corrected this equation and divided their derivations into four situations: noncollimated and collimated projection in parallel and crossed optical axes geometry [28–30]. The arrangement shown in Figure 7.6 can be used for FTP as a noncollimated projection crossed optical axes setup by leaving the out grid G_2. The scaling factor k for Figure 7.6 then becomes

$$k\left(\Delta\varphi(x,y)\right) = \frac{\ell\left(1 + \frac{y}{D}\sin^2\Omega\right)^2}{\Delta\varphi(x,y)\left(1 + \frac{y}{D}\sin^2\Omega\right)\left[1 - \left(1 - \frac{y}{D}\right)\sin^2\Omega\right] - \frac{D\cos\Omega}{pM}} \quad (7.16)$$

Alternatively, calibration can be done so that one does not have to know or determine the underlying parameters for $k(\Delta\varphi)$. In fact, the choice of a calibration method may be purposeful to avoid knowledge of these parameter values. It facilitates conversion of image coordinates to real-world coordinates and from unwrapped phase maps to absolute height values. Multiple measurements using a known object ensure that the calibration is based on the entire 3-D measurement volume [31].

7.3.2
Extensions

Often, noise is present in the recorded intensity distribution $I(x, y)$. Noise and weak frequency suppression can be achieved using a thresholding operator T_κ on $F_I(f_x, y)$ before performing the inverse Fourier transform:

$$T_\kappa(\varepsilon) = \varepsilon \text{ if } \varepsilon \geq \kappa \quad (7.17)$$
$$T_\kappa(\varepsilon) = 0 \text{ if } \varepsilon < \kappa \quad (7.18)$$

with κ the threshold value estimated from the noise in the intensity image.

Another improved version of the FTP method is gaining in popularity, namely *windowed Fourier transform profilometry* (WFTP) [32]. This method involves multiplying a movable window with $I(x, y)$, forming many local subimages of which the Fourier transform is calculated. The window function $g(x)$ can be Gaussian, Hanning, Hamming (rectangular), or any other profile, depending on one's application. To avoid leakage in the frequency domain originating from the (discontinuities at the edges of the often non-periodic) subimages, one best chooses a window going to zero at its sides. As an example, a Gaussian function $g(x)$ is used.

$$g(x) = \exp\left(-\frac{x^2}{2\sigma^2}\right) \quad (7.19)$$

$$F_{Iw}(f_x, y, u) = FT_w\left(I(x, y), u\right) = \int_{-\infty}^{+\infty} I(x, y)g(x - u)\exp\left(-i2\pi f_x x\right) dx \quad (7.20)$$

Using again half-band frequency filtering combined with zero-term and weak frequency suppression, a term $\tilde{A}\left(f_x - (1/p), y, u\right)$ is selected from $F_{Iw}(f_x, y, u)$.

Inverse Fourier transformation then yields

$$J(x, y, u) = \text{FT}^{-1}\left[\tilde{A}\left(f_x - \frac{1}{p}, y, u\right)\right] \quad (7.21)$$

Finally, the desired phase distribution is obtained by taking the following integral and arctangent:

$$\varphi(x, y) = \arctan\left(\frac{\text{Im}\left(\int_{-\infty}^{+\infty} J(x, y, u)du\right)}{\text{Re}\left(\int_{-\infty}^{+\infty} J(x, y, u)du\right)}\right) \quad (7.22)$$

Two inherent advantages of WFTP need to be emphasized [32]:

- As a WFTP is performed over a local area (determined by the width of the window $g(x)$), the signal in one position will influence the signal less in another position in the spectral analysis.
- The spectrum of a smaller (local) area is expected to be more straightforward and simpler than when using the whole-field image.

Summarizing, FTP is fast because it requires only a single image recording. Therefore, it is often used for dynamic measurements [33]. The major disadvantage is that (frequency domain) filtering is required, which inherently reduces spatial resolution. Filtering leads to smoothing of the surface and, as a consequence, sharp edges become more rounded than they are in reality. Furthermore, performing the filtering and choosing the optimal region to analyze in the spectrum is difficult. Determining the exact frequency of the projected grid at a single image point is not possible. The frequency (and eventually phase) at a certain point is always obtained through Fourier analysis of the vicinity of this point, even with WFTP. Finally, the higher the grid frequency, the better the FTP resolution, but the individual grid lines still need to be resolved by the camera. Once the (deformed) grid lines can no longer be distinguished from one another, the frequency cannot be correctly determined and phase unwrapping will certainly fail. So, a recording device with very high spatial resolution is needed to obtain modest measurement accuracy.

Box 7.1: Variations of FTP

Some authors project a two-frequency grid of vertical lines onto the object [10]. Using Fourier analysis, it is of course possible to study the high- and low-frequency (deformed) grid lines separately in the spectrum (if their respective spectra do not overlap). This then allows for more robust unwrapping (Chapter 3). The low-frequency grid has fewer phase jumps and offers information on the validity of the 2π phase differences in the result derived from the high-frequency grid. One could even project three grids of different frequency, one in red, one in green, and one in blue. Using a color acquisition device, it is easy to separate the different distorted frequency patterns by performing the Fourier analysis on each color channel of the image. Other authors have suggested projecting two grids with different frequencies convoluted with phase-shifted horizontal line grids [34]. Using this approach, they have succeeded in perfectly removing the pure zero-order term from the Fourier spectrum.

Another option is to use wavelet transformations instead of the Fourier analysis described above [11, 35]. These have the advantage that the density of spectral lines is adapted to the frequency. The main principle remains the same, but the details are beyond the scope of this text. A nice literature study referring to many other FTP-like methods is given by Gorthi and Rastogi [36].

7.3.3
Simulation Example

As an illustration of the FTP method, a simulated example is shown in Figure 7.4. An undistorted grid of vertical lines was calculated and random noise was added to simulate the output of a projector (Figure 7.4a). When these vertical grid lines are imaged by projection onto an object that is observed at an angle, one obtains deformed grid lines (Figure 7.4b). Only this image is required for all calculations leading to the 3-D height profile.

The deformation of the grid lines was created by applying a fisheye lens deformation on the image in Figure 7.4a. Figure 7.4c shows the positive half-band frequency spectrum of one horizontal line from the center of the image in Figure 7.4b. The undistorted grid with random noise would show some low noise in the spectrum, a large zero-order term, and a peak at frequency of 0.025/pixel in the spectrum. The distorted grid generates values exceeding the noise over a certain frequency range around 0.025/pixel in the spectrum, indicated by the rectangle window function in Figure 7.4c. When selecting this area of the spectrum, an inverse Fourier transform delivers Figure 7.4d, and after unwrapping one ends up with Figure 7.4e. After subtraction of the (fitted) linearly increasing phase (Eq. (7.4)), the phase map shown in Figure 7.4f is obtained. A 3-D representation can be found in Figure 7.4g. The height is not calibrated here so it corresponds to radians.

7.4
Moiré profilometry

7.4.1
Shadow Versus Projection Moiré

Moiré topography is a full-field optical technique in which the shape of object surfaces are measured by means of geometric interference between two identical

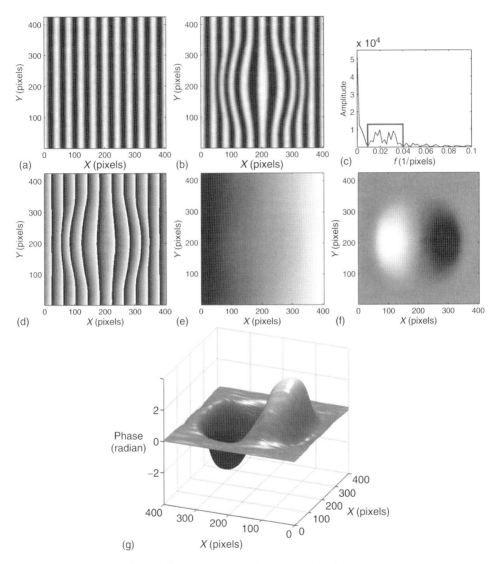

Figure 7.4 Fourier transform profilometry simulation. (a) Computer-generated grid with random noise. (b) Adding a fish-eye effect changes the grid frequency. (c) The absolute value of a Fourier transformation on horizontal line 200 as a function of spatial frequency. (d) Frequency filtering and inverse Fourier transformation delivers a wrapped phase image. (e) After phase unwrapping, a smooth steadily increasing phase is obtained. (f) After subtraction of the linearly increasing phase (Eq. (7.4)), the distorted phase related to the height (Eq. (7.13)) is obtained. (g) 3-D representation of the distorted phase (and thus height) of the simulated object.

line grids. A grid of straight lines, called the *projection grid*, is projected onto the surface of an object, and becomes deformed by the surface geometry of this object when observed at an angle Ω. When this frequency-modulated grid is viewed through a second straight grid with the same original pitch p, a geometrical interference pattern arises consisting of dark and bright zones, designated as *moiré fringes*. For specific setup geometries, these fringes represent contours of equal height and can be regarded as the intersections of the object surface with the dark and bright fringe planes (Figures 7.2b, 7.3, and 7.5). The object shape is encoded in the frequency modulation of the projected deformed grid lines, and is extracted by transmitting the image of the deformed grid through the second grid, called the *demodulation grid*. In shadow moiré topography, the projection grid and the demodulation grid are physically one and the same, while in projection moiré two separate grids are used.

In the recorded fringe image – also called *moiré interferogram* (Figure 7.5(a, b) – there still is grid noise present from the original grids on top of the moiré interference fringes. In order to apply PSAs to calculate the object height at every pixel, this grid noise needs to be removed (Figure 7.5(c)). When a high-frequency grid is used, the noise can be removed using a low-pass filter, but inevitably this will again be at the cost of spatial resolution. A better technique is to use grid averaging. Such averaging can be done continuously during the acquisition of one image or in steps with a number of moiré interferogram recorded images, but in both cases the grid is translated in its own plane and along the X-axis. In shadow moiré, grid noise can easily be removed by translating the grid while recording the moiré interferogram. As projection and demodulation grid are the same, the grid lines and their shadow move perfectly in phase, leaving the moiré interference fringes in place and unaltered while wiping out the grid lines [20, 37, 38]. In projection moiré, both grids need to shift synchronously (Section 7.4.2).

To obtain the object surface function $O(x, y) = z$, several recordings of moiré fringes *without* grid noise but *with* different fringe phases are needed. In shadow moiré, these phase changes are obtained by translating the object along the Z-axis over a fraction of the fringe plane distance [39]. In projection moiré, these phase changes are obtained by translating one of the grids in its own plane and along the X-axis over a fraction of the grid period, while the other grid remains fixed [40]. Hence, in projection moiré, the combined action of grid averaging and phase change needs a rather complicated and delicate translation system if simple Ronchi rulings are used [41]. When the grids are generated on liquid-crystal light modulators, both translation processes can be generated entirely electronically, and no moving parts are needed [42–44].

The moiré technique for topography was introduced in 1970 by Meadows *et al.* [19] and Takasaki [20], who independently developed the shadow moiré method [21, 22]. The method is simple to implement but the limitations are that the object needs to be close by and of the same size as the grid (Figure 7.3).

By using a lens to project the grid and a second lens to image the deformed grid lines through a second identical grid, it is possible to measure larger objects and to measure objects from a greater distance (compared to shadow moiré)

7.4 Moiré profilometry

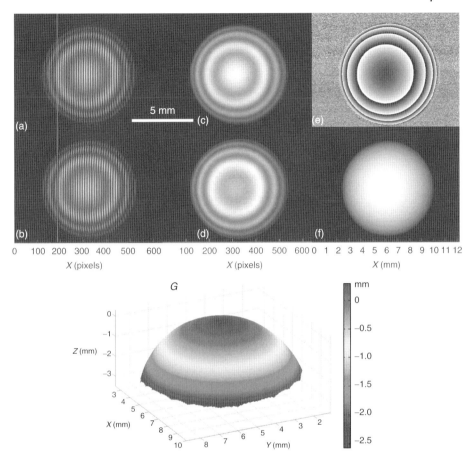

Figure 7.5 Moiré profilometry on a sphere. (a) Moiré interferogram showing moiré fringes and grid noise. (b) A similar moiré interferogram as in (a), but with both grids translated over $\Delta = p/2$. (c) After averaging a number of grid-shifted interferograms, grid noise is removed leaving a clear moiré topogram. (d) By changing the relative distance between the projection and the observation grid by $\Delta = p/4$ (in projection moiré) or by translating the object (in shadow moiré), a moiré topogram with a $\pi/2$ difference in fringe phase can be obtained. (e) From a number of (at least three) phase-shifted topograms, the wrapped fringe phase can be calculated with a phase ambiguity of 2π. (f) After removing phase jumps, the unwrapped object surface height is obtained in units of radians. The gray values in the figure represent the phase value, which corresponds to object height. (g) After multiplication with the appropriate calibration factor, the surface shape can be visualized in 3-D. (Please find a color version of this figure on the color plates.)

[17, 18]. The grids do not need to have the same size as the object: by using a suitable lens magnification, any size of object can be covered with grid lines. The technical implementation of the projection moiré technique is more complicated than shadow moiré. To avoid shadow problems (Section 7.6 and Figure 7.12b), some use a symmetrical setup with double grid projection at an angle $-\Omega$ and Ω

320 | 7 Fringe Projection Profilometry

and the camera in between. Others use one head-on orthogonal grid projection system and two cameras, observing the deformed grid patterns at angles $-\Omega$ and Ω for the same reason. The discussion is limited to the basic setup shown in Figures 7.3 and 7.6.

Notice how in Figures 7.2b, 7.3, and 7.5 the distance between fringe planes increases as a function of depth. This phenomenon is inevitable. In order to obtain

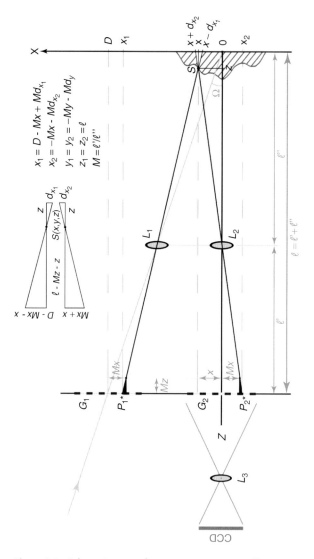

Figure 7.6 Schematic setup for projection moiré profilometry, defining the symbols used in the theoretical derivation: axes XYZ, grids G_i, lenses L_i, surface point $S(x, y, z)$, grid points $P_i(x_i, y_i, z_i)$, and magnification M.

equidistant fringe planes, the distance between the projection aperture (or the point light source in case of shadow moiré) needs to be made much larger than the object surface height variations. In that situation, the fringe plane distance can be regarded as constant and the fringe planes as equidistant to a good approximation. But even then, there still is a small nonlinear artifact, which is discussed further in Section 7.4.2.

7.4.2
Theory of Projection Moiré

7.4.2.1 Basic Principles

Figure 7.5a shows a moiré interferogram obtained on a spherical surface. When the projection and observation grid are translated simultaneously, the fringes remain unchanged, but the grid noise pattern moves along the X-direction (Figure 7.5b). If a correct number of such images are averaged, the grid noise can be wiped out from the image entirely, and a grid-noise-free moiré topogram is obtained (Figure 7.5c). The grid noise removal can be performed in a number of discrete steps or by continuously moving the grid. This is discussed in detail later in this section.

After the grid noise removal, several phase-shifted moiré topograms are needed to solve for the function $O(x, y)$. As shown in Figure 7.5c,d, one obtains moiré topograms with a similar fringe pattern, but the phase of the fringes is different. As the equation at hand has three unknown functions/variables (Eq. (7.1)), at least three topograms of different phase will be needed. As is seen in the paragraph on *phase-shifting algorithms*, Chapter 3, more phase-shifted topograms can be used to improve the measurement accuracy, but three is the minimum. From a number of phase-shifted topograms, the phase of the fringe pattern can then be calculated using an arctangent calculation. Because the arctangent has a phase ambiguity of 2π, one ends up with *phase jumps* (Figure 7.5e). To obtain the continuous surface height function $O(x, y)$, these phase jumps need to be removed by a procedure called *phase unwrapping* (Chapter 3). The result is shown in Figure 7.5f. Whenever a sudden phase change of the order of 2π is encountered in an image pixel row, this value is corrected in all subsequent pixels. Phase unwrapping is a generic problem in all phase-based techniques, among other FTP. Many different, more and less complicated schemes have been developed to remove phase jumps [45].

Finally, the phase information in radians needs to be converted to height $O(x, y) = z$. A phase of 2π corresponds to exactly one fringe plane distance, so by multiplying the phase values with the appropriate calibration factor the final surface profile is obtained (Figure 7.5g). The calibration factor can be calculated on the basis of the setup geometry and the period of the grids, or it can be obtained from one or multiple – in case of nonequidistant fringe planes, to cover the entire working volume – calibration measurements on an object with known geometry.

All the above-mentioned steps leading to a general projection moiré setup will now be discussed in detail.

7.4.2.2 Optical Geometric Interference

We first describe how a moiré fringe is formed in a general projection moiré setup with magnification M, as schematically represented in Figure 7.6. The grids G_1 and G_2 have an identical grid period or pitch p, and both have a transmission function t_i given by

$$t_i(x) = \frac{1}{2} + \frac{1}{2}\sin\left(\frac{2\pi}{p}x + \phi_i\right) \tag{7.23}$$

with ϕ_i the initial arbitrary phase.

To describe the entire interference process, a light ray is traced starting at point on grid. This point is projected by lens L_1 onto the object surface point $S(x, y, z)$. Projection by any lens introduces an insertion loss ρ, and reflection of light on a diffusely reflective object is influenced by the reflectivity coefficient $R(x, y)$. Another lens L_2 projects the object surface point S onto the point $P_2(x_2, y_2, z_2)$ on grid G_2, again with a certain insertion loss ρ_2. Incorporating all this into one equation, the observed intensity distribution $I(x, y, z)$ is given by

$$I(x, y, z) = \frac{I_0}{r^2(x, y, z)} t_1(x_1)\rho_1 R(x, y)\rho_2 t_2(x_2) \tag{7.24}$$

$$= \frac{I_0\rho_1\rho_2 R(x, y)}{4r^2(x, y, z)}\left[1 + \sin\left(\frac{2\pi x_1}{p} + \phi_1\right)\right]\left[1 + \sin\left(\frac{2\pi x_2}{p} + \phi_2\right)\right]$$

$$= a(x, y)\left[1 + \sin\left(\frac{2\pi x_1}{p} + \phi_1\right) + \sin\left(\frac{2\pi x_2}{p} + \phi_2\right)\right. \tag{7.25}$$

$$\left. - \frac{1}{2}\cos\left(\frac{2\pi(x_1 + x_2)}{p} + (\phi_1 + \phi_2)\right) + \frac{1}{2}\cos\left(\frac{2\pi(x_1 - x_2)}{p} + (\phi_1 - \phi_2)\right)\right] \tag{7.26}$$

with r the total length of the light path from the point source $P_1(x_1, y_1, z_1)$ to S and back to $P_2(x_2, y_2, z_2)$. The amplitude factor $a(x, y) = \left(I_0\rho_1\rho_2 R(x, y)/4r^2(x, y, z)\right)$ depends on the intensity I_0 of the light source, the insertion losses of both lenses, and the local reflectivity $R(x, y)$ of the object. Its actual value is of no major importance, because, by using phase-shifting algorithms later on, it will be eliminated from the final result. Expression (7.26) represents the moiré interferogram. It contains the moiré fringes and the remaining grid noise [17].

The following substitutions, derived from similar right-angled triangles in Figure 7.6, allow us to rewrite expression (7.26) with x, y, and z as its only variables:

$$x_1 = D - Mx + Md_{x_1} \text{ with } d_{x_1} = z\frac{D - (M+1)x}{\ell - (M+1)z} \tag{7.27}$$

$$x_2 = -Mx - Md_{x_2} \text{ with } d_{x_2} = z\frac{(M+1)x}{\ell - (M+1)z} \tag{7.28}$$

$$y_1 = y_2, z_1 = z_2 = \ell \text{ and } d_y = z\frac{(M+1)y}{L - (M+1)z} \tag{7.29}$$

When Eq. (7.26) is factored out and expressed in x, y, and z, one will see that all sine and cosine terms but the last one have a strong dependency of the coordinate x

in their argument, and are thus related to grid noise. Only the last term in formula (7.26) depends solely on the height z of the object surface. This is the actual moiré fringe term:

$$I_{\text{moiré}}(x, y, z) = \frac{a(x, y)}{2} \cos\left(\frac{2\pi(x_1 - x_2)}{p} + (\phi_1 - \phi_2)\right) \quad (7.30)$$

$$= \frac{a(x, y)}{2} \cos\left(\frac{2\pi}{p} D\left(1 + \frac{Mz}{\ell - (M+1)z}\right) + (\phi_1 - \phi_2)\right) \quad (7.31)$$

7.4.2.3 Grid Noise Removal

As shown in the scheme in Figure 7.1(6), there are several procedures that can be used to eliminate the grid noise terms.

The first approach is called *continuous grid (noise) averaging*. As stated before, movement or translation of both the projection and demodulation grid does not affect the position (phase) of the moiré fringes in the interferogram. The grid noise, on the other hand, shifts together with the grids. If the interferogram acquisition device integrates the resulting moiré interferograms while the grids are continuously moving along the X-axis with a speed v over exactly one or a multiple k of grid period p during an exposure time $T = (kp/v)$, the grid noise is averaged out to a constant background offset.

The steady movement of a grid can be mathematically described by adding a time-dependent part to the constant arbitrary phase ϕ in Eq. (7.23) for both grids G_i:

$$\phi_i \rightarrow \phi_i + \frac{2\pi}{p} vt \quad (7.32)$$

If the second sine term from Eq. (7.26) is studied as an example, integration over time using expression (7.32) yields

$$a(x, y) \int_0^T \sin\left(\frac{2\pi}{p}(x_2 + vt) - \phi_2\right) dt = \frac{pa(x, y)}{2\pi v}$$

$$\times \left[\cos\left(\frac{2\pi}{p} x_2 + \phi_2\right) - \cos\left(\frac{2\pi}{p}(x_2 + vT) + \phi_2\right)\right] \quad (7.33)$$

This term becomes equal to zero when filling in $T = (kp/v)$. The same result can be obtained for the other grid noise terms, thus leaving us with a moiré topogram $I_T(x, y, z)$ with the grid noise removed:

$$I_T(x, y, z) = Ta(x, y)\left[1 + \frac{1}{2} \cos\left(\frac{2\pi DMz}{p(\ell - (M+1)z)}\right)\right] + b(x, y) \quad (7.34)$$

where the constant phase offset is defined as $\phi_1 - \phi_2 = -(2\pi D/p)$, and adding b to incorporate background illumination effects, ambient light, and electrical offset in the acquisition device. This Eq. (7.34) is in essence Eq. (7.31), and corresponds to Eq. (7.3).

Tutorial Exercise 7.1

Calculate the minimal exposure time T needed to wipe out grid noise for grids with period p that are moving at speed v along the X-axis.

Suppose the grid period of the two grids in a projection moiré system is $p = 86\,\mu\text{m}$. A motor-driven translation stage continuously shifts both grids at a speed $v = 0.75\,\text{mm/s}$.

Solution:

During the exposure time, exactly one or a multiple k of grid periods should be translated, so

$$T = \frac{kp}{v} = k \times \frac{0.086\,\text{mm}}{0.75\,\text{mm/s}} = k \times 0.1146667\,\text{s} \tag{7.35}$$

An exposure time of 115 ms should remove the grid noise, as well as any multiple of this number.

A second method is called *discrete grid (noise) averaging*. Instead of continuous movement of both grids during one exposure, it is also possible to remove the grid noise by averaging the recorded images of several moiré topograms with both grids shifted over discrete distances along the X-axis in between each acquisition. Suppose one averages N interferograms (with exposure time T) each with an extra phase shift Φ for both grids. For the j th recorded topogram, the arbitrary phase in Eq. (7.23) of each grid G_i then becomes

$$\phi_i \rightarrow \phi_i + (j-1)\Phi \tag{7.36}$$

with $j \in [1, 2, \ldots, N]$. Performing the above substitution in Eq. (7.26) and averaging over all N recordings, the moiré topogram $I_N(x, y, z)$ is obtained as

$$\begin{aligned}
I_N&(x, y, z) \\
&= \frac{Ta(x, y)}{N} \sum_{j=1}^{N} \left[1 + \sin\left(\frac{2\pi x_1}{p} + \phi_1 + (j-1)\Phi\right) + \sin\left(\frac{2\pi x_2}{p} + \phi_2 + (j-1)\Phi\right) \right. \\
&\quad - \frac{1}{2}\cos\left(\frac{2\pi(x_1 - x_2)}{p} + (\phi_1 + \phi_2 + 2(j-1)\Phi)\right) \\
&\quad \left. + \frac{1}{2}\cos\left(\frac{2\pi(x_1 - x_2)}{p} + (\phi_1 - \phi_2)\right) \right]
\end{aligned} \tag{7.37}$$

$$\begin{aligned}
&= Ta(x, y) \left[1 + \frac{1}{N}\sum_{j=1}^{N} \left\{ \sin\left(\frac{2\pi x_1}{p} + \phi_1 + (j-1)\Phi\right) \right.\right. \\
&\quad + \sin\left(\frac{2\pi x_2}{p} + \phi_2 + (j-1)\Phi\right) - \frac{1}{2}\cos\left(\frac{2\pi(x_1 + x_2)}{p} + (\phi_1 + \phi_2 + 2(j-1)\Phi)\right) \right\} \\
&\quad \left. + \frac{1}{2}\cos\left(\frac{2\pi(x_1 - x_2)}{p} + (\phi_1 - \phi_2)\right) \right]
\end{aligned} \tag{7.38}$$

Note that the last term of this equation is identical to the one in expression (7.30), again. The discrete movement of both grids and averaging of these recordings deliver the same result as continuous averaging – namely Eq. (7.34) – provided one can remove the first three terms in expression (7.38). Let us focus on the first sine term as an example:

$$\sum_{j=1}^{N} \sin\left(\frac{2\pi x_1}{p} + \phi_1 + (j-1)\Phi\right)$$

$$= \sum_{k=1}^{N/2} \left\{ \sin\left(\frac{2\pi x_1}{p} + \phi_1 + (k-1)\Phi\right) + \sin\left(\frac{2\pi x_1}{p} + \phi_1 + \left(\frac{N}{2} + k - 1\right)\Phi\right) \right\} \quad (7.39)$$

$$= 2 \sum_{k=1}^{N/2} \sin\left(\frac{\pi x_1}{p} + \frac{\phi_1}{2} + \left(\frac{N}{4} + k - 1\right)\Phi\right) \cos\left(\frac{N}{4}\Phi\right) \quad (7.40)$$

Expression (7.40) amounts to zero when the phase-shift steps Φ fulfill the following relation:

$$\Phi = \frac{2\pi}{N}(2s+1) \quad (7.41)$$

with $s \in \mathrm{IN}$ and $\frac{N}{2} \in \mathrm{IN}_0$ [42]. However, to remove the first cosine term of (7.38) using the same method, it follows that N has to be a multiple of four:

$$\frac{N}{4} \in \mathrm{IN}_0 \quad (7.42)$$

According to Eqs. (7.41) and (7.42), $N = 4$ images with phase steps $\Phi = (\pi/2)$ in between every recording for both grids, or $N = 8$ and $\Phi = (\pi/4)$, or $N = 12$ and $\Phi = (\pi/6)$, and so on, have to be taken to obtain a fringe image without grid noise. In order to obtain a grid phase shift Φ, the grids need to be translated physically along the X-axis over a distance Δ:

$$\Delta = \frac{p}{2\pi}\Phi = \frac{p}{N}(2s+1) \quad (7.43)$$

Moving the grids over discrete distances becomes useful when working with LCD generated grids. In this case, the grid lines cannot be moved continuously, but only from one discrete LCD pixel column to the next [42, 43].

7.4.2.4 Digital Geometric Interference

As shown in Figure 7.1(1), the deformed grid lines can also be recorded directly without optical interference with a second physical grid. Moiré fringes can be generated afterwards by multiplication of the acquired image with a computer-generated image of the undeformed grid or with a prerecorded image of the undeformed grid (for instance an image of the grid on a flat plate). These approaches are referred to as *digital geometric interference* and *demodulation* (Figure 7.1(5)).

Apart from a simple optical setup, the method has little advantages. Just as in FTP (Section 7.3), the grid lines need to be resolved by the camera. The obtained fringes still contain grid noise, so either the projection grid needs to be translated while a number of grid images are recorded, or the grid noise needs to be removed through filtering which reduces spatial resolution. Continuous grid averaging during a single image recording is not an option since the images of the individual stationary grids are needed to generate the fringe interference images in the computer.

7.4.2.5 Phase-Shifting Algorithms

Once the grid noise has been removed, one ends up with a moiré topogram, as shown in Figure 7.5c,d. The intensity distribution seen in such a topogram is now given by Eq. (7.34). Although phase extraction in itself is quite simple to understand, namely obtaining $\varphi(x, y)$ (and thus $O(x, y)$) from a fringe image $I(x, y)$ (Eq. (7.37)), a great variety of PEMs have surfaced, each with its individual algorithms for phase calculation and with its particular needs and performance [8, 46].

The preferred method to obtain the phase encoded in interference fringes employs PSAs (Figure 7.1(4)), also called *temporal phase-shifting methods (without spatial carrier)*. The phase $\varphi(x, y)$ can be calculated from the arctangent of the ratio between two combinations of moiré topograms I_i with different moiré fringe phases ψ_i. Looking closely at Eqs. (7.31) and (7.34), one sees that the moiré fringe term depends on the difference in phase $(\phi_1 - \phi_2)$ between the two grids. If one grid is moved along the X-axis while the other one remains stationary, the location of the fringes on the surface will shift. This means that the phase of the fringe cosine term changes by a phase shift $\Delta\psi$ related to the translation of one grid with respect to the other. The relative translation between the projection G_1 and the demodulation grid G_2 thus forms the basic mechanism to obtain moiré topograms of different phases.

The first PSA dates back to 1966 [47], but in the meantime many different and complicated algorithms appeared to extract the phase φ for every recorded pixel (x, y) from Eq. (7.34). All algorithms require $n \geq 3$ fringe images I_i, each with a fringe phase ψ_i increasing with constant phase-shift steps $\Delta\psi$.

$$I_i(x, y) = a(x, y)\left[1 + \frac{1}{2}\cos\left(\varphi(x, y) + \psi_i\right)\right] + b(x, y) \quad (7.44)$$

with $i \in [1, 2, \ldots, n]$ and phase $\varphi(x, y) = (2\pi DMz/p(\ell - (M + 1)z)) \approx z = O(x, y)$

The most popular class of PSAs are described in Chapter 3 of this book and are based on the following trigonometric function [14, 48]:

$$-\varphi = \arctan\left(\frac{\sum_{i=1}^{n} I_i \sin\left(2\pi(i-1)/n\right)}{\sum_{i=1}^{n} I_i \cos\left(2\pi(i-1)/n\right)}\right) = \arctan\left(\frac{N}{D}\right) \quad (7.45)$$

using $\psi_i = (2\pi(i - 1)/n)$ in Eq. (7.44).

As there are only three unknown variables in Eq. (7.44), namely $a(x, y)$, $b(x, y)$, and $\varphi(x, y)$, three recorded moiré topograms I_1, I_2, and I_3 are sufficient to solve for the unknowns. All variables – such as M, ℓ, $R(x, y)$, $r(x, y, z)$, ρ_i, thus $a(x, y)$, $b(x, y)$, and the object shape $O(x, y)$ – are considered constant during phase-shifting. The more number of topograms that are used, the more robust the algorithm becomes against random noise. Three analytical solutions with the common phase-shift steps of $\Delta\psi = (\pi/2)$ are given below. For further information, we recommend [46].

Three-Step Method

- Recording $n = 3$ images I_i and using expressions (7.44) and (7.45), a short three-step phase-shifting algorithm is obtained [13, 48]:

$$\varphi + \frac{\pi}{4} = \arctan\left(\frac{I_3 - I_2}{I_1 - I_2}\right) \tag{7.46}$$

- which is equivalent to

$$\varphi = \arctan\left(\frac{2I_2 - I_1 - I_3}{I_3 - I_1}\right) \tag{7.47}$$

- with ψ_i : $\psi_1 = (\pi/4)$, $\psi_2 = (3\pi/4)$, and $\psi_3 = (5\pi/4)$.

Four-Step Method

- Recording $n = 4$ images I_i and using expressions (7.44) and (7.45), a popular four-step phase-shifting algorithm is obtained [13, 14]:

$$\varphi = \arctan\left(\frac{I_4 - I_2}{I_1 - I_3}\right) \tag{7.48}$$

- with ψ_i : $\psi_1 = 0$, $\psi_2 = (\pi/2)$, $\psi_3 = \pi$, and $\psi_4 = (3\pi/2)$.

Five-Step Method

- Recording $n = 5$ images I_i and using expressions (7.44) and (7.45), the so-called Hariharan method, is obtained [48, 49]:

$$\varphi(x, y) = \arctan\left(\frac{2(I_4 - I_2)}{I_1 + I_5 - 2I_3}\right) \tag{7.49}$$

with ψ_i : $\psi_1 = -\pi$, $\psi_2 = -(\pi/2)$, $\psi_3 = 0$, $\psi_4 = (\pi/2)$, and $\psi_5 = \pi$.

The use of PSAs has many advantages:

- The phase (and therefore the height) is calculated and obtained for every image pixel independently. The spatial x–y resolution is uniform and inherently high because the number of measurement points coincides with the number of detector elements.
- A high z-resolution can be obtained, as the phase is calculated on basis of gray scale variations in the phase-shifted fringe images. This is independent of the algorithm but related to the dynamic range of the recording device [18].
- The algorithm has a low sensitivity to stationary noise in the image, so nonuniformities in the background illumination $b(x, y)$ have no importance (unless they change in time).

- The method is applicable to low-contrast fringes (the signal-to-noise ratio is a more determining factor of the performance).
- With current computer power, the calculations are fast and can easily be automated.

A major disadvantage (which also exists in (W)FTP) is the need for phase unwrapping. Phase unwrapping is a fundamental and inherent requirement in all phase measuring techniques using an *arctangent*, and many simple and complex methods have been developed to deal with the problem. The result of a PSA algorithm – an arctangent calculation – is located between $-(\pi/2)$ and $\pi/2$, or, with some advanced algorithms, between $-\pi$ and π. Phase unwrapping strives to make the 2π phase discontinuities disappear, but is prone to errors. In general-purpose software packages, some good and fast unwrapping algorithms are included, and dedicated software packages with advanced methods are commercially available as well. No further details are given here, but one is referred to the vast literature that exists on the subject [45, 50], as well as Chapter 3 in this book.

Tutorial Exercise 7.2

What is the influence of square grid profiles (box-car wave) instead of the projected grid lines with a sinusoidal profile discussed before?

❊

Solution: If square grids are used, the moiré fringe profile of Eq. (7.31) is no longer sinusoidal but triangular, which leads to systematic nonlinearities in the height calculation when using the arctangent functions [46]. However, if fine, high-frequency square grids are used, the lenses tend to act as a low-pass filter, reducing the higher order components and thus creating approximate sinusoidal grid lines. Also, defocusing the lenses just a bit has the same effect.

Theoretically, when using square transmission profiles, the above derivation holds in first approximation: a square grid has a transmission function t', given by

$$t'(x) = \frac{1}{2} + \frac{2}{\pi} \sum_{k=1,\text{odd}}^{\infty} \frac{1}{n} \sin\left(\frac{2\pi k x}{p} + \phi\right) \tag{7.50}$$

The fringes originating from such a grid transmission profile are not sinusoidal but triangular. Nonetheless, applying the four-bucket (arctangent) algorithm, one gets

$$\frac{I_4 - I_2}{I_1 - I_3} = \frac{\sum_{k=1,\text{odd}}^{\infty} \frac{1}{k^2} \sin\left(k \frac{2\pi D M z}{p(\ell - (M+1)z)}\right)}{\sum_{l=1,\text{odd}}^{\infty} \frac{1}{l^2} \cos\left(l \frac{2\pi D M z}{p(\ell - (M+1)z)}\right)} \tag{7.51}$$

Because of the presence of $1/k^2$ and $1/l^2$, the higher order terms are quickly suppressed. Therefore, the approximation of using only the first-order term is often made and valid. In practice, square-wave grids therefore are useful. They are called *Ronchi rulings* and are (more) readily available in many different grid periods, as they are a standard in optical testing. Some phase-shifting or arctangent algorithms are more robust against the use of square grids than others [46].

Box 7.2: Fringe Tracing

Before the era of PSAs, the 3-D surface information was extracted from moiré topograms using a technique called *fringe tracing*, *fringe tracking*, or the *skeleton method* [14–16]. The method is based on the assumption that the fringe extrema correspond to the minima and maxima of Eq. (7.34).

The positions of these local fringe extrema were connected and used as lines of equal height. The pixels on such a line were set at the corresponding height of the virtually intersecting (approximately equidistant) dark or bright fringe plane. All other pixels in between these fringe extrema were interpolated, thus delivering a 3-D shape of the object surface, Figure 7.1(8). Needless to say, this method offered limited resolution because of the interpolation.

7.4.2.6 Nonlinearity and Fringe Plane Distance

A phase span of 2π corresponds to a fringe plane z-distance λ, which represents the height interval along the Z-axis corresponding to one entire (dark and bright) fringe pair (Figures 7.3 and 7.7). From Eqs. (7.44) and (7.48), it is clear that a full

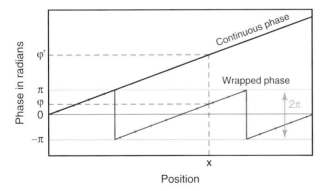

Figure 7.7 A plot of an assumed linearly climbing true continuous phase φ', and its (atan2) wrapped phase φ. To achieve correct unwrapping, phase jumps need to be detected and (multiples of) 2π need(s) to be added to the wrapped phase φ.

fringe plane distance λ equals (Eq. (7.54)):

$$\varphi(z) = \arctan\left(\frac{I_4 - I_2}{I_1 - I_3}\right) = \frac{2\pi DMz}{p(\ell - (M+1)z)} \tag{7.52}$$

$$2\pi = \Delta\varphi = \frac{2\pi DM(z+\lambda)}{p(\ell - (M+1)(z+\lambda))} - \frac{2\pi DMz}{p(\ell - (M+1)z)} \tag{7.53}$$

$$\lambda(z) = \frac{p(\ell - (M+1)z)^2}{DM\ell + p\ell(M+1) - p(M+1)^2 z} \tag{7.54}$$

As already mentioned, the fringe plane distance depends on z. The reason is that z is present in the numerator and denominator of Eq. (7.54). $\lambda(z)$ becomes smaller when the height of the object increases and vice versa (Figure 7.2). This leads to systematic errors in the moiré technique: φ is not perfectly linearly related to z (Eq. (7.52)). The moiré technique thus contains a basic nonlinearity, which is unavoidable. However, this nonlinearity can be strongly reduced when a good setup geometry is chosen [18]. If the projection distance ℓ is chosen large in comparison to the maximal object height z_{max}, the following approximation can be made using a Taylor expansion:

$$\frac{\ell z}{\ell - (M+1)z} = z\left(1 + \frac{(M+1)}{\ell}z + \frac{(M+1)^2}{\ell^2}z^2 + \cdots\right) \approx z \tag{7.55}$$

Equation (7.52) can now be rewritten as a linear relationship between z and φ:

$$z(x, y) = \frac{p\ell}{2\pi DM}\varphi(x, y) \tag{7.56}$$

and the fringe plane distance in Eq. (7.54) becomes approximately equidistant:

$$\lambda \approx \frac{p\ell}{DM} \quad \text{for } \ell \gg z_{max} \tag{7.57}$$

Tutorial Exercise 7.3

Suppose that one has constructed a projection moiré setup using two grids with $p = 10$ line pairs/mm. The separation between the grid centers is $D = 20$ cm. How large should the moiré setup be to have a fringe plane distance which can be regarded as constant within 2% error over an object depth of 5 mm? In other words, what value of projection distance ℓ is needed so that the approximation in Eq. (7.57) is valid within 2%. For simplicity, use a magnification factor M of 1.

Solution:

The height-dependent fringe plane distance $\lambda(z)$ is given by Eq. (7.54). At $z = 0$, λ equals

$$\lambda(0) = \frac{0.1\text{mm} \times \ell^2}{20\text{cm} \times \ell + 0.1\text{mm} \times 2\ell} = \frac{0.1\text{mm} \times \ell}{200.2\text{mm}} \tag{7.58}$$

At an object height of $z = 5$ mm, λ equals:

$$\lambda(5\text{ mm}) = \frac{0.1\text{mm} \times (\ell - 2 \times 5\text{mm})^2}{200.2\text{mm} \times \ell - 0.1\text{mm} \times 4 \times 5\text{mm}} = \frac{0.1\text{mm} \times (\ell - 10\text{mm})^2}{200.2\text{mm} \times \ell - 2\text{mm}^2} \tag{7.59}$$

It is demanded that the fringe plane distance varies less than 2%, so

$$\frac{\lambda(5) - \lambda(0)}{\lambda(0)} \leq 0.02 \tag{7.60}$$

Solving for ℓ, this gives

$$\ell \geq 5\text{mm} \tag{7.61}$$

The first value is impossibly small, so the second value of about 1m has to be used:

$$\frac{z_{max}}{\ell} \leq \frac{5\,\text{mm}}{1000\,\text{mm}} = 0.005 \tag{7.62}$$

7.4.3
Practical Considerations

As seen from Eq. (7.57), the inter-fringe plane distance λ decreases when the opening angle $\Omega = \arctan(D/\ell)$ between projection and observation is increased, and when the period p of the grid is decreased. A smaller fringe plane distance leads to a better height measuring resolution, but it also causes more phase-unwrapping problems. Nevertheless, the highest measurement resolution can be obtained by using a small fringe plane distance, so one would be tempted to use extremely fine grids and large angles. Both parameters are, however, limited because of practical reasons.

For any opening angle Ω, there always exists a problem of shading or occlusion, both for the viewing and for the projection direction. Shading refers to a part of the object surface that cannot be reached or viewed because an extreme surface feature is occluding it. The larger the angle, the more serious this shading problem becomes. The best choice is once again dependent on the object under study. If an object contains steep height changes, a small angle will be needed to avoid shading problems (Section 7.6). If the object contains discontinuities, sharp edges, or corners, it becomes impossible to overcome the shading problem. On a smooth object, a large angle between projection and observation direction can be used, but even then one is limited by projection lenses: even very good lenses will generally not allow larger angles than approximately 45°.

If the camera is not aligned along the optical viewing or Z-axis with the object and the observation lens, the height measurement (sensitivity) axis will not be perpendicular to the XY plane, and measurements will be obtained in a (slightly) deformed coordinate system. Hence, as shown in Figures 7.3, 7.6, and 7.8, only the projector is placed at an angle, and the camera and observation lens are aligned in front of the object along the Z-axis.

Tutorial Exercise 7.4

Suppose one has an opening angle of $\Omega = 10°$, a magnification $M = 0.2$, and a lens-to-object distance $\ell'' = 75$ cm for the setup in Figure 7.6. The grid period

$p = 200\,\mu$m for both the projection and demodulation grids creates three orders of moiré fringes on the object surface. Estimate the object height using the fringe plane distance in Eq. (7.57).

Solution:

The magnification $M = \ell'/\ell'' = \ell'/75\,\text{cm} = 0.2$ leads to $\ell' = 15\,\text{cm}$ and $\ell = \ell' + \ell'' = 90\,\text{cm}$ (Figure 7.6). From $\Omega = \arctan(D/\ell) = \arctan(D/90\,\text{cm}) = 10° = 0.1745$ rad, it follows $D = 15.87\,\text{cm}$. Now the fringe plane distance can be calculated as $\lambda \approx \frac{p\ell}{DM} = \frac{0.2\,\text{mm} \times 900\,\text{mm}}{0.2 \times 158.694\,\text{mm}} = 5.6713\,\text{mm}$.

As three entire fringe pairs cover the object, the surface is intersected by three dark and three bright planes, and the object height should be approximately 17 mm or 1.7 cm.

The other way to increase measurement resolution is to use finer grids, and therefore smaller periods p. Here also, one faces practical limitations. From a very basic point of view, the period of the grid is limited by diffraction: if the lines are so fine that the first-order diffraction maximum coincides with the next dark line, very little grid contrast will be left. In practice, however, it will only make sense to use coarser grids than predicted by the diffraction limit, because even the best projection lenses still have a limited modulation transfer function. When choosing a lens, it is therefore important to consult the manufacturer's detailed information, which will give a graph of the modulation contrast as a function of the number of lines per millimeter. It makes no sense to use an extremely fine grid when the projected lines will only have a modulation contrast of, for instance, 20%. Such low modulation contrast will also lead to moiré fringes with little contrast, while it is the

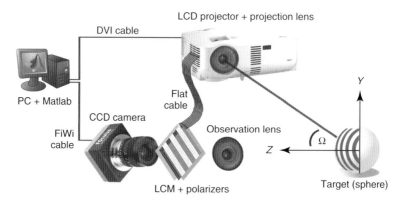

Figure 7.8 Schematic representation of an LCD moiré profilometer based on a commercial data projector. One of the LCD panels of a three-LCD data beamer is brought to the exterior of the projector. The LCD is placed between crossed polarizers, and is used as demodulation grid. One of the remaining LCDs in the projector is used to project the grid onto the object. A camera placed behind the demodulation grid records the interferograms. As each of the LCDs corresponds to one of the primary colors in an image, the period and position of the grid patterns on the LCD panels can easily and separately be controlled using a PC, a standard graphic controller, and colored grid images.

gray scale range that is used to obtain height measurement. As a rule of thumb, one should use a grid period that allows projection (and observation) with a modulation contrast of at least 80%. One should also notice that the modulation contrast of a lens differs with the angle of the light rays entering the lens, and therefore with the projection–observation angle of the setup. Finally, it is important to choose not only a high-quality lens but also a type that is specified for the purpose. If one builds a moiré setup in which the grid is projected in a 1 : 1 ratio, one cannot use a lens that is designed for a magnification ratio of 50 : 1. It makes absolutely no sense to buy a very expensive lens from a high-quality manufacturer if one uses it in the wrong way.

Several good makes of high-quality lenses are available. For magnification ratios between 50 : 1 and 10 : 1, many lenses designed for photo cameras can be used. For small magnification ratios, one needs to move to either macro lenses or, even better, lenses designed for the graphics industry.

Another important practical consideration is the setting of the lens aperture. It should not be too large, because most lenses do not perform best at their largest apertures and also because a large aperture gives a small depth of focus. If the projected lines are out of focus on the object surface, the modulation contrast degrades. On the other hand, the aperture should also not be too small, as this can also lead to less modulation contrast due to diffraction. And obviously, a small aperture also means reduced light, necessitating either very long exposure times or an excessively strong light source.

Finally, there is the choice of the camera. Obviously, more pixels mean better X–Y resolution (if the camera lens quality is sufficient). To have a good height resolution in phase-shifting moiré with optical demodulation, it is as important to have a good *gray scale resolution* – in other words, to have a camera with a broad dynamic range. It makes no sense to use a 2k × 2k charge-coupled device (CCD) camera if it only has eight bits of gray scale resolution. In recent years, much technological progress has been made in this field, and cameras with 12- or even 16-bit gray scale resolution have become affordable. Note that it is not just the number of bits of the analog-to-digital (A/D) converter that is important. The important factor is the true gray scale resolution, and so the number of gray scale steps between the noise and saturation level. Some brands tend to boast a 16-bit dynamic range, but in practice, the lowest 4 bits (or worse!) are just filled with noise.

7.4.4
Practical Implementation

In the past, optomechanical setups have been built using motors or piezo actuators to move the Ronchi ruling for grid-noise averaging and/or phase stepping. In the last decade, however, LCD technology has made huge steps in terms of resolution, contrast, and also cost, so that optomechanical setups have become outdated. In the remainder of this section, an example of a practical projection moiré setup is presented, based on LCD panels, for both the projection and the demodulation grid [42, 43]. Finally, some results and application examples used in research are shown, but it should be noted that by choosing different lenses a moiré setup can be very easily adapted to other sizes of objects and other ranges of resolution.

In a commercial LCD projector, the light of a high-pressure gas discharge lamp is collimated and divided into three spectral bands which are directed toward three liquid-crystal light modulators. After passing the modulators, the three colors are again combined in a beam recombination cube, and the final color image leaves the projector through the projection lens. The light modulators themselves consist of a polarizer (PO), a thin-film transistor (TFT), the liquid-crystal matrix (LCM), and a crossed polarizer (PO'). When a pixel on the liquid-crystal matrix is activated, it rotates the polarization direction of the incoming polarized light, so that this light now passes through the second polarizer. When inactive, the light is blocked by PO'. Intermediate activation generates intermediate light transmission. Several implementations exist, but essentially an active liquid crystal acts as a half-wave plate with its optical axis set at $45°$ to the transmission direction of the first polarizer, so that the polarization direction is turned by $90°$. Clearly, the correct rotation is only obtained for one specific wavelength. In practice, modulation contrasts of more than 500 : 1 between active (*light*) and inactive (*dark*) pixels are obtained in commercial LCD projectors. A resolution of 1024 by 768 pixels is common even on low-cost LCMs.

To construct a phase-shifting projection moiré profilometer, one liquid-crystal device needs to be brought out of the projector. In most models, for example, the NEC VT695, the recombination cube with attached liquid crystals can easily be removed from the projector, and one of the LCMs can be detached from the beam recombination cube assembly. Next, the cube with the two remaining LCMs is put back in the projector. Because the projection lamp has its peak intensity in the green-red region, it is best to isolate the blue-channel LCM for use as the demodulation grid. The best result is obtained when only the green channel of the projector is used for grid projection, as this wavelength band is closest to the optimum wavelength of the blue-channel LCM used for demodulation. The red-channel LCM is also left in place, but is not used. With an extension flat cable, the blue-channel LCM is again connected to the projector electronics, and it is mounted next to the projector. In front of and behind the LCM, crossed polarizers are installed, and their transmission direction is turned such that maximal transmission contrast is obtained between active and non-active pixels.

Figure 7.8 shows a drawing of the setup, of which the optical design follows the scheme presented earlier in Figure 7.6. Green grid lines are projected onto the object by the projector lens, and a second identical lens projects the deformed grid lines onto the demodulation LCM. A CCD camera (such as the Foculus FO442B) with an imaging lens ($f = 50$ mm) is placed behind this blue-channel demodulation light modulator. When the camera is focused on the blue-channel LCM, the moiré interference pattern of the deformed green projected grid lines through the undeformed blue grid lines can be recorded.

The standard lens of the projector (and an extra (similar) copy) can be used to measure objects with dimensions ranging from several centimeters to meters. The technique can also be used to study the shape and deformations of tiny objects such as eardrums [51]. In this case, the projection lens is removed from the LCD projector and a set of lenses with longer focal distance (Schneider-Kreuznach $f = 150$ mm) is used.

To implement phase-stepping and grid averaging, the two LCM grids need to be controlled separately. In a commercial three-LCD projector, this can be easily realized using the RGB color space, since color images have three color layers which are displayed separately. For example, an image sent to the projector consists of a 3-D matrix $1024 \times 768 \times 3$. If a certain pixel on the projection LCM and on the demodulation LCM needs to be active, it suffices to send an image that contains the color cyan (or *green* + *blue*) to that pixel. Separate matrices for projection and demodulation are thus prepared, containing grids which are then respectively stored in the second (green) layer and the third (blue) layer of a bitmap image. The projector is connected as a second monitor with the graphics card, and the display is set to *Dual View* extended desktop mode, allowing one to send the bitmap image to the second screen using custom-made software. The grids can have (discretized) sinusoidal, square, or other intensity profiles. Nonlinearity in the transmission response of liquid crystals can decrease the accuracy and resolution of the moiré profilometric method. If the sinusoidal grid transmission functions deviate from a perfect sine wave, the moiré fringe shape will deviate from a sine pattern as well. Using an arctangent method on imperfect sines or cosines introduces an error in the phase calculation [1, 26, 43]. One can easily test the response of the LCMs by measuring the projected light intensity as a function of the input value sent to the LCM. The inverse of this curve can then be used to linearize the transmission, as some sort of *gamma correction*.

As explained earlier in Section 7.4.2, grid noise needs to be removed by recording a number of interferograms each with the same relative phase between the two grids, but with both grid line patterns sequentially shifted over a number of pixels for each interferogram. As shown theoretically for a sinusoidal grid, (minimally) $N = 4$ of such images are needed to remove the grid noise (Eq. (7.42)). In reality, however, the grids are never perfectly sinusoidal or square because, between the active pixel columns, there are inactive zones that always stay dark – in current commercial devices this limited fill factor is about 80% of the pixel area. As a consequence, it may be necessary that more averaging steps are needed to remove the grid noise entirely [42]. As it is needed to move the grids N times over a phase step $\Phi = (2\pi/N)$ in an entire period p for grid noise removal (Eq. (7.41)), and as phase steps $\Delta \psi = (\pi/2)$ in PSAs correspond to a translation of one-fourth of a period, the grid period expressed in pixels, p_{pixel}, needs to be a multiple of N and 4, as N itself is a multiple of 4 (Eqs. (7.42) and (7.43)).

$$\frac{p_{pixel}}{N} \in IN_0 \tag{7.63}$$

When discrete grid averaging is used (Section 7.4.2), a moiré interferogram needs to be recorded at each grid position. The exposure time for the CCD camera depends on the intensity of the projector, but typically it will be in the range of about 100 ms. When $N = 16$ images is used for grid averaging and each grid position has to be recorded seperately, this amounts to a total recording time of at least 1.6s for one moiré topogram. The real bottleneck, however, is the time needed to transfer and store the images coming from the camera, interleaved by uploading grid images on the display board. The discrete averaging procedure works well [42], but it greatly slows down the speed of the measurement.

Using a relatively new commercial graphics board, it is possible to speed up the averaging process using semicontinuous grid averaging (Section 7.4.2) [44]. An nVidia GeForce (8800 GT) board with 512 MB of graphic memory can be used for this purpose. First, all grid images are prepared for all averaging positions (e.g., $N = 16$) of all phases (e.g., $n = 4$). Next, a custom-made program developed in C-Graphics (Cg) is used to upload all 64 images in four buffers of the graphics card memory (one for each moiré fringe phase step). The graphics processing unit (GPU) is programmed to continuously *play* the content of a certain buffer. For the first map or phase, this *movie* of $N = 16$ images is displayed at the highest frame rate the graphics board can deliver, and the first moiré topogram is recorded in one camera exposure time. Then the next *movie* is played, containing $N = 16$ images with a quarter grid period translation between the projection and the demodulation grid, and the next moiré topogram is recorded. Finally, the same procedure is repeated for the third (and fourth) topogram. In semicontinuous grid averaging, the movie of grid positions needs to be rendered on the LCMs exactly one time (or exactly a whole multiple of times) during the camera exposure. The fixed frame generation rate of the graphics board therefore determines the minimal CCD exposure time to be used (Tutorial Example 7.7).

The entire setup can be controlled by a custom-made software program and contains no physically moving components, as the projection and demodulation grid can be shifted on the liquid crystal pixel matrices. The method does have the minor limitation that shifts are confined to discrete steps of one pixel (Eq. (7.63)). This low-cost setup can also be used in combination with the FTP approach.

Tutorial Exercise 7.5

Suppose one has an LCD projector in our setup with a frame rate of 60 Hz. How long should the camera exposure time be if $N = 16$ grid images are used in semicontinuous grid averaging?

Solution:

To display one cycle of a movie with $N = 16$ shifted grids images on a 60 Hz projector, the time needed is

$$16 \times \frac{1}{60} \, \text{s} = 0.266666 \, \text{s} \qquad (7.64)$$

Hence an exposure time of 267 ms or any multiple of this value should be used. This is a huge time gain compared to the discrete averaging of the $N = 16$ images, which takes almost 2 s.

7.4.5
Demonstration Measurements

The measurement results shown in this chapter were made using $N = 8$ or 16 shifted moiré interferograms for discrete and semicontinuous grid noise removal, and $n = 3$ or 4 phase-shifted moiré topograms (Eqs. (7.47) and (7.48)). Different object sizes are measured, thus requiring different lengths ℓ' and ℓ'', magnifications M, and opening angles Ω.

Tests on an oblique flat plate have shown that the height resolution is better than 15 µm over a depth of 5 mm with $M = 1$ [1, 42], and similarly over 30 mm depth with $M = 0.16$ [43]. In the case of $M = 0.16$, the following setup parameters were used: $\ell'' = 1060$ mm, $\ell' = 170$ mm, $\Omega = 7.4°$. The calibrated scaling factors were 51 µm/pixel for X–Y and 1.54 mm/rad for Z, with a respective maximum error of 1.1 and 2.6% in an imaging volume of $71 \times 53 \times 30$ mm. A fringe plane distance $\lambda = 2\pi \times 1.54$ mm $= 9.65$ mm ≈ 1 cm is thus obtained for $p_{\text{pixel}} = 16$, and half in case of $p_{\text{pixel}} = 8$.

As a demonstration of the isometric X–Y–Z calibration, a square pyramid is measured with sides of 36 mm for the base and an apex of 18 mm, placed on a 40×40 mm block. All measured dimensions and angles correspond to reality within a maximum error of 2.7%. A moiré topogram, the height map, and the cross sections are shown in Figure 7.9, together with the resulting 3-D surface. The pixel period used on the LCMs was $p_{\text{pixel}} = 8$. As each pixel has a dimension of 12 µm, it follows from Eq. (7.57) that $\lambda = 4.8$ mm, corresponding to the presence of 3.7 fringe planes over an object height of 18 mm. This can easily be verified in Figure 7.9a,e. In Figure 7.9f, the right-angled steps on another custom-made test object are correctly reproduced as 90.0 ± 0.5.

As stated before, large objects can be measured as well as small ones. As an example of a larger object, Figure 7.10 shows a shape measurement obtained on a life-size mannequin head; on a golf ball (with diameter 42.62 mm), clearly showing its dimples; and on a small plastic figure (5cm height). To conclude this demonstration of the performance of the setup, an upscaled artificial model of a tympanic membrane is investigated. Moiré topography has become one of the established techniques for measurement of eardrum shape and deformation in research on the mechanics of hearing [1, 41, 51]. One performs point indentations on real eardrum membranes, and the shape and deformation behavior are measured with profilometry to extract material parameters – such as elasticity – by reverse engineering and finite-element modeling. For calibration purposes, an upscaled model of the eardrum was built and measured (Figure 7.11).

7.5
Noncontinuous Surfaces

Both in Fourier transform as in projection moiré profilometry, an angle is being used between the projection and the observation direction to obtain the object depth information. Consequently, there is an inherent asymmetry between the

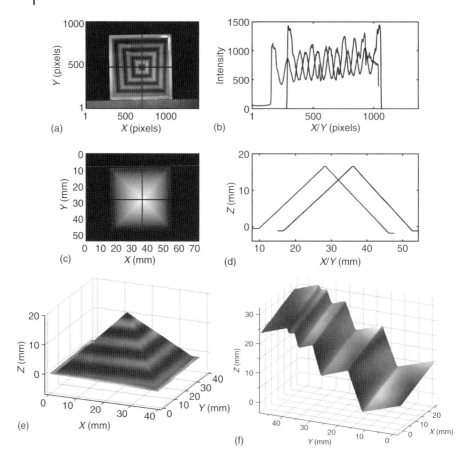

Figure 7.9 Performance measurements with a projection moiré topography setup. (a) Moiré topogram obtained on a pyramid with an apex height of 18 and 36 mm wide square base, placed on a square ground surface of 40 mm. Grid noise was removed using discrete averaging with $N = 8$ images. (b) Cross section along a vertical and a horizontal line of the moiré topogram shown in (a). (c) Gray scale representation of the calibrated phase (thus height) map calculated from four phase-shifted moiré topograms. (d) Cross section along a horizontal and a vertical line going through the apex of the pyramid in (c). (e) 3-D representation of the surface of the pyramid. The image of the first moiré fringe topogram is shown in overlay on the surface. (f) Measurement result obtained for a second test object with a surface consisting of right-angled steps of 2, 5, 10, and 15 mm.

two in-plane axes of the measurement X and Y. This can lead to different behavior when one measures a noncontinuous object or objects with very steep height variations on their surfaces.

If an object shows steep height changes along the X-axis (as defined in Figure 7.6), shading or occlusion problems might appear: either a part of the object is not seen, or it is not reached by the projected structured light pattern, and a shadow is cast on the object. In fact, this shadow area is exactly the same as missing data, like in

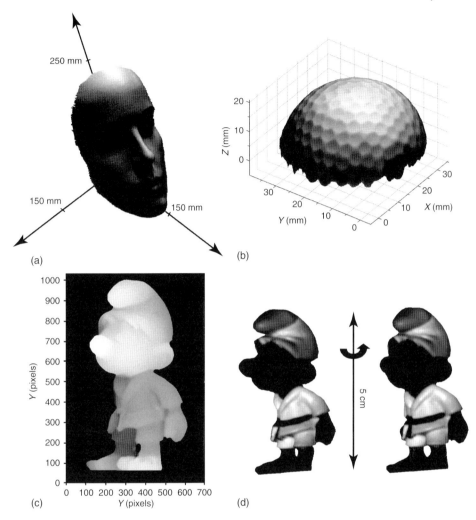

Figure 7.10 Objects of different sizes measured with a projection moiré topography setup. (a) 3-D representation of the surface of a life-sized mannequin head. (b) 3-D representation of the surface of a golf ball. Notice how the dimples in the surface are well resolved. (c) Gray scale representation of the phase map obtained for a plastic figure. (d) Two views of a 3-D reconstruction of the phase map in (c) for a rotation along the Y-axis, with a photograph of the figure overlaid.

the case of measuring an object that has holes on its surface or when measuring multiple objects which are disconnected (Figure 7.12). These difficulties are known as *noncontinuous difficulties*.

Shading problems on steep objects can only be avoided by making the angle between the projection and observation direction smaller than the smallest angle between the local surface normal and the X-axis for all points of the object surface. As seen before, measurement resolution depends on the angle between the projection

340 | *7 Fringe Projection Profilometry*

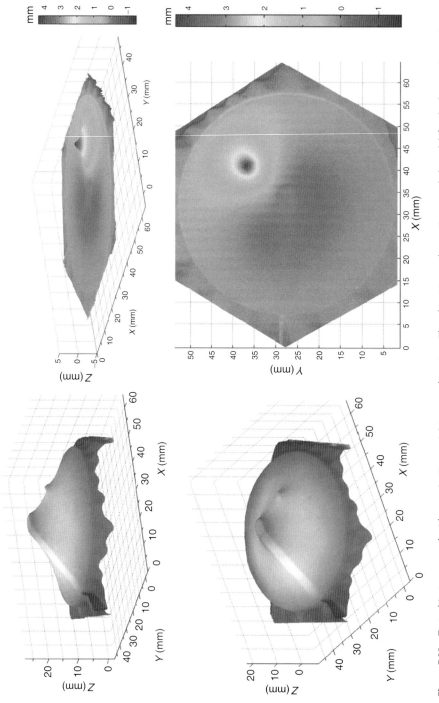

Figure 7.11 Test object measured with a projection moiré topography setup. The object is used as a 6× upscaled model for the human tympanic membrane. An oblique beam pushes the circular latex membrane outward, representing the malleus ossicle. A needle locally indents the surface during the measurement. (a,c) Two views of a 3-D surface shape reconstruction. (b,d) Two views of a 3-D reconstruction showing the indented membrane shape minus the membrane shape in resting position. Color represents deformation. (Please find a color version of this figure on the color plates.)

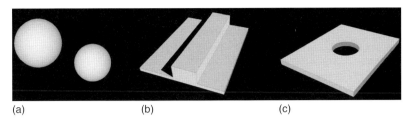

Figure 7.12 Three typical problematic situations in fringe projection profilometry. (a) Disconnected objects, such as two spheres, lose their relative Z-positioning. (b) Steep height changes or discontinuities cast shadows caused by the required (opening) angles in the setup, and thus have missing projected grid lines. (c) Objects with holes also cause missing projected grid lines.

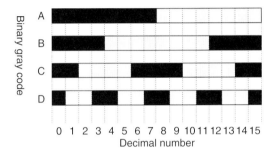

Figure 7.13 By projecting binary Gray code patterns on the object surface (sequentially A–D), every pixel along the X-axis receives a specific decimal number that is related to its fringe order number.

and observation direction, so there will always be a trade-off between accuracy and the steepest surface inclinations that can be measured. If this condition is not fulfilled, shadows occur, and data will be missing. When these regions are small, the local surface height in such zones can be obtained through (2-D) interpolation. Also, simple row- and column-based phase unwrapping will lead to errors. Either a sophisticated unwrapping algorithm is needed, or the operator will have to manually mask the zones of missing data so that the phase-unwrapping algorithm can skip these parts. In the case of FTP, it is advisable to switch to WFTP to reduce the contamination of the spectrum by noise and false data from shade.

Steep height changes along the Y-axis are not prone to the shading problem, as the projection and observation axis lie in the same XZ plane. Even discontinuities do not cause shade here. So, if an object mainly shows height variations or discontinuous steps along one direction (Figure 7.12b), it is advisable to align this direction with the Y-axis of the measurement setup.

Discontinuities or multiple nonconnected objects always pose problems (Figure 7.12a,c). In FTP, an object surface step or discontinuity can be resolved locally only if all projected fringes are resolved between the object surface points on both sides of a discontinuity. Otherwise, the Fourier analysis and filtering will reduce the sharp edge to a smoothed feature. When measuring multiple disconnected objects, FTP will again introduce noise and false data in the spectrum and both FTP and WFTP will most likely fail in correct phase unwrapping. Consequently, the multiple objects will lose their relative Z-distance. However, this can be corrected by determining the fringe (line pair) numbers from *Gray code* projections (Figure 7.13). The Gray code or *reflected binary code* is a binary numeral system where two successive values differ in only one bit, which reduces possible incorrect identification of the code. Before or after measuring the object surface, a series of black and white square grids is projected according to the Gray code principle. By matching one of these Gray code patterns with the projected grid frequency used in FTP (or in direct recording of the grid lines combined with phase-shifting), the (line pair) fringe number and thus the correct 2π interval is identified in which a pixel point should be situated [52]. Using this information in phase unwrapping, the relative height positioning of the objects can be corrected.

Phase-shifting moiré profilometry can manage discontinuities or multiple disconnected objects as long as their height variations remain smaller than the fringe plane distance. Otherwise, phase unwrapping also will lead to errors. As soon as the height difference becomes larger than a full fringe plane distance (and the discontinuity cannot be circumvented using for instance sideways 2-D phase unwrapping), the moiré method no longer resolves if one or a whole multiple of 2π radians lies between two object surface points. The solution then involves making two (or more) measurements with high- and low-frequency grid lines. The low-frequency grid needs to generate a fringe plane distance large enough to cover the two objects or the discontinuity. The resulting phase map which contains no phase jumps can then be used in the unwrapping of the second high(er) resolution phase map [9, 53–55].

7.6
Summary

Structured light projection techniques and especially fringe projection profilometry methods are important and popular methods for whole-field surface topography. Many papers on – or making use of – these methods can be found in the scientific literature. However, few give a full introductory step-by-step explanation of the theory.

In this chapter, we attempted to provide an introductory guide to fringe projection profilometry. Fourier profilometry and projection moiré profilometry were extensively discussed. Some inherent problems or choices related to fringe projection profilometry were also discussed, such as grid noise removal, phase-shifting algorithms, noncontinuous surfaces, etc.

A practical and low-cost setup was described, to implement a home-made profilometer using a commercial liquid crystal display projector. Both the Fourier and moiré approach could be implemented on this device. Finally, extensive literature references were provided to continue one's journey in the field of fringe projection profilometry.

Problems

7.1 Explain all the options and branches of the scheme in Figure 7.1 to a friend. Focus on the advantages of each branch.

7.2 Verify the formulae in Eqs. (7.19–7.21). Derive them from right-angled triangles hidden in Figure 7.6. Use the small inset at the top of Figure 7.6 as a guidance.

7.3 Suppose the setup in Figure 7.6 has an opening angle $\Omega = 20°$ and a magnification $M = 1$. The distance ℓ between the object reference plane and the plane of the projection and demodulation grid equals 70 cm. One wants to measure an object surface with height variation 5 cm. Furthermore, five dark fringes need to be present on this height range. Which grid period p does one need to use?

7.4 An optomechanical moiré setup uses long-travel piezoelectric positioning stages to wipe out grid noise. Both Ronchi rulings have a period $p = 7$ line pairs/mm. The stages on which the grids are placed move at a speed of $v = 0.85$ mm s^{-1} along the X-axis. How long does the camera need to expose its CCD?

7.5 A modern LCD projector with 120 Hz frame rate is used in semicontinuous grid averaging, and plays a movie in which eight images of a grid configuration are sequentially shifted in repeat. How long does the exposure time need to be to wash out all the grid noise and obtain a clear moiré fringe topogram?

7.6 Project a vertical square grid with a slide or LCD projector on a football, and take a picture of the deformed grid lines with a webcam or other photographic device at an angle of about 20°. Try to implement FTP as described in Section 7.2. Let Figure 7.4 guide you.

8
Thermoelastic Stress Analysis

Janice M. Dulieu-Barton

8.1
Introduction

The purpose of the chapter is to introduce the basics of the full-field, noncontact stress analysis technique known as *thermoelastic stress analysis* (TSA). The technique is based on the measurement of a small temperature change that occurs in an elastic solid as a result of a change in stress or strain. The temperature change is obtained using an infrared detector and is then calibrated in terms of stress. The chapter starts with an introduction to the theory of TSA for isotropic and orthotropic bodies. The basics of infrared thermography as applied to TSA are provided. Following on, the major assumptions in deriving the theory are discussed. The chapter has not been written as a review but as a tutorial for the topic. There are numerous references cited as background throughout the chapter. The chapter ends with an overview of progress and prospects for the future; here some important applications are described and the relevance of the technique to engineering community is demonstrated.

8.2
The Thermoelastic Effect

The relationship between mechanical deformation and thermal energy in an elastic solid is known as the *thermoelastic effect*. The first theoretical treatment of this phenomenon was published for a single material element by Weber and Thomson (Lord Kelvin) as early as 1855. Based on Kelvin's approach, it can be shown that, for a linear elastic homogeneous material, the rate of change of temperature ($\dot{T}$) is a function of the applied deformation in the form

$$\dot{T} = \frac{T_0}{\rho C_\varepsilon} \frac{\partial \sigma_{ij}}{\partial T} \dot{\varepsilon}_{ij} - \frac{\dot{Q}}{\rho C_\varepsilon} \quad \text{for} \quad i, j = 1, 2, 3 \tag{8.1}$$

where T is the temperature, T_0 is the absolute (reference) temperature, C_ε is the specific heat at constant strain, $\dot{Q}$ is the rate of heat loss per unit volume, ρ is the mass density, σ_{ij} is the stress tensor, and $\dot{\varepsilon}_{ij}$ is the rate of change of the strain tensor.

During TSA, the test specimen is dynamically loaded at a frequency high enough so that the heat transfer term ($\dot{Q}$) can be neglected, and hence the relationship given in Eq. (8.1) can be assumed to be adiabatic. The minimum frequency required is dependent on the thermal conductivity of the test material and stress gradients in the structure and is discussed in detail in a following section.

To develop Eq. (8.1) into a simple equation that can be applied experimentally, it is necessary to express the stresses in terms of strains and temperature to obtain the derivative $\partial \sigma_{ij}/\partial T$. For an isotropic, linear elastic material, the constitutive stress–strain–temperature relationships can be expressed in terms of the Lamé constants in the form

$$\sigma_{ij} = 2\mu\varepsilon_{ij} + (\lambda\varepsilon_{kk} - \beta\delta T)\delta_{ij} \tag{8.2}$$

for

$$\delta_{ij} = \begin{cases} 1 & \text{for} \quad i=j \\ 1 & \text{for} \quad i \neq j \end{cases}$$

and

$$\beta = (3\lambda + 2\mu)\alpha$$

where ε_{kk} is the first strain invariant (i.e., $\varepsilon_{11} + \varepsilon_{22} + \varepsilon_{33}$), δT is the change in temperature ($\delta T = T - T_0$), and α is the coefficient of linear thermal expansion. The Lamé constants λ and μ are functions of the Young's Modulus E and Poisson's ratio ν, as follows:

$$\mu = \frac{E}{2(1+\nu)}$$

and

$$\lambda = \frac{\nu E}{(1+\nu)(1-2\nu)} \tag{8.3}$$

The derivatives of stress with respect to T are obtained from Eq. (8.2) as follows:

$$\frac{\partial \sigma_{ij}}{\partial T} = 2\frac{\partial \mu}{\partial T}\varepsilon_{ij} + \left(\frac{\partial \lambda}{\partial T}\varepsilon_{kk} - \frac{\partial \beta}{\partial T}\delta T - \beta\right)\delta_{ij} \tag{8.4}$$

Equation (8.4) contains the *temperature derivatives of the material elastic properties*. For most engineering materials, the variations of the elastic properties with temperature are practically zero at room temperature, so in this treatment they are neglected. (The consequences of making this assumption are described in Section 8.9.) For a stress-induced temperature field, δT is in the order of millikelvin and therefore the term $(\partial \beta/\partial T)\delta T$ will be negligible compared to β, so this term is also neglected to give a simplified version of Eq. (8.4) as

$$\frac{\partial \sigma_{ij}}{\partial T} = -\beta \delta_{ij} \tag{8.5}$$

Substituting Eq. (8.5) into Eq. (8.1), and assuming *adiabatic conditions*, where $\dot{Q} = 0$, gives an expression for the rate of temperature change in terms of the material properties and the applied deformation, which can be written as

$$\dot{T} = -\frac{T_0 \beta}{\rho C_\varepsilon} \dot{\varepsilon}_{kk} \tag{8.6}$$

Expressing $\dot{\varepsilon}_{kk}$ in terms of stress using the relationship given in Eq. (8.2) yields

$$\dot{T} = -\alpha \left[\frac{T_0}{\rho C_\varepsilon} + \frac{1 - 2\nu}{3\alpha^2 E} \right] \dot{\sigma}_{kk} \tag{8.7}$$

where $\dot{\sigma}_{kk}$ is the rate of change of the sum of the first stress invariant $(\sigma_{11} + \sigma_{22} + \sigma_{33})$.

The bracketed term in Eq. (8.7) comprises mainly material constants. However, T_0 may vary from test to test and it would be convenient to express Eq. (8.7) as a linear function of the material properties. To do this, C_ε can be expressed in terms of the specific heat at constant pressure (C_p) using the relationship

$$C_\varepsilon = C_p - \frac{3 E \alpha^2 T_0}{\rho(1 - 2\nu)} \tag{8.8}$$

Therefore, Eq. (8.7) can be rewritten as

$$\dot{T} = -\frac{\alpha T_0}{\rho C_p} \dot{\sigma}_{kk} \tag{8.9}$$

Equation (8.9) directly relates the rate of change of temperature to the rate of change of stress in the specimen. Integrating this expression over a time period from the initial state to the final state of deformation provides a linear relationship between the change in temperature (ΔT) and the change in the first stress invariant ($\Delta \sigma_{kk}$) in the form

$$\Delta T = -K T_0 \Delta \sigma_{kk} \tag{8.10}$$

where K is known as the *thermoelastic constant* and is a function of the material properties as follows:

$$K = \frac{\alpha}{\rho C_p} \tag{8.11}$$

to yield the more familiar form of

$$\Delta T = -\frac{\alpha T_0}{\rho C_p} \Delta(\sigma_1 + \sigma_2) \tag{8.12}$$

Equation (8.12) is the standard form of the thermoelastic equation for an isotropic, homogeneous material loaded elastically, where it is assumed that the temperature change occurs adiabatically.

Tutorial Exercise 8.1

An image is obtained from an aluminum alloy aircraft structure during testing where the ambient temperature is measured as 27 °C. The aluminum alloy is reported to have a coefficient of thermal expansion of 23.5×10^{-6} K^{-1}, a density of 2700 kg m^{-3}, and a specific heat at constant pressure of 900 J kg^{-1} K^{-1}. During a tensile test, the yield strength of the aluminum alloy was determined as 110 MPa and Young's modulus of 68 MPa. In the image the maximum ΔT value is 0.377 K. Discuss the implications of this finding on the integrity of the component.

Solution:

From Eqs. (8.11) and (8.12)

$$K = \frac{\alpha}{\rho C_p} = \frac{23.5 \times 10^{-6}}{2700 \times 900} = 9.67 \times 10^{-6} \text{ MPa}^{-1}$$

Neglecting the negative sign in Eq. (8.12) (see Section 8.7 for explanation),

$$\frac{\Delta T}{KT} = \Delta(\alpha_1 + \alpha_2) = \frac{0.377}{9.67 \times 10^{-6} \times 300} = 130 \text{ MPa}$$

$\Delta(\alpha_1 + \alpha_2)$ is in excess of the yield stress of the material. However, this does not indicate that the component will yield, as the quantity is the sum of the principal stresses. If $\sigma_1 = \sigma_2$, then neither of the principal stresses is above the yield stress. If one of the principal stresses is negative, then this means that either may be above yield. The conclusion is that further investigation is required either by attaching a strain gauge rosette or by modeling. It should be noted that the negative sign in Eq. (8.12) has been ignored. This is because the phase of the response is set 180° out of phase to make a tension give a positive ΔT.

8.3
Infrared Thermography

A body with a temperature above absolute zero will emit energy in the form of electromagnetic radiation/thermal radiation at its surface. As the temperature increases, the quantity of heat transferred by means of thermal radiation increases. By using the amount of energy emitted in the form of electromagnetic radiation, accurate temperature measurements can be made over a wide range by means of infrared thermography. The spectral radiant emittance ($\Phi_{\lambda,b}$) for a blackbody in a hemisphere in the wavelength range from λ to $\lambda + \delta\lambda$ can be found using *Planck's law* [2], that is,

$$\Phi_{\lambda,b} = \frac{C_1}{\lambda^5 \left(\exp^{C_2/\lambda T} - 1 \right)} \tag{8.13}$$

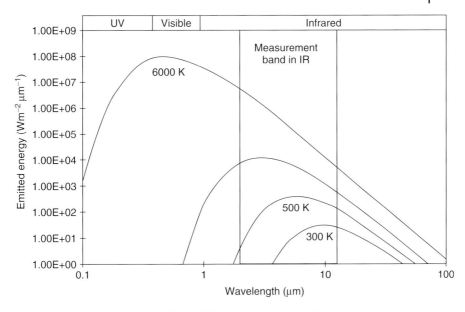

Figure 8.1 Radiant emitted energy from a blackbody according to Planck's law.

where C_1 is the first radiation constant $= 2\pi c^2 h$ (h is Planck's constant, c is the speed of light, and λ is wavelength), C_2 is the second radiation constant $= ch/k$ (k is the Boltzmann constant), and T is the absolute temperature.

The spectral radiant emittance for a blackbody at various temperatures can be plotted against wavelength as shown in Figure 8.1. It is possible to detect the radiant emittance by using an infrared detector. These are basically electronic devices that are sensitive to photon emission in the infrared spectrum, which has a peak between 2 and 12 μm at ambient temperature as shown in Figure 8.1. In modern infrared detection systems, the detector is built into a camera system with a lens and electronics, which enables the camera to have a computer interface. Low-sensitivity detectors are usually bolometers; higher sensitivity devices are usually *photon detectors* and these are used for strain measurement. The intensity of the photon collection defines the electrical signal output from the detector as the detector acts as a transducer turning a photon strike into a voltage signal.

By integrating Planck's law (Eq. (8.13)) over all wavelengths between zero and infinity, it is possible to obtain a finite integral and obtain the well-known fourth-power *Stefan–Boltzmann* relationship for evaluating the radiant emittance Φ_b as follows:

$$\Phi_b = 2\pi c^2 h \int_0^\infty \frac{d\lambda}{\lambda^5 \left(\exp\left(ch/kT\lambda \right) - 1 \right)} \tag{8.14}$$

Letting $x = ch/kT\lambda$ allows the integration to be processed by substitution with respect to x. Thus, x must be rewritten to make λ the subject and subsequently differentiated with respect to x. That is: $\lambda = ch/kTx$, $d\lambda/dx = ch/kTx^2$, and $\lambda^5 = (c^5h^5)/(k^5T^5x^5)$.

Substituting the relationship for λ^5 and the derivative of λ with respect to x into Eq. (8.14) leaves

$$\Phi_b = \frac{2\pi c^3 h^2 k^5 T^5}{c^5 h^5 kT} \int_0^\infty \frac{x^5 \, dx}{x^2 \left(\exp\left(ch/kT\lambda\right) - 1 \right)} \tag{8.15}$$

and Eq. (8.15) can be simplified to

$$\Phi_b = \frac{2\pi k^4 T^4}{c^2 h^3} \int_0^\infty \frac{x^3 \, dx}{\exp(x) - 1} \tag{8.16}$$

Using the integral in Eq. (8.16), the radiant emittance can be written as follows:

$$\Phi_b = \left(\frac{2\pi^5 k^4}{15 c^2 h^3} \right) T^4 \tag{8.17}$$

This result can be simplified by rewriting the right-hand side of Eq. (8.20) through the introduction of the Stefan–Boltzmann constant B as the bracketed term to give the fourth-power Stefan–Boltzmann relationship for evaluating the radiant emittance over all wavelengths λ_b as follows:

$$\Phi_{\lambda,b} = BT^4 \tag{8.18}$$

It can be shown that $\Phi_{\lambda,b}$ has a maximum. The wavelength at which this occurs is defined as λ_{max}, and can be calculated from Wien's displacement law [2].

In a similar manner, it is possible to obtain a discrete equation for the number of photons (N_b) emitted by an object at a specific temperature by dividing the energy in each wavelength interval by the energy carried by each photon. To evaluate the relationship for a general case, the photon flux can be derived for the total number of photons per unit area and time by producing a closed-form integral of the equation for spectral radiant emittance; again, this is only possible by considering the wavelength range between zero and infinity, as

$$N_b = \int_0^\infty \frac{2\pi c}{\lambda^4 \left(\exp\left(hc/\lambda kT\right) - 1 \right)} \, d\lambda \tag{8.19}$$

The integration is possible using the substitution method and the subsequent derivatives presented for Eq. (8.15). Noting that here λ is raised to the power 4, in this case the λ terms can be rewritten for this derivation as

$$\lambda^4 = \frac{c^4 h^4}{k^4 T^4 x^4} \tag{8.20}$$

Substituting into Eq. (8.19) leaves

$$N_b = \frac{2\pi c k^4 T^4}{c^4 h^4} \frac{ch}{kT} \int_0^\infty \frac{x^4}{x^2 (\exp((x) - 1)} dx \tag{8.21}$$

which again can be simplified as

$$N_b = \frac{2\pi k^3 T^3}{c^2 h^3} \int_0^\infty \frac{x^2 \, dx}{\exp(x) - 1} \qquad (8.22)$$

The integral in Eq. (8.22) may be evaluated in a similar manner to that used previously to give

$$N_b = \frac{2\pi k^3 T^3}{c^2 h^3}(2.4041) \qquad (8.23)$$

This formulation again allows the substitution of the *Stefan–Boltzmann* constant to provide the following relationship between the number of photons incident and the surface temperature:

$$N_b = \frac{0.370 B}{k} T^3 \qquad (8.24)$$

The quantity $0.370 B/k = 1.52 \times 10^{15}$ photons s^{-1} m^{-3} sr^{-1} K^{-3} can be regarded as the *Stefan–Boltzmann constant for photodetectors*. Denoting the constant for photodetectors as B', Eq. (8.24) simplifies to

$$N_b = B' T^3 \qquad (8.25)$$

The relationship presented in Eq. (8.25) shows that, when considering the entire electromagnetic spectrum, the total number of photons increases with the cube of the absolute temperature, whereas the radiant emittance over the entire spectrum increases with the fourth-power of absolute temperature. The relationship for the number of photons (N_b) emitted from a surface as derived above and presented in Eq. (8.25) has been used to relate the photon output from an infrared detector to the temperature, but these should be regarded as an approximation. For infrared thermography, there are two atmospheric windows of interest: one is located from 3 to 5 μm (most bolometers and photodetectors) and the other is from 8 to 12 μm (some photodetectors). The number of photons per unit area and time is obtained by integrating the spectral photon emittance over the operating wavelength band of the photodetector instead of the range $0 - \infty$. It is impossible to derive a closed-form relationship, as presented for the radiant emittance or photon flux above, for practical narrow-band IR detectors. Therefore, it is necessary to consider other mathematical methods to express the relationship between the spectral radiant power and the absolute temperature between the wavelength limits of the infrared detectors of interest. Clearly, this will not yield the same result given by the formulation of Eq. (8.25). For detectors where the operating wavelength is much less than λ_{max}, then the response of the detector to temperature changes follows an approximate power law, that is, $N_b \propto T^n$ [3]. Therefore, it is possible to propose an expression that relates the surface temperature of a body to the total number of photons emitted over a particular wavelength range as follows:

$$N_{b_\lambda} = B' T^n \qquad (8.26)$$

where B' is a constant that is dependent on the detector. The index n can be evaluated by determining N_{b_λ} from Eq. (8.19) by numerical integration over the wavelength of interest for a variety of temperatures. Taking logs of Eq. (8.26) will yield a simple linear form, that is,

$$\ln N_{b_\lambda} = \ln B' + n \ln T \tag{8.27}$$

where n can be determined from the slope of a plot of $\ln N_{b_\lambda}$ against $\ln T$ and B can be derived from the intercept of the plot. Equation (8.30) can be used as a basis for calibration; however, most instruments are supplied with empirical calibration curves from the manufacturer.

When radiation impinges on a body, it is either transmitted through the body, absorbed by the body, or reflected away from the body, so that

$$\tau + a + r = 1 \tag{8.28}$$

where τ is the transmissibility, a is the absorptivity, and r is the reflectivity.

Engineering materials are usually opaque in the infrared region, even if they are transparent to visible light, for example, glass. The transmitted energy is therefore zero. So Eq. (8.28) can be rewritten as

$$a + r = 1 \tag{8.29}$$

The reflectivity is then $1 - a$, so a portion of incident radiation is reflected back to the detector. Care must be taken to avoid anomalous readings from reflections from heat sources. The absorbed energy is therefore equal to the emitted energy, and so

$$a = e \tag{8.30}$$

where e is the emissivity.

The variability in emissivity means that care is required to interpret the measured temperature and to take account of background radiation. Often, a layer of matt black paint is applied to the surface of a material to create an enhanced and uniform emissivity. It is interesting that this is not usually necessary for polymeric materials. Infrared systems are usually calibrated using a blackbody; therefore, it is necessary to input a surface emissivity before calibrating the response into a temperature value.

8.4
Obtaining Thermoelastic Measurements from an Infrared System

In early work, the total radiant flux emitted from a surface was used to develop a working relationship for thermoelastic studies [4]. It follows by differentiation of the standard Stefan–Boltzmann relationship that the flux change $\Delta\Phi$ resulting from a small change in the surface temperature ΔT is given by

$$\Delta\Phi = 4eBT^3 \Delta T \tag{8.31}$$

The surface emissivity e is included, which is important to consider in TSA, as it is probable that the surface will not behave like a blackbody.

If the flux change is recorded by a linear detecting system, the detector voltage output (S) will be proportional to the change in temperature, and therefore it follows from Eq. (8.31) that the change in the principal material stresses is given by

$$S = -\frac{4R^* e B \rho C_p T^4}{\alpha}(\Delta\sigma_1 + \Delta\sigma_2) \tag{8.32}$$

where R^* is some detector response factor for the system.

Grouping the variables before the bracket on the right-hand side of Eq. (8.32) as those dependent on the material under test and the settings of the detector system, a calibration constant A is defined. The general thermoelastic relationship is therefore as follows [4]:

$$AS = (\Delta\sigma_1 + \Delta\sigma_2) \tag{8.33}$$

This treatment neatly obviates the inaccuracies associated with using the simple Stefan–Boltzmann relationship. However, the calibration constant has to be defined for every surface temperature, as there is a very strong detector dependence on surface temperature particularly for the detectors that operate on the 2–5 μm wavelength range. Equation (8.33) has been used as the basis for numerous studies spanning almost three decades. Since the initial validation in [4], it has been demonstrated that the approach can be used in a wide range of applications; some example applications are provided throughout the chapter, particularly in the final section.

Orthotropic Materials

The simple thermoelastic theory devised for an isotropic body in Eq. (8.36) is not valid for orthotropic materials [5]. For orthotropic materials, the following equation is used [5]:

$$(\alpha_1 \Delta\sigma_1 + \alpha_2 \Delta\sigma_2) = A^* S \tag{8.34}$$

where $\Delta\sigma$ is the change in the direct surface stress, A^* is a further calibration constant, and the subscripts 1 and 2 denote the principal material directions of the surface lamina.

Stanley and Chan [5] validated Eq. (8.34) using two types of composite component. Potter [6] proposed a thermoelastic theory relating the thermoelastic output to that of the surface strains and demonstrated its validity on a carbon fiber/epoxy resin laminate. Bakis and Reifsnider [7] investigated the influence of material inhomogeneity and anisotropy using carbon-fiber-reinforced plastics. It was found that the thermoelastic response was affected by a number of factors, which included the volume fraction, the thermoelastic properties of the microconstituent materials, the orientations of the laminae within the laminate, and the orientation of the lamina on the surface.

8.5
Temperature Dependence of Thermoelastic Response

To derive a surface temperature correction factor that could be realistically inserted into Eqs. (8.33) and (8.34), an expression that is just a function of temperature is desirable. An approximate approach has been suggested [8] for detectors where the operating wavelength is less than λ_{max}. Here, the response of the detector to temperature changes follows an approximate power law, that is, $N_b \propto T^n$. Equations (8.26) and (8.27) can be used to provide a relationship [8] for the surface temperature of a body to the total number of photons emitted over a particular wavelength range. Equation (8.27) shows that the index n and B' can be obtained from a log–log plot of $\ln N_{b_\lambda}$ against $\ln T$; B can be derived from the intercept of the plot.

For a specimen with an absolute temperature of 293 K, λ_{max} occurs at 9.89 μm. Therefore, as the power law is valid only for cases where the operating wavelength of the detector is less than λ_{max}, at around room temperature the relationship given by Eq. (8.26) is valid for the 2–5 μm range but not in the 8–12 μm range. A numerical integration of Planck's law over the 2–5 μm range for temperatures of 293–323 K using MATLAB and application of the above procedure [8] yielded $B = 6.322 \times 10^6$ photons s^{-1} m^{-3} sr^{-1} K$^{-10.47}$ and $n = 10.47$; the correlation coefficient for the curve fit was 0.99. Therefore, a temperature correction factor can be introduced as $(T_0/T)^{10.47}$. There is a huge difference between the standard approach using the photon detector law and calculating over the correct wavelength range. Hence Eqs. (8.33) and (8.34) should be used with caution. In view of this, it is much better to have an empirical radiometric calibration against a blackbody and apply Eq. (8.10) directly. The validations carried out in [5] used a detector that worked in the 8–12 μm range. Here, the temperature sensitivity is less and the temperature correction follows approximately a cube law.

8.6
Derivation of the Thermoelastic Constant

Until recently, the approach in TSA was to derive the calibration factor (A in Eq. (8.33)). The three principal calibration techniques used to derive a value for A, are as follows [9]:

1) Direct calibration, using properties of the infrared detector, system variables, specimen surface emissivity, and the thermoelastic constant of the specimen material;
2) Calibration against measured stress;
3) Calibration against calculated stress.

The direct calibration (method (1)) is based on solving Eq. (8.32) and utilizes both material and detector properties to theoretically derive a value for A. This method is not regarded as the most accurate. The second method utilizes an independent measure of strain (typically an electrical resistance strain gauge). The sum of the principal stresses can be determined using the strain measurements and Hooke's law; however, values for the material's Young's modulus and Poisson's ratio are required. The third technique used for calibration requires a known stress field. Normally, a specimen constructed from the material in question is loaded in simple tension. (A beam in four-point bending and a disc in two-point diametral compression are other recommended arrangements.) The stress can be calculated theoretically using the cross-sectional area and the applied load. Equation (8.33) is simplified to find a value for A. This method has the least potential for error sources. Techniques based on these calibration approaches are currently being developed into a standard for TSA measurements [10].

Identical approaches can be used to find the thermoelastic constant K. Here, Eq. (8.12) is used, and a known stress is applied to a specimen; ΔT and T are obtained from the measurement and therefore K is obtained.

Tutorial Exercise 8.2

An independent verification of the thermoelastic constant K of the aluminum alloy used in Tutorial Exercise 8.1 is required. This is to be done with a strip of the alloy 15 mm wide by 5 mm thick loaded in uniaxial cyclic tension with a load cycle of 4 $\pm$2.5 kN at 10 Hz. During the test, room temperature is measured as 21 °C and the average ΔT values over the uniform area of the strip is obtained at 0.180 $\pm$0.005 K. Derive the value of K and discuss this in the context of the value of K obtained in Tutorial Exercise 8.1.

Solution:

For an uniaxial test, the stress is obtained from the force and geometrical values:

$$\Delta \sigma_1 = \frac{5 \times 10^3}{15 \times 5} = 66.7 \text{ MPa}$$

$$K = \frac{\Delta T}{T \Delta (\sigma_1 + \sigma_2)} = \frac{0.180}{294 \times 66.7} = 9.20 \times 10^{-6} \text{ MPa}^{-1}$$

The K value obtained from the experiment is 5% less than that calculated from the literature. This is typical, as the exact grade of aluminum alloy is not known and even so properties vary with the grade. It is essential that a calibration test piece is made from the same billet of aluminum alloy from the same rolling direction.

A number of quantitative studies have been carried out on orthotropic composite structures, for example, by deriving a calibration constant [11–13]. A generalized calibration routine was suggested in [14] but has not been fully developed. The routine was based on the response from the resin-rich surface layer and assumes that the response is that of the resin alone acting as strain witness. This may only be true for certain material combinations but has also been observed by other researchers [15, 16]. A general conclusion is that it is not possible at present to apply TSA in a straightforward manner to a general composite structure and obtain quantitative stress or strain values.

8.7
Nonadiabatic Conditions

Nonadiabatic conditions exist when the rate of heat flux per unit volume ($\dot{Q}$) is different from zero. It is impossible to achieve fully adiabatic conditions experimentally; however, for the purposes of TSA a *pseudo-adiabatic* state may be achieved, where there is no measurable attenuation of the thermoelastic signal because of heat transfer. To understand the limitations of the pseudo-adiabatic assumption, it is useful to use the following expression:

$$\dot{T} = -\frac{\alpha T_0}{\rho C_p}\dot{\sigma}_{kk} - \frac{\dot{Q}}{\rho C_\varepsilon} \tag{8.35}$$

The magnitude of $\dot{\sigma}_{kk}$ can be amplified by increasing the loading frequency and thereby making it much greater than the $\dot{Q}$ term. This is usually the approach taken to obtain a pseudo-adiabatic state, with a loading frequency of 10–20 Hz being sufficient for most applications.

$\dot{Q}$ is a function of the thermal conductivity of the material and the stress-induced thermal gradients in the component, and can be expressed in the form

$$\dot{Q} = k\nabla^2 T \tag{8.36}$$

where k is the material thermal conductivity and $\nabla^2 T$ is the thermal flux defined as

$$\nabla^2 T = \frac{\partial^2 T}{\partial x^2} + \frac{\partial^2 T}{\partial y^2} + \frac{\partial^2 T}{\partial z^2} \tag{8.37}$$

Analysis of Eq. (8.35) shows that, for a specimen with a uniform stress field or a low k value, the $\dot{Q}$ term will be small. In both cases, a pseudo-adiabatic state is achieved at low loading frequencies. However, most engineering components have stress gradients and are made from metals that have a high thermal conductivity.

Finite element analysis (FEA) may be used to analyze nonadiabatic affects at loading frequencies that cannot be achieved experimentally. It may be used to determine the loading frequency where $\dot{Q}$ can be neglected for a given stress field and a given set of material properties. In close proximity of stress raisers, attenuation of the thermoelastic signal is an important consideration, even at loading frequencies of 20 Hz and above. Often, the attenuation effects are localized and TSA can still be used to obtain quantitative data from most specimens. This is verified by the large number of publications that have successfully used TSA to obtain quantitative data at loading frequencies less than 15 Hz.

Nonadiabatic behavior can be investigated experimentally by examining the thermoelastic response over a range of frequencies. If the response is independent of frequency, it can be assumed that a pseudo-adiabatic state has been achieved. Furthermore, it is possible with the lock-in thermography system required for TSA not only to obtain the magnitude of the thermoelastic response but also its phase relative to the applied cyclic load. As a positive stress change results in a decrease in response (i.e., a cooling analogous to the expansion of a gas and denoted by the minus sign in Eq. (8.12)), the stress change is always 180° out of phase with the temperature change. However, in standard practice the measurement system is usually set so that a tensile stress change is in phase with the temperature change and hence the minus sign in Eq. (8.12) is often neglected. The phase provides a convenient tool for assessing nonadiabatic behavior, as any phase lead or lag indicates that the stress change is not in phase with the temperature change and Eq. (8.12) is not valid. Heat conduction in areas of high stress gradient causes changes in the magnitude of the thermoelastic response as well as a phase shift. Discontinuities in the phase image can be used to identify nonadiabatic regions in a specimen. This is evident in Figure 8.2, which shows the magnitude and phase images from a strip of material loaded in uniaxial tension. Here, the area above and below the hole, which are in compression, are out of phase with the rest of the plate. Moreover, close to edge of the hole (right and left in the image), areas of out-of-phase signal are apparent, where the stress gradient is the greatest.

The resolution of the detection system is also a consideration when obtaining pseudo-adiabatic conditions. TSA of small-scale structures requires a detection system with a high spatial resolution to ensure that an adequate number of measurements are obtained across the structure to build an accurate representation of the signal distribution. However, at high resolutions each detector element is projected over a very small area on the test specimen. The average temperature in the projected area is used to derive the thermoelastic signal. The smaller the measurement area, the greater effect the $\dot{Q}$ term on the thermoelastic measurements. At high detector resolutions, acute stress gradients are not averaged out over large measurement areas and therefore the sensitivity of the signal to nonadiabatic effects is increased. A treatment for establishing whether the response is adiabatic has recently been proposed [17]. Here, small discs were examined, and it was shown that by using a combination of the thermal diffusion properties and the specimen geometry a procedure could be applied to identify whether adiabatic conditions had been achieved.

Bakis and Reifsnider [7] investigated the limitations of Eq. (8.34) in terms of the adiabatic thermoelastic assumption made in its development. For composite materials, it was suggested that the nonadiabatic behavior in carbon-fiber-reinforced plastic (CFRP) laminates could be due to heat transfer between the fiber and matrix or caused by viscoelastic effects. The former was discounted [19] for fibers of diameter ≈ 7 μm, which is typical for carbon fibers. Wong et al. [19] discussed the effects of nonadiabatic conditions on the thermoelastic signal recorded from the specimen surface due to heat transfer characteristics at large stress gradients, such as those experienced between plies orientated at different angles in a laminate.

TSA is widely considered as a technique for establishing surface strains or stresses. However, it was suggested in [20] that the nonadiabatic response could be used to establish subsurface stresses. Some initial work on identifying subsurface damage is described in [21], which demonstrates the potential of the technique for materials with high thermal conductivities.

8 Thermoelastic Stress Analysis

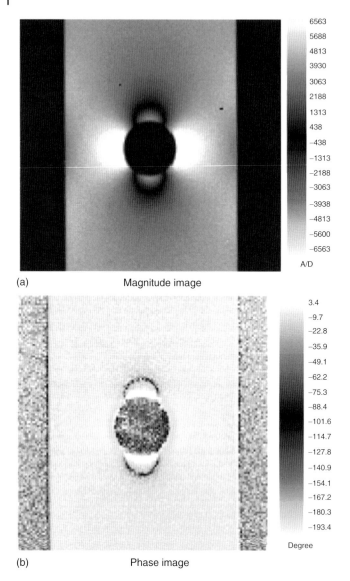

(a) Magnitude image

(b) Phase image

Figure 8.2 TSA data from hole in a plate specimen. (Please find a color version of this figure on the color plates.)

8.8
Paint Coatings

It is a standard practice when taking thermographic measurements to coat the specimen surface with a thin layer of matt black paint to improve emissivity and ensure that the response is uniform. However, this coating is thermally inert in comparison with the specimen and may, in certain instances, give rise to nonadiabatic effects. Thermal lag [22] is caused by the insulating effect of the paint coating. Even with no heat loss to the environment, a surface coating will still have a capacitance and resistance associated with it, which will result in a temperature drop across it. In short, an increase in paint coating thickness will result in a smaller ΔT at the paint surface than at the substrate surface. This is significant for small-scale specimens where it can be difficult to apply an even paint coating, especially in areas where there are sharp changes in section. In general, most practitioners will recommend a minimal amount of paint [23]. Two short passes of paint supplied using an aerosol is deemed suitable for most applications.

Thermal drag-down [22] is a phenomenon relating to both the paint thickness and the loading frequency. Since no heat is being generated by the coating, its temperature can only change because of the heat transferred to it from the substrate. As the loading frequency increases, the spatial wavelength decreases; as a result, there is less heat input into the coating. In brief, the heat will not flow into the surface coating sufficiently quickly to maintain the temperature change for accurate thermoelastic measurements to be recorded.

A theoretical treatment for covering the effects of paint coating has been developed [24], where four distinct regions of coating response were identified:

- *High-emissivity coating region*, where the temperature change on the surface of the coating is identical to that on the surface of the material;
- *Opacity-limited region*, where the coating is so thin that the surface emissivity is a combination of the paint coating and the specimen surface;
- *Coating diagnostic region*, where thermal drag-down is experienced as the loading frequency is increased;
- *Strain witness region*, where the coating is so thick, or loading frequency so high, that the thermoelastic response is from the coating alone.

A detailed experimental study of the effect of paint coating on a variety of materials is provided in [25]. A typical plot showing contours of ΔT for changing frequency and paint thickness is shown in Figure 8.3. Here, it can be seen clearly that the desirable high-emissivity region occurs to the top left of the plot and it is likely, except for very thin coatings, that the measurements will be taken in the coating diagnostic region where the response is dependent on frequency because of thermal dragdown.

8.9
Temperature Dependence of the Material Elastic Properties

For the small changes in temperatures associated with the thermoelastic effect, the thermoelastic constant K is assumed to be a material constant that is independent of the stress field. However, K is a function of the coefficient of linear expansion α, which has been shown to be stress dependent as follows:

$$\left(\frac{\partial \alpha}{\partial \sigma}\right)_T = \frac{\partial}{\partial \sigma}\left[\left(\frac{\partial \varepsilon}{\partial T}\right)_\sigma\right]_T = \frac{\partial}{\partial T}\left[\left(\frac{\partial \varepsilon}{\partial \sigma}\right)_T\right]_\sigma \tag{8.38}$$

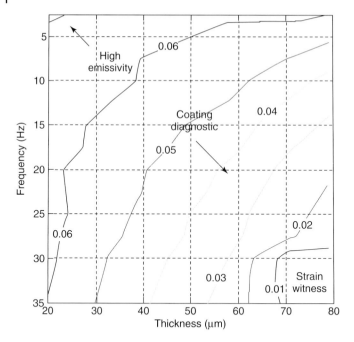

Figure 8.3 Contour plot of experimental response with changing frequency (2.5–35 Hz) and paint thickness (20–80 μm).

Therefore, if $E = (\partial \sigma / \partial \varepsilon)_T$, differentiating by parts gives

$$\left(\frac{\partial \alpha}{\partial \sigma}\right)_T = -\frac{1}{E^2}\left(\frac{\partial E}{\partial T}\right)_\sigma \tag{8.39}$$

Equation (8.39) shows that α will be stress dependent for materials whose elastic properties are sensitive to changes in temperature. This also implies that the thermoelastic response of such materials will be dependent on the mean stress of the applied stress cycle and cannot be assumed to be constant.

The theoretical approach presented previously assumes that the dependence is a higher order effect and can be neglected for most engineering materials. Although this hypothesis has been shown to be valid for most applications, it may not be valid for some materials, particularly those with high-temperature sensitivity used in components where a high mean stress is applied, in which case the following equation must be used [26]:

$$\dot{T} = \frac{T_0}{\rho C_\varepsilon}\left[\left(-\beta - \frac{\partial \beta}{\partial T}\delta T + \frac{\partial \lambda}{\partial T}\varepsilon_{kk}\right)\delta_{ij} + 2\frac{\partial \mu}{\partial T}\varepsilon_{ij}\right]\dot{\varepsilon}_{ij} - \frac{\dot{Q}}{\rho C_\varepsilon} \tag{8.40}$$

The $(\partial \beta / \partial T)\delta T$ term will be negligible compared to β and can be neglected. However, for some materials the terms $(\partial \mu / \partial T)\varepsilon_{ij}$ and $(\partial \lambda / \partial T)\varepsilon_{kk}$ can be of a

8.9 Temperature Dependence of the Material Elastic Properties

significant order. Therefore, retaining these terms and omitting higher order components allows Eq. (8.42) to be written in terms of stresses for adiabatic conditions as follows:

$$\dot{T} = \frac{T_0}{\rho C_p} \left\{ -\left(\alpha + \left(\frac{\nu}{E^2} \frac{\partial E}{\partial T} - \frac{1}{E} \frac{\partial \nu}{\partial T} \right) \sigma_{kk} \right) \dot{\sigma}_{kk} \right.$$
$$\left. + \left(\frac{(1+\nu)}{E^2} \frac{\partial E}{\partial T} - \frac{1}{E} \frac{\partial \nu}{\partial T} \right) \sigma_{ij} \dot{\sigma}_{ij} \right\} \quad (8.41)$$

Equation (8.41) is known as the *revised version of the thermoelastic equation*, where the rate of temperature change is both a function of the stresses and their rate of change. It should be noted that C_p is used in Eq. (8.41). The higher order terms that constitute the difference between C_ε and C_p are neglected to allow Eq. (8.41) to be presented in a concise form. The difference between C_ε and C_p can be calculated using Eq. (8.8) and is negligible for most metals, and hence C_p is used here for consistency.

To illustrate more clearly the effects of the stresses in Eq. (8.41), a uniaxial stress system is considered, where $\sigma_{11} = \sigma_{kk}$ and $\sigma_{22} = \sigma_{33} = \sigma_{12} = \sigma_{23} = \sigma_{13} = 0$. Under these conditions, Eq. (8.41) can be written as follows [27]:

$$\dot{T} = -\frac{T_0}{\rho C_p} \left(\alpha - \frac{1}{E^2} \frac{dE}{dT} \sigma_{11} \right) \dot{\sigma}_{11} \quad (8.42)$$

Equation (8.42) shows that, for the revised treatment, the rate of temperature change is a function of the applied stress and its rate of change. It is now possible to integrate Eq. (8.42) over the time period corresponding to the deformation in a similar manner to the approach with Eq. (8.9). However, in a simple deformation process it is difficult to see what value should be used for the σ_{11}, that is, a static stress, which is independent of time. In a cyclic load, as with that used in TSA, σ_{11} could be regarded as the mean stress of the cycle σ_m, as this does not vary with time, and therefore Eq. (8.42) can be integrated to give

$$\Delta T = -\frac{T_0}{\rho C_\varepsilon} \left(\alpha - \frac{1}{E^2} \frac{dE}{dT} \sigma_m \right) \Delta \sigma_{11} \quad (8.43)$$

so that

$$\Delta T = -K^* \Delta \sigma_{11} \quad (8.44)$$

where K^* is the revised version of the thermoelastic constant, which is a function of the Young's modulus and the mean stress and can be written as

$$K^* = \frac{1}{\rho C_p} \left(\alpha - \frac{1}{E^2} \frac{dE}{dT} \sigma_m \right) \quad (8.45)$$

Equation (8.45) provides an expression for the thermoelastic constant that takes the stress dependence of α into account by incorporating the quantity $(1/E^2)dE/dT$ (Eq. (8.43)) and the σ_m term. If the material's Young's modulus is high and the dE/dT term is small, then the mean stress will have a negligible effect on the thermoelastic response. The stress dependence of the thermoelastic constant for

Table 8.1 Material properties for three high-strength alloys.

Material	E (MPa)	dE/dT (MPa K^{-1})	σ_y (MPa)
4340 steel	210×10^3	−56.7	304
Al-2024	72×10^3	−36.0	197
Ti–6Al–4V	120×10^3	−61.8	430

a variety of materials is given in Table 8.1. The percentage difference between K^* values calculated at zero stress and at the yield stress was evaluated for each material. The greatest dependence (40%) was calculated for Ti–6Al–4V, with values of 10 and 14% obtained for Al-2024 and 4340 steel, respectively. Analysis of the material properties given in Table 8.1 shows that relative to the other materials Ti–6Al–4V has a relatively low E value combined with a high dE/dT and yield strength values.

Tutorial Exercise 8.3

It is known that the aluminum alloy used in Tutorial Exercise 8.1 has a dE/dT value of 0.7×10^{-4} MPa K^{-1}. Assuming uniaxial conditions, discuss how this will influence the TSA data.

Solution:

From Eq. (8.45)

$$\Delta T = -\frac{T_0}{\rho C_\varepsilon}\left(\alpha - \frac{1}{E^2}\frac{dE}{dT}\sigma_m\right)\Delta\sigma$$

The worst case scenario is that $\sigma_m = (\Delta(\sigma_1+\sigma_2)/2)65$ MPa. The error in ΔT by neglecting the effect of σ_m is

$$\left(\frac{\Delta(\sigma_1+\sigma_2)}{2}\frac{1}{\alpha E^2}\frac{dE}{dT}\sigma_m\right) = 65 \times \frac{1}{23.5 \times 10^6 \times 60} \times 0.7 \times 10^{-4} = 0.05 \text{ or } = 5\%$$

The practical significance of the mean stress effect for the materials given in Table 8.1 can be established by calculating the percentage of the material's yield stress required to introduce an error of 5% by neglecting the mean stress dependence. Calculations were carried out for steel, an aluminum alloy, and a titanium alloy, with values of 29, 17, and 5% obtained, respectively. In practical terms, this analysis shows that, at least in the case of steel and aluminum, the mean stress effect will be effectively hidden within the ambient noise of the measurements (approximately 5%). The effect is shown to be significant in the case of the titanium alloys where a mean stress that is just 5% of the yield stress is required to introduce an error of 5% in the thermoelastic measurements. It has been shown that for nickel–titanium alloys (shape memory alloys) the effect can be significant depending on the heat treatment the material has received this rendered quantitative analysis impossible because of the coupling of the mean stress and the stress change in Eq. (8.41).

It has been suggested [26] that the mean stress effect can be used to detect residual stresses. As residual stress is essentially a mean stress, it is accepted that the linear form of the TSA relationship given in Eq. (8.9) does not allow its evaluation. As described above, the changes in the thermoelastic response resulting from the inclusion of residual stress produce temperature change differences of a few millikelvin, which are significantly less than those expected to be resolved in standard TSA. At present, three approaches have been investigated as potential candidates for residual stress measurement using the thermoelastic response. Two are based on the mean stress effect and the revised higher order thermoelastic effect [26]. One utilizes the thermoelastic response at the second harmonic of the loading frequency [29], and the other directly relates the change in the thermoelastic response to the principal stresses [30]. The major limitation of these two approaches is that they are not suitable for steel components since the temperature dependence of the elastic properties of steel is negligible at room temperature. The third approach [31] is based on Eq. (8.9) and the change in the thermoelastic constant K resulting from plastic deformation during manufacture or assembly. In the third approach, the main disadvantage is that plastic deformation must have taken place, but it has the advantage that it may be valid for a larger range of materials, not just those with temperature-dependent elastic properties. A significant disadvantage common to all three approaches is that any change in the thermoelastic response resulting from either the mean stress σ_m or from the modification of K will be small. In actual components, the changes in the response are around the noise threshold of the detectors. Success in detecting these changes has been achieved by applying very large residual stress or plastic strain, by using materials that are very sensitive to the mean stress effect, or from investigation of specifically designed specimens. Recently, the sensitivity of infrared detectors has improved to the extent where it may be possible to accurately measure changes representative of those in actual components, thereby leading to a renewed interest using TSA for residual stress analysis. Since the variations in thermoelastic response are small, it is important to minimize sources of signal attenuation and to understand the possible sources of error within the measurement. The major factors known to influence the change in thermoelastic response (and subsequently K) are the high-emissivity coating, background temperature, applied stress, and the infrared detector settings; these effects must always be considered in parallel to any evaluations of the significance of changes in response due to mean stress or from the modification to K.

8.10 Progress, Applications, and Prospects

In 1990, TSA was the subject of a book edited by Harwood and Cummings [32]. It captured the then state of the art and covered aspects such as the mean stress effect, applications to composite materials, paint coatings, and nonadiabatic behavior. It is an excellent publication, and most of the material is relevant today. Following on from this, there have been two good review papers: one in 1998 [33] that covered the development and applications on the technique, and another in 2001 [34] that brought together the theory for isotropic materials. In 2008, a special issue of the *Journal of Strain Analysis* (Vol. 43, No. 6), edited by the author, was published on the topic, which demonstrated the range of diversity of applications of TSA.

One of the most focused applications has been on crack-tip stress studies, with the intention of deriving the stress intensity factors (SIFs) from thermoelastic data alone (see, for example, [35–41]). The original treatment was based on the first-order Westergaard equations [35, 36], or equivalently, the singular terms in the eigenfunction expansion of Williams. This was demonstrated in [36], where the SIFs were obtained for simulated cracks (in the form of spark eroded slots) in mode I and mixed mode opening. The analysis in [36] made use of the cardioid geometry of the isopachics by deriving expressions for the SIFs in terms of the cardioid area and the positions of certain tangents to the curve. This treatment was fully exploited by developing algorithms and software that automatically obtained the SIFs from the thermoelastic data and demonstrated that the technique could be applied to actual cracks. Others have developed successful techniques for crack-tip stress studies, for example, [37, 38], based on the Mushkelishvili stress function. Recent work [39] directly fitted the cardioid form to extracted isopachics to estimate the crack-tip SIFs. The fitting was performed using a genetic algorithm (GA). An added advantage of the approach is that the crack tip is located in the data field, which can be difficult to detect by visual inspection. The technique has been applied to fatigue crack growth [40]. Most recently, attempts have been made to derive the higher order terms in the Westergaard equations [41]. The work is continuing, and the prospect of deriving very accurate SIFs from growing cracks in real structure is a distinct future possibility.

It is evident from Eq. (8.9) that, in the general case, the individual surface stress components cannot be directly derived from the thermoelastic response, and, consequently, the development of a means of determining individual surface stresses from the stress sum data, that is, stress separation, has become a topic of considerable importance. Work on stress separation, principally numerical, up to the mid-1990s is reviewed in [42]. Further studies over the last decade include hybrid photoelastic–thermoelastic techniques [43, 44] and other numerical approaches based on finite difference [45, 46] and finite element techniques [47]. It may be surmised that the additional effort and cost involved in implementing any one of these proposed approaches may be considerable. In [42], an approach is described that does not rely on the use of other techniques or additional computational effort but is based on the use of devices referred to as *thermoelastic strain gauges*. These can be bonded to the surface of a stressed specimen to provide a supplementary thermoelastic signal; from this signal and that from the adjacent surface of the specimen, individual stresses at the gauge position on the specimen surface can be derived. Clearly, the benefit of being able to deliver individual stresses from TSA is clear. However, Eq. (8.9) provides a very important stress metric that can provide the basis for validation of numerical studies or the assessment of damage.

TSA has been applied extensively to composite materials, and some examples have been given earlier in the chapter. Studies relating to stress analysis of composite structure are few. An example for aircraft sandwich structures is given in [48]. Growth areas are high-resolution studies, damage assessments, and evaluation of stress in complex components and assemblies. Residual stress assessment using TSA is a topic of current interest, and with the growth of more sensitive detectors this is becoming a real possibility. However, the major bar to progress is the need for a cyclic load. Recent work [18] has shown that for composite materials it is possible to obtain quantitative data from a component subjected to a transient load. This represents a real step forward in progress and opens the technique for exploitation as a nondestructive examination (NDE) methodology.

Acknowledgments

The author would like to acknowledge the support of her coworkers and students in the production of this chapter, in particular Dr Trystan Emery and Dr James Eaton-Evans. She is extremely grateful to her former Ph.D. supervisor and coworker Emeritus Professor Peter Stanley from the University of Manchester, who introduced her to the technique and provided unrelenting encouragement to build on the research conducted in the early 1990s in a small laboratory in Manchester.

Problems

8.1 A further validation of the K value of Tutorial Exercise 8.2 is carried out at different frequencies. At 15 Hz, ΔT is obtained as 178 $\pm$0.006 K and at 30 Hz, $\Delta T 150 \pm 0.010$ K. Explain the cause of the reduction in the response and devise a means of verifying K.

8.2 Describe the procedure and derive the equations to obtain K using a two-gauge orthogonal strain gauge rosette bonded to the component under investigation.

8.3 If the component in Tutorial Exercise 8.1 was made for Al-2024, explain how this would modify the influence of the mean stress value.

8.4 Is it possible to obtain residual stresses from TSA (i) using Eq. (8.12) and (ii) using Eq. (8.43). Describe an experiment procedure using only TSA to determine the material's sensitivity to the mean stress.

8.5 Explain why it is necessary to use imaging systems based on photon detectors for TSA.

8.6 In determining the sensitivity of a photon detector, why is it not sufficient to apply the standard Stefan–Boltzman relationship.

8.7 Explain how a Brazilian disc may be used to calibrate the thermoelastic response (see [9]).

8.8 Explain why in the vicinity of cracks and stress concentrations the thermoelastic response may not give an accurate representation of the stress field.

8.9 The image and plot shown below are obtained from a crack in an aluminum alloy plate loaded in uniaxial tension in the vertical direction. The plot shows ΔT and the phase. Explain why there is a step change in phase to the left of the plot and why there is an increase in phase to the right of the plot, in the vicinity of the crack tip.

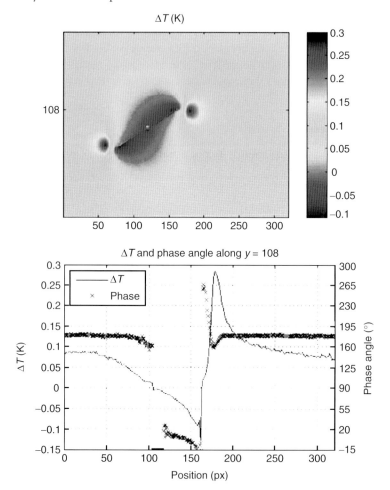

9
Photoelasticity

Eann A. Patterson

9.1
Introduction

Photoelasticity is a technique involving the analysis of fringe patterns produced by polarized light transmitted through transparent materials, and is used to determine the state of stress and strain in a material. It is utilized extensively in the glass industry to evaluate residual stresses during manufacture as well as in fatigue and fracture research using materials such as polycarbonate. Transparent coatings bonded to complex engineering components by a reflective adhesive are used as strain witnesses to determine surface stresses in the components, and models made from materials, such as epoxy resins, are employed to evaluate the three-dimensional stress system in engineering components in order to validate numerical stress analyses.

Photoelasticity arises due to the property of temporary birefringence exhibited by many transparent materials; that is, when subjected to strain, these materials become optically anisotropic and exhibit refractive indices that vary with direction. When the strain is removed, the materials revert to being optically isotropic. This property of temporary birefringence or temporary double refraction was first observed by Brewster in 1816. In 1841, Neumann investigated the changes in refractive index as a function of strain, and in 1853 Maxwell observed that the changes in refractive indices were proportional to the stresses in the material.

When polarized light enters a birefringent material subject to strain, it is resolved into two components with vectors parallel to the directions of principal strains of the material. The two components of the light propagate through the material at speeds proportional to the refractive indices, and hence strains, parallel to their vectors. Consequently, for nonhydrostatic strain states, one light component will be retarded relative to the other, and a common-path interferometer can be used to combine the components as they exit the material in order to generate fringes that are contours of the retardation, or difference in principal strains. Figure 9.1 shows an example of such a fringe pattern for a two-dimensional model of a crane hook.

Most photoelastic stress analysis involves an assumption of two-dimensionality: either plane stress as in two-dimensional photoelasticity or plane strain as in

Optical Methods for Solid Mechanics: A Full-Field Approach, First Edition. Edited by Pramod Rastogi and Erwin Hack.
© 2012 Wiley-VCH Verlag GmbH & Co. KGaA. Published 2012 by Wiley-VCH Verlag GmbH & Co. KGaA.

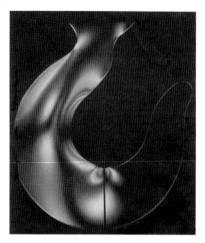

Figure 9.1 Photoelastic fringe pattern in two-dimensional model of a loaded crane hook viewed in a dark-field plane polariscope with both isoclinic and isochromatic fringes visible (a color version of this photograph can be found on the front cover of *Experimental Techniques*, 33 (1), January 2009, and on-line at *http://www3.interscience.wiley.com/journal/122200412/issue*). (Please find a color version of this figure on the color plates.)

reflection photoelasticity (see Box 9.2). Three-dimensional stress states can be handled using the frozen stress technique (see Box 9.1), to lock strains into a model prior to sectioning into planar slices for two-dimensional analysis; or, using integrated photoelasticity (see Box 9.3) to "optically slice" a three-dimensional component. Two-dimensional analysis is considered throughout this chapter, except in Boxes 9.1 and 9.3 and where stated explicitly in examples.

The theory underpinning the explanation provided above is outlined in Section 9.2, including the design of a common-path interferometer, or polariscope as it is more commonly known. Subsequent sections describe the interpretation of fringes (Section 9.3) and their analysis by both manual (Section 9.4) and automated (Section 9.5) methodologies, the selection of materials and loads for photoelastic models (Section 9.6), and the determination and characterization of stress fields from photoelastic fringe data (Section 9.7).

Tutorial Exercise 9.1

Obtain a pair of Polaroid sunglasses to act as your polariscope. You can use them in reflection, with the sun as your light source. Sunlight striking an appropriate angle on glass will be reflected from the bottom surface, and, when viewed with the Polaroid sunglasses, will reveal isochromatic fringe patterns related to the residual stresses locked into the glass during manufacture; you can see these fringes in a car windshield. Alternatively, you can look at the residual stresses in the transparent CD-ROM that is supplied on the top of a packet of blank CD-ROMs, using your Polaroid sunglasses and a CD-ROM as a mirror; if you do not have a transparent CD-ROM, then look at the transparent area in the center of any CD-ROM.

9.2
Polariscope Theory and Design

Maxwell deduced that the changes in refractive indices that occur in temporary birefringent materials subject to stress could be described by [1]

$$n_1 - n_0 = C_1\sigma_1 + C_2(\sigma_2 + \sigma_3)$$
$$n_2 - n_0 = C_1\sigma_2 + C_2(\sigma_3 + \sigma_1) \quad (9.1)$$
$$n_3 - n_0 = C_1\sigma_3 + C_2(\sigma_1 + \sigma_2)$$

where $\sigma_{1,2,3}$ are the principal strains, n_0 is the refractive index in the unstrained state, $n_{1,2,3}$ are the indices of refraction coincident with the directions of principal strain, and $C_{1,2}$ are the stress-optic coefficients. In general, experimental stress analysis using photoelasticity is restricted to two-dimensional cases, that is, $\sigma_3 = 0$, so these equations reduce to

$$n_1 - n_0 = C_1\sigma_1 + C_2\sigma_2$$
$$n_2 - n_0 = C_1\sigma_2 + C_2\sigma_1 \quad (9.2)$$

In photoelasticity, of prime interest is the relative speed, or retardation, of the two light components propagating through the material with their vectors in the directions of the principal stress directions. Thus, it is convenient to eliminate n_0, the refractive index in the unstrained state, from the above equations: that is,

$$n_2 - n_1 = (C_2 - C_1)(\sigma_1 - \sigma_2) \quad (9.3)$$

$(C_2 - C_1)$ is known as the *stress-optic coefficient* C_σ, and has units of Brewsters (1 Brewster = 10^{-12} m²/N). *Refractive index* is defined as the ratio of the velocity of the light in free space to the velocity in a material; consequentially, in the time that light travels over a path length h through the thickness of the material it will travel hn_0 in free space. So it can be deduced that, at exit from a temporary birefringent material of thickness h, one component of the light will have developed a relative retardation or path difference compared to the other of

$$\delta = h(n_2 - n_1) = C_\sigma h(\sigma_1 - \sigma_2) \quad (9.4)$$

This can be expressed in terms of a relative angular phase shift Δ, which is dependent on the wavelength of the light λ: that is,

$$\Delta = \frac{2\pi}{\lambda}\delta = \frac{2\pi h}{\lambda}(n_2 - n_1) = \frac{2\pi C_\sigma h}{\lambda}(\sigma_1 - \sigma_2) \quad (9.5)$$

which can be simplified to the well-known stress-optic law

$$\sigma_1 - \sigma_2 = \frac{Nf_\sigma}{h} \quad (9.6)$$

where

$$N = \frac{\Delta}{2\pi} \quad (9.7)$$

where N is known as the *fringe order*. In essence, the fringe order is a count of the number of cycles of angular retardation Δ. The factor f_σ is known as the *material fringe constant* and is defined as

$$f_\sigma = \frac{\lambda}{C_\sigma} \qquad (9.8)$$

To generate the fringes implied by the stress-optic law, the two components of the light need to be generated, and then interfered with one another when they leave the material. This is achieved using a common-path interferometer, which is more commonly known as a *polariscope*.

In photoelasticity, a polariscope is constructed using plane or linear polarizers and quarter-wave plates, both of which exhibit a form of permanent birefringence. Light incident on these elements is resolved into two components, with mutually perpendicular vectors; in the case of a plane polarizer, one component is absorbed and the other is transmitted, while in a wave plate the two components are transmitted at different speeds.

A simple polariscope can be constructed using two plane polarizers positioned on either side of a stressed plate, as shown in Figure 9.2. The plane polarizer in front of the light source is known as the *polarizer* and the other one is known as the *analyzer*. When the axis of polarization of the polarizer and that of analyzer are crossed or perpendicular, as shown in the figure, then the field of view of the instrument will appear black because of the complete absorption of the light by the two polarizers, except at the location of the stressed specimen where fringe orders defined by the stress-optic law (Eq. (9.6)) will be visible (e.g., Figure 9.1). A more sophisticated polariscope can be constructed by the inclusion of quarter-wave plates; that is wave plates that introduce a uniform relative retardation of quarter of a wavelength, that is, $\delta = \lambda/4$ (or $\Delta = \pi/2$), as shown in Figure 9.3. A pair of quarter-wave plates placed each side of the specimen and inside the polarizers create what is known as a *circular polariscope* when their optical axes are perpendicular to one another and at $45°$ to the optical axes of the polarizer and the analyzer.

There are a number of methods for analyzing the optical system of a polariscope [2], including Jones calculus and Mueller matrices [3]. The latter are based on

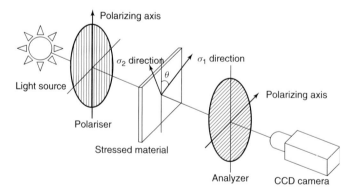

Figure 9.2 Schematic diagram of plane polariscope with polarizers crossed to generate a dark-field image, such as in Figure 9.1, with both isochromatic and isoclinic fringes.

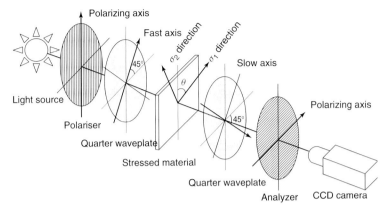

Figure 9.3 Schematic diagram of a dark-field circular polariscope with the polarizers crossed and the optical axes of the quarter-wave plates crossed and at 45° to axes of the polarizers.

a phenomenological perspective and are employed here because of their relative simplicity. In such a description, the Stokes vector **S** is utilized to describe the intensity and state of polarization of the light:

$$\mathbf{S} = \begin{bmatrix} 2I_0 \\ 2I_1 - 2I_0 \\ 2I_2 - 2I_0 \\ 2I_3 - 2I_0 \end{bmatrix} \tag{9.9}$$

where I_0 represents the total intensity of the light, while I_1, I_2, I_3 represent the intensities of the light components with horizontal, 45°, and left-hand circular polarizations, respectively. The total intensity is of interest in photoelasticity because it is the only component that can be readily observed and measured. In this notation, light that is unpolarized or randomly polarized is described by

$$\mathbf{S} = \begin{bmatrix} 1 \\ 0 \\ 0 \\ 0 \end{bmatrix} \tag{9.10}$$

The Mueller matrix $\mathbf{P}_\beta$ for a linear polarizer at an arbitrary angle β is

$$\mathbf{P}_\beta = \begin{bmatrix} 1 & \cos 2\beta & \sin 2\beta & 0 \\ \cos 2\beta & \cos^2 2\beta & \sin 2\beta \cos 2\beta & 0 \\ \sin 2\beta & \sin 2\beta \cos 2\beta & \sin^2 2\beta & 0 \\ 0 & 0 & 0 & 0 \end{bmatrix} \tag{9.11}$$

and, for a linear retarder with an optical fast axis at an angle θ generating an angular relative retardation of α, the Mueller matrix $\mathbf{R}_\alpha(\theta)$ is

$$\mathbf{R}_\alpha(\theta) = \begin{bmatrix} 1 & 0 & 0 & 0 \\ 0 & \cos^2 2\theta + \sin^2 2\theta \cos\alpha & (1-\cos\alpha)\sin 2\theta \cos 2\theta & -\sin 2\theta \sin 2\alpha \\ 0 & (1-\cos\alpha)\sin 2\theta \cos 2\theta & \sin^2 2\theta + \cos^2 2\theta \cos\alpha & \cos 2\theta \sin 2\alpha \\ 0 & \sin 2\theta \sin 2\alpha & -\cos 2\theta \sin 2\alpha & \cos\alpha \end{bmatrix} \quad (9.12)$$

For a quarter-wave plate ($\Delta = \alpha = \pi/2$) with its optical axis at an arbitrary angle β, this reduces to

$$\mathbf{R}_{\pi/2}(\beta) = \mathbf{Q}_\beta = \begin{bmatrix} 1 & 0 & 0 & 0 \\ 0 & \cos^2 2\beta & \sin 2\beta \cos 2\beta & -\sin 2\beta \\ 0 & \sin 2\beta \cos 2\beta & \sin^2 2\beta & \cos 2\beta \\ 0 & \sin 2\beta & -\cos 2\beta & 0 \end{bmatrix} \quad (9.13)$$

A dark-field circular polariscope can be modeled as

$$\mathbf{S}' = \mathbf{P}_0 \mathbf{Q}_{-\pi/4} \mathbf{R}_\alpha(\theta) \mathbf{Q}_{\pi/4} \mathbf{P}_{\pi/2} \mathbf{S} \quad (9.14)$$

and a dark-field plane polariscope as

$$\mathbf{S}' = \mathbf{P}_0 \mathbf{R}_\alpha(\theta) \mathbf{P}_{\pi/2} \mathbf{S} \quad (9.15)$$

It can be shown from this analysis (see Tutorial Example 9.2) that the intensity component of the Stokes vector emitted from a dark-field plane polariscope is given by

$$2I_0 = \sin^2 2\theta \sin^2 \frac{\alpha}{2} \quad (9.16)$$

where the $\sin^2(\alpha/2)$ term represents the fringes defined by the stress-optic law (Eq. (9.6)); they are known as *isochromatic fringes* because they appear in white light as contours of constant color as a result of the material fringe constant f_σ being a function of wavelength. In monochromatic light, isochromatic fringes appear as a series of black fringes when $\sin^2(\alpha/2) = 0$. The $\sin^2(2\theta)$ term in Eq. (9.16) generates a second set of fringes, known as *isoclinics*, which are black in both white and monochromatic light. Isoclinic fringes occur when $\sin^2(2\theta) = 0$, that is, when $\theta = 0$ or $\pi/2$, which coincides with the direction of the principal stress in the specimen being aligned with the optical axes of the polarizer and analyzer.

The same effect can be produced in a circular polariscope arrangement by rotating the quarter-wave plates so that their optical axes are parallel to the polarizer with which they are paired, thereby eliminating them optically.

Tutorial Exercise 9.2

Use Mueller matrices to determine the expression for the intensity of light from a plane dark-field polariscope containing a specimen with a uniform retardation of α at an angle θ.

Solution:

"Plane" implies the use of two polarizers and no quarter-wave plates, and "dark-field" implies that the polarizers are crossed to produce total extinction except in the presence of stressed areas of the specimen: that is,

$$S' = P_0 R_\alpha(\theta) P_{\pi/2} S$$

Or breaking it down into stages, the light emitted from the polarizer is given by

$$S_{P\pi/2} = P_{\pi/2} S = \begin{bmatrix} 1 & -1 & 0 & 0 \\ -1 & 1 & 0 & 0 \\ 0 & 0 & 0 & 0 \\ 0 & 0 & 0 & 0 \end{bmatrix} \begin{bmatrix} 1 \\ 0 \\ 0 \\ 0 \end{bmatrix} = \begin{bmatrix} 1 \\ -1 \\ 0 \\ 0 \end{bmatrix}$$

Then, light emitted from the specimen is given by

$$S_R = R_\alpha(\theta) S_{P\pi/2}$$

$$= \begin{bmatrix} 1 & 0 & 0 & 0 \\ 0 & \cos^2 2\theta + \sin^2 2\theta \cos\alpha & (1-\cos\alpha)\sin 2\theta \cos 2\theta & -\sin 2\theta \sin 2\alpha \\ 0 & (1-\cos\alpha)\sin 2\theta \cos 2\theta & \sin^2 2\theta + \cos^2 2\theta \cos\alpha & \cos 2\theta \sin 2\alpha \\ 0 & \sin 2\theta \sin 2\alpha & -\cos 2\theta \sin 2\alpha & \cos\alpha \end{bmatrix} \begin{bmatrix} 1 \\ -1 \\ 0 \\ 0 \end{bmatrix}$$

$$= \begin{bmatrix} 1 \\ -\cos^2 2\theta - \sin^2 2\theta \cos\alpha \\ -(1-\cos\alpha)\sin 2\theta \cos 2\theta \\ -\sin 2\theta \sin 2\alpha \end{bmatrix}$$

And finally, the light emitted from the analyzer is given by

$$S' = P_0 S_R$$

$$S' = \begin{bmatrix} 1 & 1 & 0 & 0 \\ 1 & 1 & 0 & 0 \\ 0 & 0 & 0 & 0 \\ 0 & 0 & 0 & 0 \end{bmatrix} \begin{bmatrix} 1 \\ -\cos^2 2\theta - \sin^2 2\theta \cos\alpha \\ -(1-\cos\alpha)\sin 2\theta \cos 2\theta \\ -\sin 2\theta \sin 2\alpha \end{bmatrix}$$

$$= \begin{bmatrix} 1 - \cos^2 2\theta - \sin^2 2\theta \cos\alpha \\ 1 - \cos^2 2\theta - \sin^2 2\theta \cos\alpha \\ 0 \\ 0 \end{bmatrix}$$

The intensity is obtained from the first term in the Stokes vector: that is,

$$2I_0 = 1 - \cos^2 2\theta - \sin^2 2\theta \cos\alpha = \sin^2 2\theta (1 - \cos\alpha) = 2\sin^2 2\theta \sin^2 \frac{\alpha}{2}$$

9.3 Isoclinic and Isochromatic Fringes

For a stressed specimen, isoclinic and isochromatic fringes can be observed simultaneously in the plane polariscope, as seen in Figure 9.1. In white light, the isoclinic fringes will appear as black fringes and the isochromatic fringes as color fringes, including black for zero-order fringes [$(\sigma_1 - \sigma_2) = N = 0$]. Simultaneous rotation of the polarizer and analyzer relative to the specimen will cause the isoclinic fringes to rotate, because these fringes occur at locations where the directions of the principal stresses in the specimen coincide with the axes of the polarization of the polarizer and analyzer. In monochromatic light, isochromatic fringes also appear black; however, they will not move with rotation of the polarizer and analyzer, so that they can be readily identified from isoclinic fringes. If the magnitude of the stress inducing the temporary birefringence is increased or decreased, then the isochromatic fringes will tend to grow or recede from an area of zero stress, whereas isoclinic fringes will not move in the case of linear elastic loading. The movement of isochromatic fringes with applied load provides an alternative way of distinguishing isoclinic and isochromatic fringes. The use of a circular polariscope provides information on the isochromatic fringes only by removing the isoclinic fringes that often mask the isochromatic fringes.

The basic form of the stress-optic law (Eq. (9.6)) dictates that, in monochromatic light, neighboring fringes either increase or decrease by one fringe order, and so together with the location of zero-order fringes, this allows the fringe order throughout a field of view to be identified. In white light, within a fringe order, each wavelength is extinguished in turn from violet, blue through green to yellow, orange, and red; and the extinguished color causes the isochromatic fringe to change color from lemon-yellow, orange, deep red, purple, deep blue, to green. The purple fringe is called the *tint of passage* and corresponds to the approximate center of the fringe order. The colors described above are present in a first-order fringe and become paler in subsequent higher order fringes, so that an experienced observer can identify a fringe order by the combination of colors present.

Zero-order fringes occur when $(\sigma_1 - \sigma_2) = 0$, and points where this occurs as a result of $\sigma_1 = \sigma_2$ are known as *isotropic points*. In common with all zero-order fringes, isotropic points are black in a white-light circular polariscope and hence are easily identifiable. It is important to note that isoclinic fringes of all values must pass through isotropic points as well as singular points, such as locations of

Figure 9.4 Photograph of a disc subject to compression along its vertical diameter viewed in a dark-field circular polariscope (see Figure 9.3).

concentrated loads. The value of the isoclinic angle on a free boundary is equal to the slope of the boundary relative to the axis of the polarizer. In any specimen, at a sharp corner the boundary rapidly rotates through many angles; and thus, the sharp corner must be a meeting point for many isoclinic fringes, so that it is always an isotropic point with a zero-order fringe. This latter information is often critical in the interpretation of the isochromatic fringe patterns, particularly from photographs. Further extensive information on the interpretation of photoelastic fringe patterns is given in the classic text by Frocht [4] as well as other sources such as Durelli and Shukla [5].

Tutorial Exercise 9.3

Identify the fringe orders in the photograph of the disc subjected to compression across its vertical diameter in a dark-field circular polariscope shown in Figure 9.4; plot the fringe order along the horizontal and vertical diameters.

Solution:

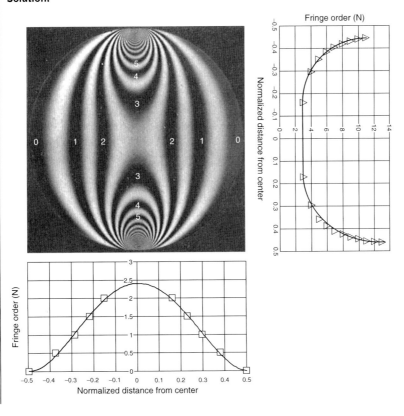

9.4
Fractional Fringe Analysis Using Compensation Techniques

In Section 9.3, strategies for identifying the integer-order fringes were outlined using the polariscope set up with the polarizer and analyzer axes crossed so that no light is transmitted when the polariscope is empty. If the analyzer and its associated quarter-wave plate are rotated, so that their axes are perpendicular to those in the polarizer and its associated quarter-wave plate, then a light-field polariscope is created in which there is maximum transmission of the light when the polariscope is empty. In this arrangement, half-order fringes can be identified using strategies similar to those described in Section 9.3 for whole-order fringes. However, other fractions of a fringe require more sophisticated techniques for their measurement. One such method is to employ a device known as a *compensator* with an adjustable and precisely known value of birefringence which is inserted into the polariscope over the point in the specimen where it is required to make the measurement. The device is adjusted until the fringe order at the point of interest reduces to zero or an integer fringe value. Many such devices have been invented, of which the null-balance and Babinet–Soleil compensators have been the most widely used [6].

An alternative approach is to apply a set of techniques known as *goniometric compensation*, of which the Tardy compensation technique is the most popular and is simple to apply. In step one, a plane polariscope is used and the polarizer and analyzer rotated together in the dark-field configuration until an isoclinic fringe is exactly over the point of interest P. In step two, a circular polariscope is created without moving the polarizer and analyzer and by inserting the quarter-wave plates to create a dark-field circular polariscope, that is, with their optical axes crossed and at 45° to the axes of the polarizer and analyzer. The point P may lie between two integer fringes whose orders N_L and N_H must be determined using the strategies outlined in Section 9.3. In the third step, the analyzer is rotated to move the lower order fringe N_L until it is exactly over the point P. The exact fringe order N_P will be given by

$$N_P = N_L + \frac{\phi_L}{180} \tag{9.17}$$

where ϕ_L is the angle of rotation of the analyzer required to move the integer fringe to the location P. If you chose to move the higher order fringe N_H, then the analyzer will need to be rotated in the opposite direction by $\phi_H (= 180 - \phi_L)$ and

$$N_P = N_H - \frac{\phi_H}{180} \tag{9.18}$$

The Senarmont method of goniometric compensation is less straightforward to describe and also to employ. Step 1 is identical to Tardy compensation; however, in step 2, only one quarter-wave plate is inserted in the same orientation as before and the polarizer is rotated so that its axis is parallel to that of the quarter-wave plate and hence at 45° to the axis of the analyzer. Step 3 is executed as for Tardy compensation by rotating the analyzer except that the point P is not neatly enclosed by fringes of high and lower orders; instead, a jumble of partial isochromatic

9.4 Fractional Fringe Analysis Using Compensation Techniques

fringes is observed until on rotation of the analyzer a solid black fringe appears at the location of interest.

These goniometric techniques have become the basis of phase-stepping techniques in digital photoelasticity, with Tardy compensation being the basis of most, if not all, techniques based on circular polariscopes [7, 8] and Senarmont compensation being the basis of the commercially available gray-field polariscope [9].

Tutorial Exercise 9.4

Derive, using Mueller calculus, an expression for the light intensity as a function of the rotation of the analyzer during Tardy compensation, and hence find the condition for extinction during Tardy compensation.

Solution:

The Stokes vector for the light emitted by a circular polariscope set up for Tardy compensation can be described by

$$S' = P_\phi Q_{-\pi/4} R_\alpha \left(\frac{\pi}{2}\right) Q_{\pi/4} P_{\pi/2} S$$

The specimen is aligned to the polarizer so that $R_\alpha(\theta)$ becomes $R_\alpha(\pi/2)$ and the quarter-wave plates are aligned for a dark-field circular polariscope, that is, with their axes mutually perpendicular at $45°$ to the axis of the polarizer, so $\beta = \pm\pi/4$ in Eq. (9.13).

The circularly polarized light entering the specimen can be obtained as (using $S_{P\pi/2}$ from Tutorial Example 9.3)

$$S_{Q\pi/4} = Q_{\pi/4} P_{\pi/2} S = Q_{\pi/4} S_{P\pi/2} \begin{bmatrix} 1 & 0 & 0 & 0 \\ 0 & 0 & 0 & 1 \\ 0 & 0 & 1 & 0 \\ 0 & 1 & 0 & 0 \end{bmatrix} \begin{bmatrix} 1 \\ -1 \\ 0 \\ 0 \end{bmatrix} = \begin{bmatrix} 1 \\ 0 \\ 0 \\ -1 \end{bmatrix}$$

Then

$$S_R = R_\alpha(\pi/2) S_{Q\pi/4} \begin{bmatrix} 1 & 0 & 0 & 0 \\ 0 & -1 & 0 & 0 \\ 0 & 0 & -\cos\alpha & -\sin\alpha \\ 0 & 1 & \sin\alpha & \cos\alpha \end{bmatrix} \begin{bmatrix} 1 \\ 0 \\ 0 \\ -1 \end{bmatrix} = \begin{bmatrix} 1 \\ 0 \\ -\sin\alpha \\ \cos\alpha \end{bmatrix}$$

and

$$S_{Q-\pi/4} = Q_{-\pi/4} S_R = \begin{bmatrix} 1 & 0 & 0 & 0 \\ 0 & 0 & 0 & 1 \\ 0 & 0 & 1 & 0 \\ 0 & 1 & 0 & 0 \end{bmatrix} \begin{bmatrix} 1 \\ 0 \\ -\sin\alpha \\ \cos\alpha \end{bmatrix} = \begin{bmatrix} 1 \\ \cos\alpha \\ -\sin\alpha \\ 0 \end{bmatrix}$$

Finally,

$$S' = P_\phi S_{Q-\pi/4}$$

$$= \begin{bmatrix} 1 & \cos 2\phi & \sin 2\phi & 0 \\ \cos 2\phi & \cos^2 2\phi & \sin 2\phi \cos 2\phi & 0 \\ \sin 2\phi & \sin 2\phi \cos 2\phi & \sin^2 2\phi & 0 \\ 0 & 0 & 0 & 0 \end{bmatrix} \begin{bmatrix} 1 \\ \cos \alpha \\ -\sin \alpha \\ 0 \end{bmatrix}$$

$$= \begin{bmatrix} 1 + \cos 2\phi \cos \alpha - \sin 2\phi \sin \alpha \\ \cos 2\phi (1 + \cos 2\phi \cos \alpha - \sin 2\phi \sin \alpha) \\ \sin 2\phi (1 + \cos 2\phi \cos \alpha - \sin 2\phi \sin \alpha) \\ 0 \end{bmatrix}$$

thus using the first term in the resultant Stokes vector to obtain the light intensity as

$$2I_0 = 1 + \cos 2\phi \cos \alpha - \sin 2\phi \sin \alpha = 1 + \cos(2\phi - \alpha)$$

And for extinction $I_0 = 0$, so $\cos(2\phi - \alpha) = 1$ and $(2\phi - \alpha) = 0$, that is, for extinction $\phi = \alpha/2$.

Box 9.1: Three-Dimensional Photoelasticity Using Stress-Freezing

The three-dimensional stress state in a component can be determined from a three-dimensional epoxy model in which the strains due to the applied loading are locked in, via a thermal cycle, thus allowing subsequent sectioning into two-dimensional slices for analysis of the fringe patterns. Subslicing of the planar slices permits the complete three-dimensional stress state in the model to be established, and, hence, the stress state in the component can be calculated using the scaling techniques outlined in Section 9.6.

Epoxy resin has a biphase chemical structure of primary and secondary bonds, which is analogous to a collection of springs surrounded by wax and is set hard at room temperature, so when a load is applied it is borne by the wax. On heating to above the glass transition temperature ($\approx 140°$C in the case of epoxy), the wax becomes soft and the springs carry the load so that, when cooled without removing the load, the wax sets around the springs, holding them in their deformed state. In epoxy, once this process is complete, birefringence due to the strains induced at the elevated temperature is locked into the material, which can be sectioned providing no heating occurs. Sections in the form of planar slices are removed, usually to coincide with the expected orientations of principal planes. Slices should be thin enough to ensure constant stress through their thickness, but sufficiently thick to present adequate fringes for analysis. The latter requirement can be alleviated by the use of a fringe-multiplying polariscope [23] (see inset in Figure 9.6) in which the light is reflected back and forth through the slice, using partial mirrors, thereby increasing the path length and the number of fringe orders with each passage. Alternatively, automated photoelasticity allows the analysis of subfringe orders, permitting the use of thin slices and low strain values. Often, to enhance clarity, slices are viewed using glass-walled cells filled with a fluid of matched refractive index.

Three-dimensional photoelasticity involves work by skilled craftsmen with appropriate facilities, which makes it expensive to perform; however, it is practically the only technique available that can provide the complete state of stress in a complex three-dimensional component; and, hence, is very valuable in the validation of complex numerical stress analyses [24, 25].

9.5
Digital Fringe Analysis

The quantitative analysis of photoelastic fringe patterns has been enhanced considerably by the advent of digitally based automation, both in terms of the quality and comprehensiveness of the available data and the speed of processing. Automated photoelasticity, or *digital photoelasticity* as it has become known, was a popular topic of research at the end of the last century, with many approaches being developed and demonstrated. A short review is provided by Patterson [7], while a more comprehensive one is provided by Ramesh [8]. In general, three classes of approach can be identified for full-field analysis of a photoelastic fringe pattern, namely, Fourier analysis, phase-stepping methods, and spectral analysis.

Fourier transforms have been used in a number of ways for automated photoelastic fringe analysis. The isoclinic angle can be calculated from the ratio of the real and imaginary parts of the Fourier transform of the intensity captured as a function of the angular rotation of the analyzer in a plane polariscope. In a circular polariscope, if output intensity is considered as a function of a spatial coordinate, then the fringe order can be found from the ratio of the real and imaginary parts of its Fourier transform when the fringe order is modulated with a high-frequency signal by the addition of a quartz wedge in the field of view. These Fourier-based methods are relatively free of noise, but require a large number of images (typically 100) and, as a consequence, are not widely utilized. A variant on these methods, which is commercially available, is the gray-field polariscope [9], which contains only one quarter-wave plate as in the Senarmont compensation method (Section 9.4), and for which the intensity is taken as a function of the orientation of a rapidly rotating analyzer. Typically 16 images are captured per revolution of the analyzer, and spatially periodic relative retardation and isoclinic angle are obtained. A disadvantage of this and other Fourier techniques is that the periodic relative retardation cannot be unwrapped so that the technique is restricted to subinteger fringe analyses.

Many phase-stepping algorithms have been proposed, though few have been used for the analysis of engineering components, as opposed to laboratory specimens. If methods employing a Fourier transform can be considered phase-shifting, that is, a continuous modulation of the phase, then phase-stepping is discrete changes in the phase brought about by careful selection of a small number of angular orientations of the optical elements. Typically, the ambient light intensity, intensity of the light source, the isochromatic fringe order, and isoclinic angle are taken as unknowns, and so at least four images, described by four simultaneous equations, are needed for a solution. Equations, in terms of the angular positions of the optical elements, can be derived using one of a number of the matrix theories of photoelasticity [2], including the Mueller matrices described in Section 9.2. The nature of the expression for the light intensity implies that the solutions will always be periodic functions of the spatial coordinates, and hence will need unwrapping to generate a continuous map of isochromatic fringe order. Most phase-stepping methods are closely related to Tardy compensation, and hence identification of the fringe order for, at least, one point is required to generate a map of absolute fringe order. The latter is one of the disadvantages of phase-stepping methods, which also include the lack of robustness of the unwrapping methods, partly due to the poor quality of isoclinic data obtained and the interaction between isochromatic and isoclinic fringes. Ramesh and Ganapathy [10] have reviewed phase-stepping methods with particular attention to their effectiveness, and more recently digital photoelasticity in general [11], while Ghiglia and Pritt [12] provide many unwrapping techniques that can be adapted to photoelasticity. The principal advantages of phase-stepping photoelasticity are the small number of images required, which allows the option of simultaneous image capture for dynamic experiments, and the sophisticated range of algorithms available, which have been employed in a number of industrial case studies (see, for example, Figure 9.5 and [13]).

Spectral analysis requires measurement of the light intensity in a circular polariscope as a function of wavelength, which is possible using a standard instrument for a single point but is not viable for a full field of view. For analysis of a single point, a theoretical description of the light intensity as a function of wavelength is fitted to the measured spectral signature [14]. For full-field analysis, RGB cameras can be used to obtain a coarse spectral signature to which either a theoretical description can be fitted, or a "look-up" table used to identify fringe orders [15]. The coarseness of the data restricts RGB photoelasticity to fringe orders less than 3; otherwise, the uniqueness of the spectral signature is lost. The disadvantages of spectral analysis are that the results are strongly dependent on the spectrum of the light source and the absorption spectrum of the photoelastic specimen, particular for RGB photoelasticity, and that isoclinic information cannot be obtained because it is independent of wavelength. The advantages include the simplicity of the setup for RGB photoelasticity and that absolute fringe order is obtained without the need for any postprocessing or *a priori* knowledge of the stress distribution.

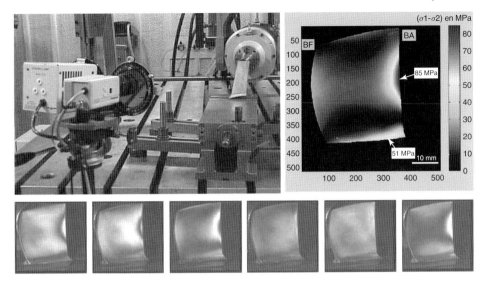

Figure 9.5 Experimental arrangement (top left) for reflection photoelasticity (see Box 9.2) applied to a compressor blade excited at around 7 kHz and illuminated using a strobe light source, in order to capture six phase-stepping images (bottom) that were processed to generate a map of principal stress difference (top right) for use in the validation of finite element analyses (see [13] for more details). (Please find a color version of this figure on the color plates.)

Tutorial Exercise 9.5

Derive, using Mueller matrices, the expression for light intensity as a function of the angular position of the analyzer and associated quarter-wave plate in a circular polariscope and then design a simple phase-stepping algorithm to evaluate the isoclinic angle and relative retardation.

Solution:

The Stokes vector for the light emitted by a circular polariscope set up for phase-stepping can be described by

$$\mathbf{S}' = \mathbf{P}_\beta \mathbf{Q}_\phi \mathbf{R}_\alpha(\theta) \mathbf{Q}_{\pi/4} \mathbf{P}_{\pi/2} \mathbf{S}$$

The process is similar to that shown in Tutorial Example 9.4 and results in

$$I_0 = 1 + \sin 2(\beta - \phi) \cos \alpha - \sin 2(\theta - \phi) \cos 2(\beta - \phi) \sin \alpha$$

To render the unwrapping processes viable, it is preferable to obtain the isoclinic angle θ and the relative retardation α as functions with distinct discontinuities between periods when plotted against the spatial coordinates; both tangent and cotangent satisfy this requirement. Further, a practical phase-stepping scheme requires stepping angles that are easy to set in regular increments, particular if the rotation of the optical elements of the polariscope is not automated. So, starting with $\phi = \beta = 0$,

$$I_0 = 1 - \sin 2(\theta - \phi) \sin \alpha$$

and rotating both elements together in $\pi/4$ steps yields,

from $\phi = \beta = 0$ $\quad I_{0a} = 1 - \sin 2\theta \sin \alpha$

from $\phi = \beta = \dfrac{\pi}{4}$ $\quad I_{0b} = 1 + \cos 2\theta \sin \alpha$

from $\phi = \beta = \dfrac{\pi}{2}$ $\quad I_{0c} = 1 + \sin 2\theta \sin \alpha$

from $\phi = \beta = \dfrac{3\pi}{4}$ $\quad I_{0d} = 1 - \cos 2\theta \sin \alpha$

and, by observation $\dfrac{I_{0c} - I_{0a}}{I_{0b} - I_{0d}} = 2 \tan 2\theta \quad$ so, $\theta = \dfrac{1}{2} \tan^{-1} \left[\dfrac{I_{0c} - I_{0a}}{I_{0b} - I_{0d}} \right]$

To obtain the relative retardation, the isoclinic angle needs to be eliminated, so rotating the quarter-wave plate by $\pi/4$ relative to the analyzer yields

from $\phi = 0$ and $\beta = \dfrac{\pi}{4}$ $\quad I_{0e} = 1 + \cos \alpha$

from $\phi = 0$ and $\beta = -\dfrac{\pi}{4}$ $\quad I_{0f} = 1 - \cos \alpha$

and, by observation $\dfrac{I_{0b} - I_{0d}}{(I_{0e} - I_{0f}) \cos 2\theta} = \tan \alpha \quad$ so, $\alpha = \cot^{-1} \left[\dfrac{(I_{0e} - I_{0f}) \cos 2\theta}{I_{0b} - I_{0d}} \right]$

Using the cotangent function, rather than a tangent function, for the relative retardation changes its periodicity, so that the period ends occur at $N = 0, 1/2, 1, 1 1/2, \ldots$ rather than $N = 1/4, 3/4, 1 1/4, 1 3/4, \ldots$, which slightly reduces the need for unwrapping because, for example, a fringe-order distribution, with a minimum of zero and a maximum of 1, would have a discontinuity only along $N = 1/2$, instead of along $N = 1/4$ and $3/4$.

Box 9.2: Reflection Photoelasticity

The stresses on the surface of complex engineering components can be determined using reflection photoelasticity, which involves bonding a thin transparent coating to the component surface using a reflective adhesive. The coating acts as a strain witness, by reproducing the strains experienced by the surface of the component. Polarized light is transmitted through the coating, reflected by the adhesive layer, and transmitted back through the coating. A dark-field circular polariscope, combined with a null-balance compensator, is the common instrument used to analyze the resultant fringe patterns.

It is assumed that a state of plane stress exists in the coating, that is, $\sigma_3 = \sigma_{zz} = 0$, and that the surface strain in the component is transmitted to the coating without distortion, that is, $\varepsilon_1^{\text{coating}} = \varepsilon_1^{\text{component}}$ and $\varepsilon_2^{\text{coating}} = \varepsilon_2^{\text{component}}$; thus using Hooke's Law and the stress-optic law yields

$$\sigma_1^{\text{component}} - \sigma_2^{\text{component}} = \frac{Nf_\varepsilon E^{\text{component}}}{1+\nu}$$

A wide range of coatings are available commercially and are usually supplied with a sensitivity value K where $f_\varepsilon = K/\lambda$. In general, coatings should be selected to provide a large number of fringes for analysis, for which high values of coating sensitivity and thickness are required; however, the coating needs to be thin, in order to reproduce the component strain and avoid reinforcement effects, particularly in bending. This conundrum is relieved, to a large extent, by digital photoelasticity, which allows analysis of subinteger fringes with ease. This also reduces the impact of Poisson ratio mismatch between the coating and component, which causes errors in the fringe data in a border (4h wide) around the edge of the coating.

The disadvantages of reflection photoelasticity are that detailed surface preparation is required, it has a limited temperature range, and detailed analysis is difficult especially on curved surfaces. The advantages are that it applies to real components, a full-field view of surface strain is readily seen, and the equipment is well established and simple to use. See [19] for more details.

An example of the application of the technique to a compressor blade from a jet engine is shown in Figure 9.5.

9.6
Material and Load Selection

In two- and three-dimensional (see Box 9.1) photoelasticity, a model is used to represent the component of interest, whereas in reflection photoelasticity a coating is bonded to the component and acts as a strain witness. Coating selection is a specialized topic that is dealt with in Box 9.2. For photoelastic models, the material is usually selected to be isotropic and homogeneous, in order to the render the analysis of the fringe patterns viable. Since many transparent materials exhibit temporary birefringence, the choice can be difficult and a high figure of merit is often defined as a criterion, along with good machinability, low creep, and low time-edge effects. The *figure of merit* Q is defined as the ratio of Young's modulus to material fringe constant: that is,

$$Q = \frac{E}{f_\sigma} \tag{9.19}$$

Time-edge effects refer to the tendency for the material to absorb moisture from the atmosphere, especially through machined edges. The resultant gradient of moisture in the specimen creates a differential expansion and strains that produce a fringe order along the edge, which has the appearance of a rind.

Glass, epoxy, and polycarbonate have figures of merits of 175, 292, and 354, respectively, which are high. While they all exhibit low creep and time-edge effects, only epoxy has a high machinability. Epoxy is widely used as a photoelastic material, though at room temperature it exhibits brittle failure, and hence polycarbonate has been used to model fatigue processes [16].

In the elastic state, based on considerations of equilibrium and compatibility, the distribution of stress is independent of the loads and the scale of the model employed. Thus, it is appropriate to select the loads and scale as convenient to the experimental setup and the facilities available. However, it is important to maintain geometric similarity before and after loading; otherwise, a change in component shape will cause a change in the distribution of stress. Thus, it is desirable for

$$\frac{\varepsilon_m}{\varepsilon_c} = 1 \tag{9.20}$$

where subscript m refers to the model and c to the component. And,

$$\frac{\varepsilon_m}{\varepsilon_c} = \frac{\sigma_m}{\sigma_c} \cdot \frac{E_c}{E_m} = \frac{W_m}{W_p} \cdot \frac{L_c h_c}{L_m h_m} \cdot \frac{E_c}{E_m} \tag{9.21}$$

where E is Young's modulus, W is the applied load, and L is a characteristic length; and it is assumed that the Poisson ratio of the model and component are the same. Combining expressions (9.20) and (9.21) results in

$$W_m = W_c \cdot \frac{L_m h_m}{L_c h_c} \cdot \frac{E_m}{E_c} \tag{9.22}$$

So, an appropriate selection of load and scale can be made to maintain dimensional similitude between the model and component being analyzed, and using Eq. (9.6) to ensure that sufficient fringes are present for analysis.

It is important to note that similitude will not hold adjacent to loading points because of nonlinear behavior; so, it is important to invoke St. Venant's principle, by not collecting data closer to the loaded area than the largest dimension of the loaded area.

Tutorial Exercise 9.6

The fringe pattern in Figure 9.6 was obtained from a slice taken from a model of a full-scale M30 bolt that was stress-frozen in an oven [17]. At stress-freezing temperature (see Box 9.1), the ratio of Young's moduli of the epoxy resin and steel is 1.60×10^{-3}, and the material fringe constant is 0.463 MPa/fringe per mm. Calculate a reasonable model load to apply during stress-freezing, if the applied load for a steel M30 bolt is 210 kN.

Solution:

An M30 bolt has a nominal diameter of 30 mm which can be taken as a characteristic dimension, so it would be appropriate to replace h by D; but it is

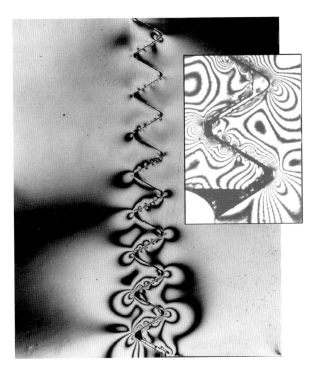

Figure 9.6 Photograph from a dark-field circular polariscope (see Figure 9.3) of the photoelastic fringe pattern from a slice (thickness 1.5 mm) removed from a stress-frozen model (see Box 9.1) of a nut and bolt assembly with the bolt axis on the left edge of the picture and the clamped face of the nut along the right end of the bottom edge; inset is a slice from different meridional position viewed in a fringe-multiplying polariscope at ×5 multiplication (see [17] for more details).

a full-scale model, so the dimensional scaling factor is 1.

$$W_m = W_c \cdot \frac{D_m^2}{D_c^2} \cdot \frac{E_m}{E_c} = (210 \times 10^3) \times 1 \times (1.6 \times 10^{-3}) = 336 \, \text{N}$$

And so, the tensile stress along the shank of the bolt will be

$$\sigma_m = \frac{W_m}{(\pi D_m^2)/4} = \frac{336}{\pi (30 \times 10^{-3})^2 / 4} = 0.475 \times 10^6 \, \text{Pa}$$

And, assuming that this is a principal stress and that the minimum principal stress is zero, then substituting into Eq. (9.6),

$$N = \frac{(\sigma_1 - \sigma_2) h}{f_\sigma} = \frac{0.475 \times 1.5}{0.463} = 1.53 \, \text{fringes}$$

for a slice thickness of 1.5 mm, which is small enough to assume that the stress distribution is approximately constant through its thickness. So, in the absence of stress concentrations, the fringe order in the slice would be just more than

1½ fringes; and this is a reasonable number to allow an accurate evaluation of the stress distribution.

9.7
Stress Analysis

The calculation of stress (or strain) values from fringe order data initially requires knowledge of the material fringe constant in Eq. (9.6). A calibration of the material can be performed using a specimen for which the stress field is known. The most common approach is to use a tension specimen for which, in two-dimensional photoelasticity, the fringe order can be plotted as a function of applied load. It is often more convenient to plot load as a function of fringe order, at every half-order fringe, by switching between dark field (for integer fringes) and light field (for half-order fringes) in a circular polariscope. In a stress-freezing experiment, a loaded tensile specimen can be subjected to the same thermal cycle as the model, and then can be cut in two using a single cut at about 45° to the loading axis, to form a wedge in which the fringe order increases with the thickness of the wedge. A precise determination of the through-thickness fringe order in the full section of the specimen can be made to find the material fringe constant. Alternative geometries for calibration specimens include a disc compressed along a diameter (Figure 9.4) or a beam subject to four-point bending. For the disc, the principal stress difference as a function of x along the radius perpendicular to the load is given by

$$(\sigma_1 - \sigma_2) = \frac{8P}{\pi Dh} \cdot \frac{D^4 - 4D^2 x^2}{(D^2 + 4x^2)^2} \tag{9.23}$$

And at the center, where $x = 0$,

$$(\sigma_1 - \sigma_2) = \frac{8P}{\pi Dh} = \frac{Nf_\sigma}{h}, \quad \text{that is,} \quad f_\sigma = \frac{8P}{\pi DN} \tag{9.24}$$

For the beam loaded in four-point bending, in the mid-section the stress is uniform and equal to

$$\sigma_1 = \frac{My}{I} \quad \text{and} \quad \sigma_2 = 0 \tag{9.25}$$

where M is the applied moment and y is the distance from the neutral axis parallel to the applied loads. So given $I = (hb^3/12)$, for loads P applied at L and $2L$ from the center of the beam

$$(\sigma_1 - \sigma_2) = \frac{12PLy}{hb^3} = \frac{Nf_\sigma}{h}, \quad \text{that is,} \quad N = \frac{12PL}{b^3 f_\sigma} y \tag{9.26}$$

And the fringe order can be plotted as a function of y, in order to obtain the material fringe constant f_σ. The four-point bending arrangement is more complicated to set up, but provides an accurate value of material fringe constant from a single load application and, hence, is well-suited to use with photography in two-dimensional photoelasticity and three-dimensional stress-freezing.

Photoelasticity readily provides the difference in the principal stresses; however, separating the stresses is an involved process and rarely undertaken in practice. The classical approach is known as the *shear difference method* [4] and is based on the equilibrium equations for plane stress in the absence of body forces [1]:

$$\frac{\partial \sigma_x}{\partial x} + \frac{\partial \tau_{yx}}{\partial y} = 0 \quad \text{and} \quad \frac{\partial \sigma_y}{\partial y} + \frac{\partial \tau_{xy}}{\partial x} = 0 \tag{9.27}$$

Integration of these expressions leads to

$$\sigma_x = (\sigma_x)_0 - \int \frac{\partial \tau_{yx}}{\partial y} dx \quad \text{and} \quad \sigma_y = (\sigma_y)_0 - \int \frac{\partial \tau_{xy}}{\partial x} dy \tag{9.28}$$

Or in terms of finite differences,

$$\sigma_x = (\sigma_x)_0 - \sum \frac{\Delta \tau_{yx}}{\Delta y} \quad \text{and} \quad \sigma_y = (\sigma_y)_0 - \sum \frac{\Delta \tau_{xy}}{\Delta x} \tag{9.29}$$

where $(\sigma_x)_0$ and $(\sigma_y)_0$ are initial values that can be found at a free boundary, where the fringe order is directly proportional to single principal stress. Subsequent values of σ_x or σ_y can be obtained by implementing the finite difference operation over a rectangular grid, given that from a Mohr's circle of stress

$$2\tau_{xy} = (\sigma_2 - \sigma_1)\sin 2\theta = \frac{Nf_\sigma}{2h}\sin 2\theta \tag{9.30}$$

Modern techniques of automated photoelastic fringe analysis have made this and more complex methods of stress separation more practical, though stress separation remains unreliable for engineering components because of the dependence on isoclinic fringes for which it is difficult to obtain high-quality data [18]. In reflection photoelasticity (Box 9.2), the finite difference operation must be applied along the isoclinics. Stress separation at a point can be performed using measurements taken at oblique incidence but is tedious to perform for an area [19]. To avoid these issues, reflection photoelasticity can be combined with thermoelastic stress analysis [20] which provides maps of the sum of the principal stresses $(\sigma_1 + \sigma_2)$, and so the combination by addition and subtraction readily yields separated principal stresses. In addition, photoelastic coatings are generally black in the infrared spectrum used by thermoelastic stress analysis, so that the two techniques can be used simultaneously.

Detailed fringe order data can be used to obtain stress maps, if there is *a priori* knowledge of the form of the stress field in the model or component; in these circumstances, the stress field equations can be fitted to the fringe order data by varying the coefficients in the equations. It is important that there are substantially more data points than the number of coefficients in the equations. The most common application of this approach is in fracture mechanics where the stress field around a crack tip is assumed to be described by one of a number of theoretical descriptions, such those due to Westergaard or Muskhelishvili [21]. In these cases, the form of the stress field equation implies an iterative fit to an array of fringe order data, which is often performed using a Newton–Raphson scheme. Overall, the approach is known as the *multipoint over-deterministic method* and is widely used in fatigue and fracture studies using two-dimensional, three-dimensional [22], and reflection photoelasticity. The approach can be used in other applications where the form of the stress field is well defined, such as in contact mechanics.

Tutorial Exercise 9.7

For the crack (blunt notch) in Figure 9.7, plot the fringe orders along the vertical line through the crack tip; hence, calculate the stress intensity factor for the crack, if the far-field stress σ_{ox} is assumed to be zero and given that the material fringe constant is $f_\sigma = 14$ MPa fringe^{-1} mm^{-1}.

Solution:

The very-near-field equations for the stress field around the tip of a mode I crack are [1]

$$\sigma_{xx} = \frac{K_I}{\sqrt{2\pi r}} \cos\frac{\theta}{2} \left(1 - \sin\frac{\theta}{2} \sin\frac{3\theta}{2}\right) - \sigma_{ox}$$

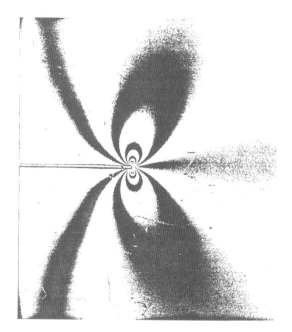

Figure 9.7 Photograph of the photoelastic fringe pattern in a two-dimensional plate (thickness 6 mm) with a machined blunt notch of length $a = 20$ mm subject to uniaxial loading perpendicular to the notch, that is, pure mode I loading.

$$\sigma_{yy} = \frac{K_I}{\sqrt{2\pi r}} \cos\frac{\theta}{2}\left(1 + \sin\frac{\theta}{2}\sin\frac{3\theta}{2}\right)$$

$$\tau_{xy} = \frac{K_I}{\sqrt{2\pi r}} \sin\frac{\theta}{2}\cos\frac{\theta}{2}\cos\frac{3\theta}{2}$$

and the maximum shear stress $\hat{\tau}$ is given by

$$(2\hat{\tau})^2 = (\sigma_{yy} - \sigma_{xx})^2 + 2\tau_{xy}^2$$

So,

$$(2\hat{\tau})^2 = \frac{K_I^2}{2\pi r}\sin^2\theta + \frac{2\sigma_{ox}K_I}{\sqrt{2\pi r}}\sin\theta\sin\frac{3\theta}{2} + \sigma_{ox}^2$$

For analysis along a vertical line, $\theta = 90°$, and assuming $\sigma_{ox} = 0$, then this reduces to

$$2\hat{\tau} = \frac{K_I}{\sqrt{2\pi r}} \quad \text{and} \quad 2\hat{\tau} = (\sigma_1 - \sigma_2) = \frac{Nf_\sigma}{h}$$

So,

$$\frac{Nf_\sigma}{h} = \frac{K_I}{\sqrt{2\pi r}} \quad \text{or} \quad N = \frac{h}{\sqrt{2\pi f_\sigma}} K_I \frac{1}{\sqrt{r}} = 0.1709 K_I \frac{1}{\sqrt{r}}$$

Now, measure the location of the fringe orders along a vertical line through the crack tip and plot them as a function of $1/\sqrt{r}$. Note that the photograph was taken in a light-field circular polariscope because the background is white and there are no isoclinics. Hence, the light fringes are integer values and the dark fringes are half-order values. There are zero-order fringes at the corners formed by the plate and the start of the notch, so counting fringe orders toward the notch tip, the complete fringe loop furthest from the crack tip is of order $1\frac{1}{2}$. The scale can be obtained by measuring the length of notch.

Fringe order (N)	Mean r (mm)	$1/\sqrt{r}$
1.5	12.4	0.3
2.0	6.5	0.4
2.5	3.4	0.5
3.0	2.3	0.7
3.5	1.6	0.8
4.0	1.2	0.9
4.5	0.9	1.1

Mean r refers to the mean of the measured values above and below the notch tip.

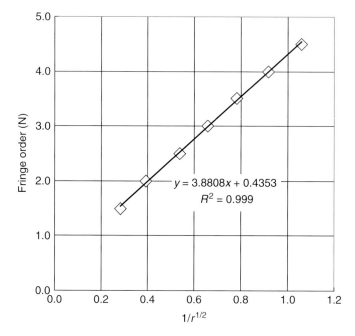

The data is linear as close to the crack tip as it is possible to measure, that is, there is no appreciable plastic zone, so a linear regression line can be fitted to all of the data. The gradient is $3.8808 = 0.1709 K_I$ so $K_I = 22.7$ MPa$\sqrt{\text{m}}$.

9.8
Conclusions

The principal techniques of photoelastic stress analysis have been described, namely, transmission, reflection (Box 9.2), and three-dimensional photoelasticity (Boxes 9.1 and 9.3). The stress-optic law, which governs the relationship between the observed isochromatic fringes and the stress inducing them, is

$$\sigma_1 - \sigma_2 = \frac{N f_\sigma}{h}$$

Methods for interpretation and evaluation for fringes have been outlined, together with comments on converting fringe order data into stress distributions. Finally, recent developments in automated or digital photoelasticity have been summarized, including phase-stepping, spectral contents analysis, and RGB photoelasticity.

Although the popularity of photoelastic stress analysis is fading, it retains some key advantages over other techniques, including the ability to evaluate the full stress tensor throughout a three-dimensional prototype.

Box 9.3: Integrated Photoelasticity

The photoelastic analysis of certain three-dimensional shapes can be performed by optical, rather than physical, sectioning, which allows the use of live-loading and the nondestructive evaluation of residual stresses in glass objects.

The direction and magnitudes of the principal stresses usually vary continuously along the light path when a three-dimensional component is viewed in a polariscope. In integrated photoelasticity, this situation is represented by an optically equivalent model that consists of a linear retarder δ at a certain angle θ, and a pure rotator χ. These quantities, known as *characteristic parameters*, can be determined by the collection of tomographic images of the component combined with the phase-stepping method, and are related to the principal stresses in the component. The method is used mainly for the nondestructive determination of residual stresses in glass [26] and, in some circumstances, has to be augmented by magneto-photoelasticity in order to handle stress distributions that are symmetric through the thickness of the glass.

Problems

9.1 Explain to a friend, without using technical terms, the mechanics underlying the generation of a photoelastic fringe pattern in a two-dimensional material subject to strain.

9.2 Using Mueller matrices, or otherwise, derive an expression for the light intensity emitted by a circular dark-field polariscope containing a specimen with a relative retardation α orientated at angle γ to the optical axis of the polarizer.

9.3 Identify the fringe orders in the nut in the photograph of a slice from a nut and bolt in a dark-field circular polariscope, as shown in Figure 9.6, and plot the distribution of stress concentration in the root of the nut thread, as a function of distance from the clamped face of the nut.

9.4 Derive, using Mueller matrices or by another means, an expression for the light intensity as function of the rotation of the analyzer during Senarmont compensation. Hence, establish the condition for extinction in the Senarmont compensation method.

9.5 Using the solution to Question 9.2 or another method, obtain an expression describing the light intensity emitted from a dark-field circular polariscope as a function of wavelength, assuming the polariscope contains a two-dimensional photoelastic specimen with a uniform relative retardation and isoclinic angle. Hence, plot the spectral signature for a number of values of fringe order between 0 and 6; finally, integrate these signatures in three intervals representing the red, green, and blue channels in an RGB camera.

9.6 Design the photoelastic experiment required to generate the photoelastic fringe pattern in a two-dimensional disc subject to compression across its vertical diameter in Figure 9.4. You should select an appropriate material, dimensions, and loads for the specimen and demonstrate that your design will generate the elastic fringes in the photograph.

9.7 Using Westergaard's equations for the distribution around a crack tip, or otherwise, calculate the relative retardation and isoclinic angle in the region of a mixed-mode crack where the ratio of $(K_{II}/K_{I}) = 0.1$; then simulate the images that would be observed in the polariscope using the phase-steps in Tutorial Example 9.5; finally, obtain the relative retardation and isoclinic angle from the phase-stepping algorithm. Comment on how these simulated images might differ from those obtained in an experiment carried out using a crack initiated from a notch in a polycarbonate plate subject to cyclic loading.

Color Plates

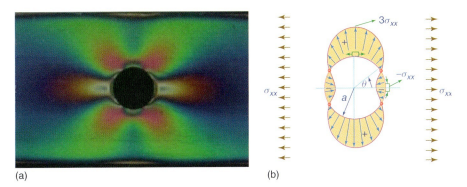

Figure 4.6 (a) Photoelastic fringe patterns in the vicinity of a hole in a finite plate subjected to axial tension. (b) Blowup of the nature of stress distribution in the boundary of a hole in an infinite plate subjected to uniaxial tension (Adapted from [3]). (This figure also appears on page 152.)

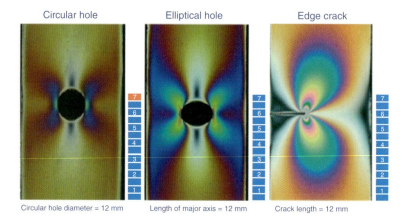

Figure 4.7 Isochromatics (contours of difference in principal stresses, see Chapter 9) observed near various stress raisers (Courtesy [3]). (This figure also appears on page 154.)

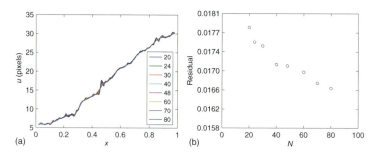

Figure 5.6 (a) Effect of the number of degrees of freedom N on the determined displacement field versus dimensionless abscissa. The natural splines basis is chosen and different discretizations N (given in the caption) are compared. The gray-level corrections both set to $N' = 2$. (b) Corresponding change of the global correlation residual. (This figure also appears on page 196.)

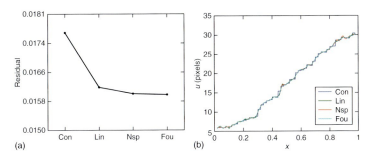

Figure 5.7 Comparison between (a) residuals and (b) displacement fields for different interpolation bases when $N = 48$ and $N' = 2$ (This figure also appears on page 196.)

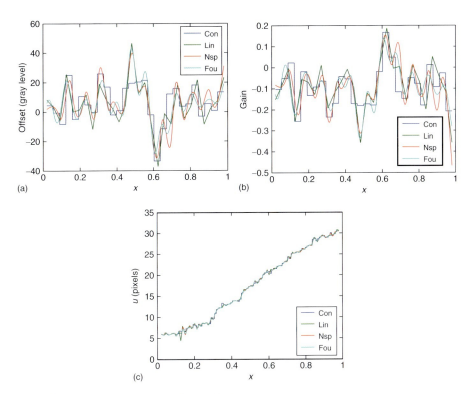

Figure 5.8 Comparison between (a,b) gray level corrections and (c) displacement fields for different bases when $N = 48$ and $N' = 28$. (This figure also appears on page 197.)

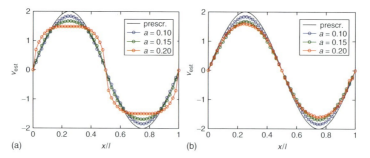

Figure 5.14 Prescribed (prescr.) and estimated displacements as functions of the ZOI center coordinate for different ratios $a = \ell/\lambda$ when (a) a piecewise constant and (b) linear interpolation is used. (This figure also appears on page 207.)

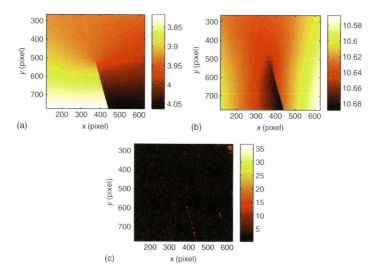

Figure 5.16 (a) Horizontal and (b) vertical components of the displacement field expressed in pixels and identified by using the integrated approach (1 pixel ↔ 1.85 μm). (c) Residual error map. The spatial integral of the square of this field gives the objective function to be minimized. The mean value of correlation residuals is about 1.5 gray level. (This figure also appears on page 212.)

Color Plates | 397

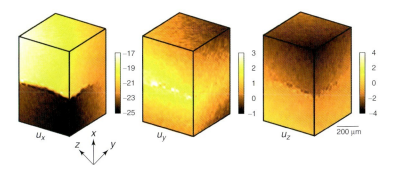

Figure 5.22 3-D rendition of the displacement field components inside the cracked sample expressed in voxels (1 voxel ↔ 3.5 µm). (This figure also appears on page 223.)

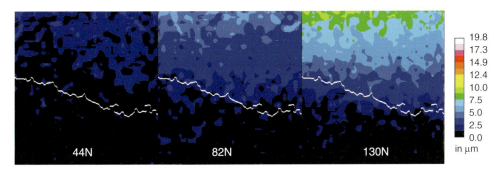

Figure 5.23 Maps of crack opening displacement Δu_x extracted from the jump in u_x displacement across the crack surface for different values of applied load. The white line indicates the position of the crack front [57]. (This figure also appears on page 223.)

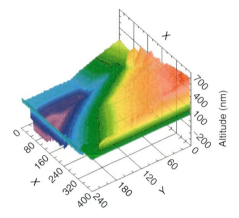

Figure 6.11 Displacement map. (This figure also appears on page 243.)

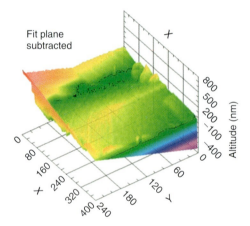

Figure 6.12 Tilt correction. (This figure also appears on page 243.)

Figure 6.44 (a) R image. (b) G image. (c) Two-color image. (This figure also appears on page 291.)

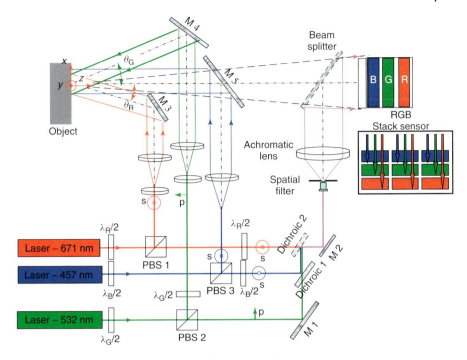

Figure 6.48 Experimental setup (M1–M5: flat mirrors, object–sensor distance is 1630 mm). (This figure also appears on page 290.)

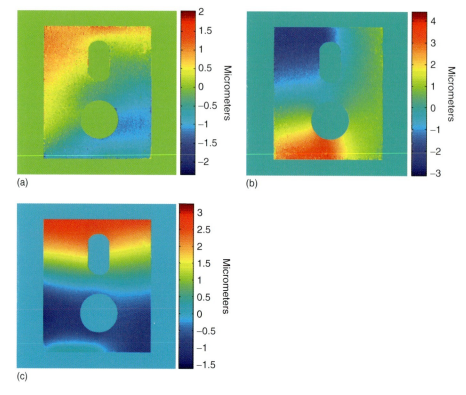

Figure 6.52 (a) u_x, (b) u_y, (c) and u_z – displacement field obtained from the set of data. (This figure also appears on page 292.)

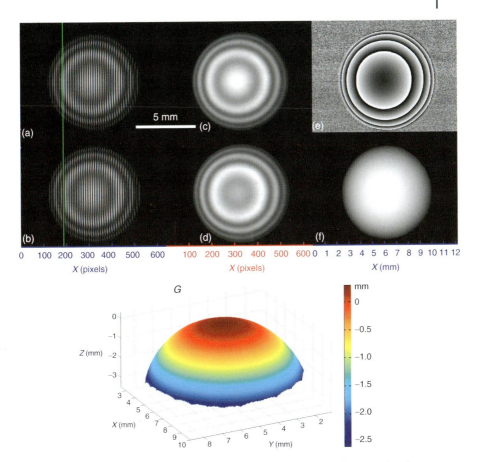

Figure 7.5 Moiré profilometry on a sphere. (a) Moiré interferogram showing moiré fringes and grid noise. (b) A similar moiré interferogram as in (a), but with both grids translated over $\Delta = p/2$. (c) After averaging a number of grid-shifted interferograms, grid noise is removed leaving a clear moiré topogram. (d) By changing the relative distance between the projection and the observation grid by $\Delta = p/4$ (in projection moiré) or by translating the object (in shadow moiré), a moiré topogram with a $\pi/2$ difference in fringe phase can be obtained. (e) From a number of (at least three) phase-shifted topograms, the wrapped fringe phase can be calculated with a phase ambiguity of 2π. (f) After removing phase jumps, the unwrapped object surface height is obtained in units of radians. The gray values in the figure represent the phase value, which corresponds to object height. (g) After multiplication with the appropriate calibration factor, the surface shape can be visualized in 3-D. (This figure also appears on page 319.)

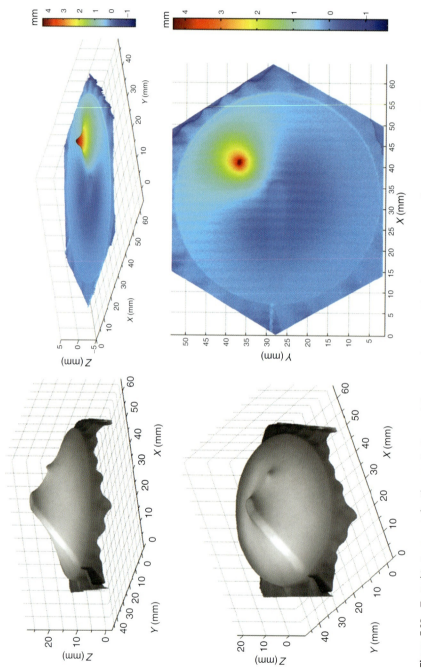

Figure 7.11 Test object measured with a projection moiré topography setup. The object is used as a 6× upscaled model for the human tympanic membrane. An oblique beam pushes the circular latex membrane outward, representing the malleus ossicle. A needle locally indents the surface during the measurement. (a,c) Two views of a 3-D surface shape reconstruction. (b,d) Two views of a 3-D reconstruction showing the indented membrane shape minus the membrane shape in resting position. Color represents deformation. (This figure also appears on page 340.)

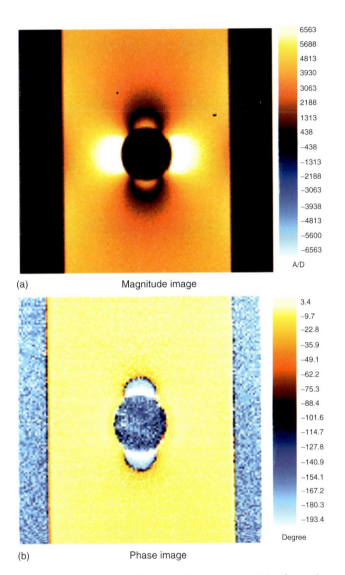

Figure 8.2 TSA data from hole in a plate specimen. (This figure also appears on page 358.)

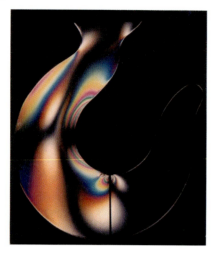

Figure 9.1 Photoelastic fringe pattern in two-dimensional model of a loaded crane hook viewed in a dark-field plane polariscope with both isoclinic and isochromatic fringes visible (a color version of this photograph can be found on the front cover of *Experimental Techniques*, 33 (1), January 2009, and on-line at *http://www3.interscience.wiley.com/journal/122200412/issue*). (This figure also appears on page 368.)

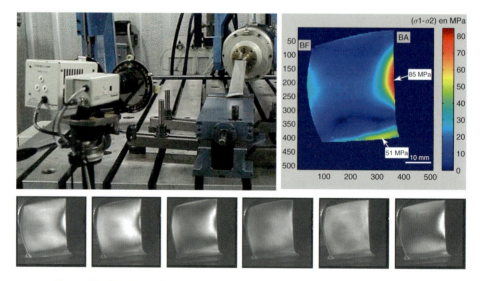

Figure 9.5 Experimental arrangement (top left) for reflection photoelasticity (see Box 9.2) applied to a compressor blade excited at around 7 kHz and illuminated using a strobe light source, in order to capture six phase-stepping images (bottom) that were processed to generate a map of principal stress difference (top right) for use in the validation of finite element analyses (see [13] for more details). (This figure also appears on page 381.)

References

Chapter 1

1. Ghatak, A. and Thyagarajan, K. (1978) *Contemporary Optics*, Plenum Publishing Corporation, New York.
2. Ghatak, A. (2009) *Optics*, McGraw-Hill, New York.
3. Born, M. and Wolf, E. (1999) *Principles of Optics*, Cambridge University Press, New York.
4. Longhurst, R.S. (1973) *Geometrical and Physical Optics*, Longman, London.
5. Mayer-Arendt, J.R. (1995) *Introduction to Classical and Modern Optics*, Prentice Hall, New York.
6. Pedrotti, F.L. and Pedrotti, L.S. (1993) *Introduction to Optics*, 2nd edn, Prentice Hall, New York.
7. Nussbaum, A. (1998) *Optical System Design*, Prentice Hall, New York.
8. Chi, K.R. (2009) Ever-increasing resolution. *Nature*, **462**, 675.
9. Thyagarajan, K. and Ghatak, A. (2010) *Lasers: Fundamentals and Applications*, Springer, New York.
10. Seigman, A.E. (1986) *Lasers*, University Science Books, Mill Valley, CA.
11. Ghatak, A. and Thyagarajan, K. (1998) *Introduction to Fiber Optics*, Cambridge University Press, New York.
12. Marcuse D. (1977) Loss analysis of single mode fiber splices, *Bell Syst. Tech. J.*, **56**, 703.

Chapter 2

1. Saleh, B.E.A. and Teich, M.C. (2007) *Fundamentals of Photonics*, 2nd edn, John Wiley & Sons, Inc.
2. Hecht, E. (2001) *Optics*, 4th edn, Addison-Wesley.
3. Boreman, G.D. (1998) *Basic Electro-Optics for Electrical Engineers*, Tutorial Texts in Optical Engineering, Vol. **TT31**, SPIE Optical Engineering Press.
4. RCA (Radio Coorporation of America) (1974) *RCA Electro-Optics Handbook*, Burle Technologies Inc.
5. Holst, G.C. (1996) *CCD Arrays Cameras and Displays*, SPIE Press.
6. ZEMAX ZEMAX–Software for Optical Simulation and System Design, www.zemax.com.
7. Code V -Optical Design, Analysis, and Illumination Calculations, www.opticalres.com.
8. OSLO Optical Design Software, www.sinopt.com.
9. Sze, S.M. (1981) *Physics of Semiconductor Devices*, 2nd edn, John Wiley & Sons, Inc.
10. Sze, S.M. (2002) *Semiconductor Devices – Physics and Technology*, 2nd edn, John Wiley & Sons, Inc.
11. Theuwissen, A.J.P. (1995) *Solid-State Imaging with Charge-Coupled Devices*, Kluwer Academic Publishers.

12. Seitz, P. (1999) in *Handbook of Computer Vision and Applications* (eds B. Jähne, H. Haussecker, and P. Geissler), Academic Press, New York, pp. 165–222.
13. Otha, J. (2008) *Smart CMOS Image Sensors and Applications*, CRC Press.
14. Lotto, C. (2010) Synchronous and asynchronous detection of ultra-low light levels using CMOS-compatible semiconductor technologies. PhD thesis. University of Neuchâtel.
15. Seitz, P. and Theuwissen, A.J.P. (eds) (2011) *Single-Photon Imaging*, Springer Series in Optical Design.
16. EMVA 1288 Standard. *Standard for Characterization and Presentation of Specification Data for Image Sensors and Cameras (Most Complete Document for Comparison of Imager and Camera Performances, but Still Missing the Photon Transfer Curve)*, EMVA, www.standard1288.org.
17. Photonfocus (High-Speed) CMOS Imager and Camera Producer, www.photonfocus.com.
18. Wikipedia Image Sensor Format, http://en.wikipedia.org/wiki/Image_sensor_format.
19. Bovik, A. (2009) *The Essential Guide to Image Processing*, 2nd edn, Academic Press.
20. Jain, A.K. (1988) *Fundamentals of Digital Image Processing*, Prentice Hall.
21. Baechler, T. et al. (2009) Single-photon resolution CMOS integrating image sensors. Proceedings EuroSensors XIII, Lausanne.

Chapter 3

1. Bothe, T. and Burke, J. (1998) Minimierung störender phasenfluktuationen beim einsatz des phasenshiftverfahrens in der elektronischen speckelmuster-interferometrie. 2nd edn, Diploma thesis. Carl von Ossietzky University Oldenburg.
2. Ichioka, Y. and Inuiya, M. (1972) Direct phase detecting system. *Appl. Opt.*, 11 (7), 1507–1514.
3. Womack, K. (1984) Interferometric phase measurement using spatial synchronous detection. *Opt. Eng.*, 23 (4), 391–395.
4. Ströbel, B. (1996) Processing of interferometric phase maps as complex-valued phasor images. *Appl. Opt.*, 35 (13), 2192–2198.
5. Burke, J. and Helmers, H. (1998) Complex division as a common basis for calculating phase differences in ESPI in one step -- Technical note. *Appl. Opt.*, 37 (13), 2589–2590.
6. Bracewell, R. (1987) *The Fourier Transform and its Applications*, 2nd edn, McGraw-Hill, New York.
7. Surrel, Y. (1996) Design of algorithms for phase measurements by the use of phase stepping. *Appl. Opt.*, 35 (1), 51–60.
8. Schwider, J., Burow, R., Elßner, K., Grzanna, J., Spolaczyk, R., and Merkel, K. (1983) Digital wave-front measuring interferometry: some systematic error sources. *Appl. Opt.*, 22 (21), 3421–3432.
9. Hariharan, P., Oreb, B., and Eiju, T. (1987) Digital phase-shifting interferometry: a simple error-compensating phase calculation algorithm. *Appl. Opt.*, 26 (13), 2504–2505.
10. Surrel, Y. (1998) Phase-shifting algorithms for nonlinear and spatially nonuniform phase shifts: comment. *J. Opt. Soc. Am. A*, 15 (5), 1227–1233.
11. Schwider, J., Falkenstörfer, O., Schreiber, H., Zöller, A., and Streibl, N. (1993) New compensating four-phase algorithm for phase-shift interferometry. *Opt. Eng.*, 32 (8), 1883–1885.
12. Schmit, J. and Creath, K. (1995) Extended averaging technique for derivation of error-compensating algorithms in phase-shifting interferometry. *Appl. Opt.*, 34 (19), 3610–3619.
13. de Groot, P. (1995) Derivation of algorithms for phase-shifting interferometry using the concept of a data-sampling window. *Appl. Opt.*, 34 (22), 4723–4730.
14. Schmit, J. and Creath, K. (1996) Window function influence on phase error in phase-shifting algorithms. *Appl. Opt.*, 35 (28), 5642–5649.
15. Surrel, Y. (1998) Extended averaging and data windowing techniques in phase-stepping measurements: an approach using the characteristic polynomial theory. *Opt. Eng.*, 37 (8), 2314–2319.
16. Schreiber, H. and Bruning, J. (2007) Phase shifting interferometry, in *Optical*

Shop Testing, 3rd edn (ed. D. Malacara), John Wiley & Sons, Inc. pp. 547–666.

17. de Groot, P. and Deck, L. (2008) New algorithms and error analysis for sinusoidal phase shifting interferometry. Proc. SPIE, **7063**, 70630k. (Interferometry XIV: Techniques and Analysis).

18. Onodera, R. and Ishii, Y. (1996) Phase-extraction analysis of laser-diode phase-shifting interferometry that is insensitive to changes in laser power. J. Opt. Soc. Am. A, **13** (1), 139–146.

19. Deck, L. (2003) Fourier-transform phase-shifting interferometry. Appl. Opt., **42** (13), 2354–2365.

20. Burke, J., Hibino, K., Hanayama, R., and Oreb, B. (2007) Simultaneous measurement of several near-parallel surfaces with wavelength-shifting interferometry and a tunable phase-shifting method. Opt. Lasers Eng., **45** (2), 326–341.

21. (a) Mertz, L. (1983) Complex interferometry. Appl. Opt., **22** (10), 1530–1534; (b) Mertz, L. (1983) Real-time fringe-pattern analysis. Appl. Opt., **22** (10), 1535–1539.

22. Küchel, M. (1991) The new Zeiss interferometer. Proc. SPIE, **1332**, 655–663. (Optical Testing and Metrology III: Recent Advances in Industrial Optical Inspection).

23. Bothe, T., Burke, J., and Helmers, H. (1997) Spatial phase shifting in ESPI: minimization of phase reconstruction errors. Appl. Opt., **36** (22), 5310–5316.

24. Fricke-Begemann, T. and Burke, J. (2001) Speckle interferometry: three-dimensional deformation field measurement with a single interferogram. Appl. Opt., **40** (28), 5011–5022.

25. Küchel, M. (1994) Method and apparatus for phase evaluation of pattern images used in optical measurement. US Patent 5361312.

26. Andrä, P., Jüptner, W., Kebbel, V., and Osten, W. (1997) General approach for the description of optical 3-D measuring systems. Proc. SPIE, **3174**, 207–215. (Videometrics V).

27. Freischlad, K. and Koliopoulos, C. (1990) Fourier description of digital phase-measuring interferometry. J. Opt. Soc. Am. A, **7** (4), 542–551.

28. Larkin, K. and Oreb, B. (1992) Design and assessment of symmetrical phase-shifting algorithms. J. Opt. Soc. Am. A, **9** (10), 1740–1748.

29. Smith, S. (1997) The Scientist and Engineer's Guide to Digital Signal Processing, California Technical Publishing, http://dspguide.com,http://www.analog.com/static/imported-files/tech_docs/dsp_book_Ch33.pdf. (accessed January 2012).

30. Gonzalez, A., Servin, M., Estrada, J.C., and Quiroga, J.A. (2011) Design of phase-shifting algorithms by fine-tuning spectral shaping. Opt. Express, **19** (11), 10692–10697.

31. Servin, M., Estrada, J.C., and Quiroga, J.A. (2009) Spectral analysis of phase-shifting algorithms. Opt. Express, **17** (19), 16423–16428.

32. Lynn, P. (1972) Recursive digital filters with linear-phase characteristics. Comput. J., **15** (4), 337–342.

33. Hibino, K., Hanayama, R., Burke, J., and Oreb, B. (2004) Tunable phase-extraction formulae for simultaneous shape measurement of multiple surfaces with wavelength-shifting interferometry. Opt. Express, **12** (23), 5579–5594.

34. Hibino, K., Larkin, K., Oreb, B., and Farrant, D. (1998) Phase-shifting algorithms for nonlinear and spatially nonuniform phase shifts: reply to comment. J. Opt. Soc. Am. A, **15** (5), 1234–1235.

35. Larkin, K. and Oreb, B. (1992) Propagation of errors in different phase-shifting algorithms: a special property of the arctangent function. Proc. SPIE, **1755** 219–227. (Interferometry: Techniques and Analysis).

36. Helmers, H. and Schellenberg, M. (2003) CMOS vs. CCD sensors in speckle interferometry. Opt. Laser Technol., **35**, 587–595.

37. Guo, H., Yu, Y., and Chen, M. (2007) Blind phase shift estimation in phase-shifting interferometry. J. Opt. Soc. Am. A, **24** (1), 25–33.

38. Xu, J., Xu, Q., and Chai, L. (2008) Iterative algorithm for phase extraction from interferograms with random and spatially nonuniform phase shifts. Appl. Opt., **47** (3), 480–485.

39. Brophy, C. (1990) Effect of intensity error correlation on the computed phase of phase-shifting interferometry. *J. Opt. Soc. Am. A*, **7** (4), 537–541.
40. Ghiglia, D. and Pritt, M. (1998) *Two-Dimensional Phase Unwrapping: Theory, Algorithms, and Software*, John Wiley & Sons, Inc.
41. Basistiy, I., Soskin, M., and Vasnetsov, M. (1995) Optical wavefront dislocations and their properties. *Opt. Commun.*, **119**, 604–612.
42. Burke, J. (2001) Application and optimisation of the spatial phase shifting technique in digital speckle interferometry. Doctoral thesis. Carl von Ossietzky University Oldenburg, Shaker, Aachen. http://aop.physik.uni-oldenburg.de/download/paper/Dissertation_Jan_Burke_2000.pdf. (accessed January 2012).
43. Aebischer, H. and Waldner, S. (1999) A simple and effective method for filtering speckle-interferometric phase fringe patterns. *Opt. Commun.*, **162**, 205–210.
44. Burke, J. and Helmers, H. (2000) Matched data storage in ESPI by combination of spatial phase shifting with temporal phase unwrapping. *Opt. Laser Technol.*, **32** (4), 235–240.
45. Huntley, J. (2001) Three-dimensional noise-immune phase unwrapping algorithm. *Appl. Opt.*, **40** (23), 3901–3908.
46. Falaggis, K., Towers, D., and Towers, C. (2009) Multiwavelength interferometry: extended range metrology. *Opt. Lett.*, **34** (7), 950–952.
47. Burke, J., Bothe, T., Osten, W., and Hess, C. (2002) Reverse engineering by fringe projection. *Proc. SPIE*, **4778**, 312–324. (Interferometry XI: Applications).

Chapter 4

1. Srinath, L.S. (2009) *Advanced Mechanics of Solids*, 3rd edn, Tata McGraw Hill.
2. Agarwal, B.D. and Broutman, L.J. (1990) *Analysis and Performance of Fiber Composites*, 2nd edn, John Wiley & Sons, Inc.
3. Ramesh, K. (2007) *e-Book on Engineering Fracture Mechanics*, IIT Madras, http://apm.iitm.ac.in/smlab/kramesh/book_4.htm.
4. Ramesh, K. (2009) *e-Book on Experimental Stress Analysis*, IIT Madras, http://apm.iitm.ac.in/smlab/kramesh/book_5.htm.
5. Broek, D. (1986) *Elementary Engineering Fracture Mechanics*, Kluwer Academic Publishers, Dordrecht.
6. Dally, J.W. and Riley, W.F. (1991) *Experimental Stress Analysis*, McGraw Hill, New York.
7. Sharpe, W.F. (ed.) (2008) *Springer Handbook of Experimental Solid Mechanics*, Springer, New York.
8. Post, D. and Han, B. (2008) Moiré Interferometry, in *Springer Handbook of Experimental Solid Mechanics* (ed. W.F. Sharpe), Springer, New York.
9. Durelli, A.J. (1977) The difficult choice: evaluation of methods used to determine experimentally displacements, strains and stresses. *Appl. Mech. Rev.*, **30**, 1167–1178.
10. Zheng, T. and Danyluk, S. (2002) Study of stresses in thin silicon wafers with near-infrared phase stepping photoelasticity. *J. Mater. Res.*, **17** (1), 36–42.
11. Tippur, H.V. (2004) Simultaneous and real-time measurement of slope and curvature fringes in thin structures using shearing interferometry. *Opt. Eng.*, **43** (12), 1–7.
12. John, L.W. (2006) *Standard Handbook of Chains: Chains for Power Transmission and Material Handling*, 2nd edn, American Chain Association, Taylor and Francis, Boca Raton.
13. Anand V., Dasari N., and Ramesh K. (2011) Innovative use of transmission and reflection photoelastic techniques to solve complex industrial problems, *Exp. Tech.*, **35** (5), 71–75.
14. Ramesh, K. (2000) *Digital Photoelasticity: Advanced Techniques and Applications*, Springer-Verlag, Berlin.
15. He, X., Zou, D., and Liu, S. (1998) Phase-shifting analysis in moiré interferometry and its applications in electronic packaging. *Opt. Eng.*, **37** (5), 1410–1419.
16. Francis, D., Tatam, R.P., and Groves, R.M. (2010) Shearography technology and applications: a review. *Meas. Sci. Technol.*, **21**, 102001–102029, doi: 10.1088/0957-0233/21/10/102001

17. Paul Kumar, U. (2010) Microscopic TV holography and interferometry for Microsystems metrology, PhD thesis. IIT Madras.
18. Paul Kumar, U., Bhaduri, B., Krishna Mohan, N., Kothiyal, M.P., and Asundi, A.K. (2008) Microscopic TV holography for MEMS deflection and 3-D surface profile characterization. *Opt. Lasers Eng.*, **46**, 687–694.
19. Paul Kumar, U., Kothiyal, M.P., and Krishna Mohan, N. (2009) Microscopic TV shearography for characterization of microsystems. *Opt. Lett.*, **34**, 1612–1614.
20. Paul Kumar, U., Bhaduri, B., Kothiyal, M.P., and Krishna Mohan, N. (2009) Two wavelength microinterferometry for 3-D surface profiling. *Opt. Lasers Eng.*, **47**, 223–229.

Chapter 5

1. Lucas, B.D. and Kanade, T. (1981) An iterative image registration technique with an application to stereo vision. Proceedings 1981 DARPA Imaging Understanding Workshop, pp. 121–130.
2. Burt, P.J., Yen, C., and Xu, X. (1982) Local correlation measures for motion analysis: a comparative study. Proceedings IEEE Conference on Pattern Recognition and Image Processing, pp. 269–274.
3. Sutton, M.A., Wolters, W.J., Peters, W.H., Ranson, W.F., and McNeill, S.R. (1983) Determination of displacements using an improved digital correlation method. *Im. Vis. Comput.*, **1** (3), 133–139.
4. Sutton, M.A., Cheng, M., Peters, W.H., Chao, Y.J., and McNeill, S.R. (1986) Application of an optimized digital correlation method to planar deformation analysis. *Im. Vis. Comput.*, **4** (3), 143–150.
5. Barker, D.B. and Fourney, M.E. (1977) Measuring fluid velocities with speckle patterns. *Opt. Lett.*, **1**, 135–137.
6. Dudderar, T.D. and Simpkins, P.G. (1977) Laser speckle photography in a fluid medium. *Nature*, **270**, 45–47.
7. Grousson, R. and Mallick, S. (1977) Study of flow pattern in a fluid by scattered laser light. *Appl. Opt.*, **16**, 2334–2336.
8. Adrian, R.J. (1984) Scattering particle characteristics and their effect on pulsed laser measurements of fluid flow: speckle velocimetry vs. particle image velocimetry. *Appl. Opt.*, **23**, 1690–1691.
9. Pickering, C.J.D. and Halliwell, N.A. (1984) Speckle laser in fluid flows: signal recovery with two-step processing. *Appl. Opt.*, **23**, 1128–1129.
10. Sutton, M.A., Zhao, W., McNeill, S.R., Helm, J.D., Piascik, R.S., and Riddel, W.T. (1999) Local crack closure measurements: Development of a measurement system using computer vision and a far-field microscope, in *Advances in Fatigue Crack Closure Measurement and Analysis: STP 1343*, 2nd edn (eds R.C. McClung and J.C. Newman Jr.), ASTM, pp. 145–156.
11. Forquin, P., Rota, L., Charles, Y., and Hild, F. (2004) A method to determine the toughness scatter of brittle materials. *Int. J. Fract.*, **125** (1), 171–187.
12. Elnasri, I., Pattofatto, S., Zhao, H., Tsitsiris, H., Hild, F. and Girard, Y. (2007) Shock enhancement of cellular structures under impact loading: Part I experiments. *J. Mech. Phys. Solids*, **55**, 2652–2671.
13. Tarigopula, V., Hopperstad, O.S., Langseth, M., Clausen, A.H., and Hild, F. (2008) A study of localisation in dual phase high-strength steels under dynamic loading using digital image correlation and FE analysis. *Int. J. Solids Struct.*, **45** (2), 601–619.
14. Besnard, G., Lagrange, J.-M., Hild, F., Roux, S., and Voltz, C. (2010) Characterization of necking phenomena in high speed experiments by using a single camera. *EURASIP J. Im. Video. Proc.*, **2010** (215956), 15.
15. Chasiotis, I. and Knauss, W.G. (2002) A new microtensile tester for the study of MEMS materials with the aid of atomic force microscopy. *Exp. Mech.*, **42** (1), 51–57.
16. Chasiotis, I. (ed.) (2007) Special issue on nanoscale measurements in mechanics. *Exp. Mech.*, **47** (1), 5–183.
17. Han, K., Ciccotti, M., and Roux, S. (2010) Measuring nanoscale stress intensity factors with an atomic force microscope. *EuroPhys. Lett.*, **89** (6), 66003.
18. Soppa, E., Doumalin, P., Binkele, P., Wiesendanger, T., Bornert, M., and Schmauder, S. (2001) Experimental and

numerical characterisation of in-plane deformation in two-phase materials. *Comput. Mater. Sci.*, **21** (3), 261–275.

19. (a) Sutton, M.A., Li, N., Joy, D.C., Reynolds, A.P., and Li, X. (2007) Scanning electron microscopy for quantitative small and large deformation measurements Part I: SEM imaging at magnifications from 200 to 10,000. *Exp. Mech.*, **47** (6), 775–787; (b) See also Sutton, M.A., Li, N., Garcia, D., Cornille, N., Orteu, J.J., McNeill, S.R., Schreier, H.W., Li, X., and Reynolds, A.P. (2007) Scanning electron microscopy for quantitative small and large deformation measurements Part II: experimental validation for magnifications from 200 to 10,000. *Exp. Mech.*, **47** (6), 789–804.

20. Sutton, M.A., Orteu, J.-J., and Schreier, H. (2009) *Image Correlation for Shape, Motion and Deformation Measurements: Basic Concepts, Theory and Applications*, Springer, New York.

21. Besnard, G., Guérard, S., Roux, S., and Hild, F. (2011) A space-time approach in digital image correlation: movie-DIC. *Opt. Lasers Eng.*, **49**, 71–81.

22. Baruchel, J., Buffière, J.-Y., Maire, E., Merle, P., and Peix, G. (2000) *X-ray tomography in material sciences*, Hermes Science, Paris (France).

23. Bay, B.K., Smith, T.S., Fyhrie, D.P., and Saad, M. (1999) Digital volume correlation: three-dimensional strain mapping using X-ray tomography. *Exp. Mech.*, **39**, 217–226.

24. Verhulp, E., van Rietbergen, B., and Huiskes, R. (2004) A three-dimensional digital image correlation technique for strain measurements in microstructures. *J. Biomech.*, **37** (9), 1313–1320.

25. Bornert, M., Chaix, J.-M., Doumalin, P., Dupré, J.-C., Fournel, T., Jeulin, D., Maire, E., Moreaud, M., and Moulinec, H. (2004) Mesure tridimensionnelle de champs cinématiques par imagerie volumique pour l'analyse des matériaux et des structures. *Inst. Mes. Métrol.*, **4**, 43–88.

26. Roux, S., Hild, F., Viot, P., and Bernard, D. (2008) Three dimensional image correlation from X-Ray computed tomography of solid foam. *Compos. Part A*, **39** (8), 1253–1265.

27. Réthoré, J., Tinnes, J.-P., Roux, S., Buffière, J.-Y., and Hild, F. (2008) Extended three-dimensional digital image correlation (X3D-DIC). *C. R. Mécanique*, **336**, 643–649.

28. Fayolle, X., Calloch, S., and Hild, F. (2007) Controlling testing machines with digital image correlation. *Exp. Tech.*, **31** (3), 57–63.

29. Fayolle, X. (2008) Correli^{FIC} : programme de pilotage d'essais asservis sur un facteur d'intensité des contraintes. MSc Dissertation. CNAM Paris.

30. Avril, S., Bonnet, M., Bretelle, A.-S., Grédiac, M., Hild, F., Ienny, P., Latourte, F., Lemosse, D., Pagano, S., Pagnacco, E., and Pierron, F. (2008) Overview of identification methods of mechanical parameters based on full-field measurements. *Exp. Mech.*, **48** (4), 381–402.

31. Rannou, J., Limodin, N., Réthoré, J., Gravouil, A., Ludwig, W., Baïetto-Dubourg, M.-C., Buffière, J.-Y., Combescure, A., Hild, F., and Roux, S. (2010) Three dimensional experimental and numerical multiscale analysis of a fatigue crack. *Comp. Meth. Appl. Mech. Eng.*, **199**, 1307–1325.

32. Maynadier, A., Poncelet, M., Lavernhe, K., and Roux, S. (2011) One-shot measurement of thermal *and* kinematic fields: Infra-Red Image Correlation (IRIC). *Exp. Mech.*. doi: 10.1007/s11340-011-9483-2.

33. Cooley, J.W. and Tukey, J.W. (1965) An algorithm for the machine calculation of complex fourier series. *Math. Comput.*, **19** (90), 297–301.

34. Schreier, H.W., Braasch, J.R., and Sutton, M.A. (2000) Systematic errors in digital image correlation caused by intensity interpolation. *Opt. Eng.*, **39** (11), 2915–2921.

35. Hild, F. and Roux, S. (2006) Digital image correlation: from measurement to identification of elastic properties - A review. *Strain*, **42**, 69–80.

36. ISO (1993) International Vocabulary of Basic and General Terms in Metrology (VIM). International Organization for Standardization, Geneva (Switzerland).

37. Hild, F., Raka, B., Baudequin, M., Roux, S., and Cantelaube, F. (2002) Multi-scale displacement field measurements of

compressed mineral wool samples by digital image correlation. *Appl. Opt.*, **IP 41** (32), 6815–6828.

38. Besnard, G., Hild, F., and Roux, S. (2006) Finite-element'' displacement fields analysis from digital images: application to Portevin-Le Chatelier bands. *Exp. Mech.*, **46**, 789–803.

39. Réthoré, J., Hild, F., and Roux, S. (2007) Shear-band capturing using a multiscale extended digital image correlation technique. *Comput. Meth. Appl. Mech. Eng.*, **196** (49–52), 5016–5030.

40. Chevalier, L., Calloch, S., Hild, F., and Marco, Y. (2001) Digital image correlation used to analyze the multiaxial behavior of rubber-like materials. *Eur. J. Mech. A/Solids*, **20**, 169–187.

41. Bornert, M., Brémand, F., Doumalin, P., Dupré, J.-C., Fazzini, M., Grédiac, M., Hild, F., Mistou, S., Molimard, J., Orteu, J.-J., Robert, L., Surrel, Y., Vacher, P., and Wattrisse, B. (2009) Assessment of digital image correlation measurement errors: methodology and results. *Exp. Mech.*, **49** (3), 353–370.

42. Bergonnier, S., Hild, F., and Roux, S. (2005) Digital image correlation used for mechanical tests on crimped glass wool samples. *J. Strain Anal.*, **40** (2), 185–197.

43. Triconnet, K., Derrien, K., Hild, F., and Baptiste, D. (2009) Parameter choice for optimized digital image correlation. *Opt. Lasers Eng.*, **47**, 728–737.

44. Williams, M.L. (1957) On the stress distribution at the base of a stationary crack. *ASME J. Appl. Mech.*, **24**, 109–114.

45. Roux, S. and Hild, F. (2006) Stress intensity factor measurements from digital image correlation: post-processing and integrated approaches. *Int. J. Fract.*, **140** (1–4), 141–157.

46. Roux, S., Réthoré, J., and Hild, F. (2009) Digital image correlation and fracture: an advanced technique for estimating stress intensity factors of 2D and 3D cracks. *J. Phys. D: Appl. Phys.*, **42**, 214004.

47. Leclerc, H., Périé, J.-N., Roux, S., and Hild, F. (2009) Integrated digital image correlation for the identification of mechanical properties, in *MIRAGE 2009*, LNCS 5496 (eds A. Gagalowicz and W. Philips), Springer-Verlag, Berlin (Germany), pp. 161–171.

48. Tikhonov, A.N. and Arsenin, V.Y. (1977) *Solutions of Ill-posed Problems*, John Wiley & Sons, Inc, New York.

49. Engl, H.W., Hanke, M., and Neubauer, A. (2000) *Regularization of Inverse Problems*, Kluwer Acad. Pub., Dordrecht (The Netherlands).

50. Leclerc, H., Périé, J.-N., Roux, S., and Hild, F. (2011) Voxel-scale digital volume correlation. *Exp. Mech.*, **51** (4), 479–490.

51. Fayolle, X., Calloch, S., and Hild, F. (2008) Contrôler une machine d'essai avec une caméra. *Méc. & Ind.*, **9** (5), 447–457.

52. Sutton, M.A., Yan, J.H., Tiwari, V., Schreier, H.W., and Orteu, J.-J. (2008) The effect of out of plane motion on 2D and 3D digital image correlation measurements. *Optics Lasers Eng.*, **46** (10), 746–757.

53. Gruen, A. and Huang, T.S. (2001) *Calibration and Orientation of Cameras in Computer Vision*, Springer Series in Information Sciences.

54. Hartley, R. and Zisserman, A. (2004) *Multiple View Geometry in Computer Vision*, Cambridge University Press, Cambridge.

55. Benoit, A., Guérard, S., Gillet, B., Guillot, G., Hild, F., Mitton, D., Périé, J.-N., and Roux, S. (2009) 3D analysis from micro-MRI during in situ compression on cancellous bone. *J. Biomech.*, **42**, 2381–2386.

56. Hild, F., Maire, E., Roux, S., and Witz, J.-F. (2009) Three dimensional analysis of a compression test on stone wool. *Acta Mater.*, **57**, 3310–3320.

57. Limodin, N., Réthoré, J., Buffière, J.-Y., Gravouil, A., Hild, F., and Roux, S. (2009) Crack closure and stress intensity factor measurements in nodular graphite cast iron using 3D correlation of laboratory X ray microtomography images. *Acta Mater.*, **57** (14), 4090–4101.

58. Limodin, N., Réthoré, J., Buffière, J.-Y., Hild, F., Roux, S., Ludwig, W., Rannou, J., and Gravouil, A. (2010) Influence of closure on the 3D propagation of fatigue cracks in a nodular cast iron investigated by X-ray tomography and 3D Volume Correlation. *Acta Mater.*, **58** (8), 2957–2967.

59. Besnard, G., Leclerc, H., Roux, S., and Hild, F. (2012) Analysis of image series

through digital image correlation, (submitted).
60. Hild, F., Roux, S., Gras, R., Guerrero, N., Marante, M.E., and Flórez-López, J. (2009) Displacement measurement technique for beam kinematics. *Opt. Lasers Eng.*, **47**, 495–503.
61. Hild, F., Roux, S., Guerrero, N., Marante, M.E., and Florez-Lopez, J. (2011) Calibration of constitutive models of steel beams subject to local buckling by using Digital Image Correlation. *Eur. J. Mech. A/Solids*, **30**, 1–10.
62. Kanninen, M.F. and Popelar, C.H. (1985) *Advanced Fracture Mechanics*, Oxford University Press, Oxford.
63. Black, T. and Belytschko, T. (1999) Elastic crack growth in finite elements with minimal remeshing. *Int. J. Num. Methods Eng.*, **45**, 601–620.
64. Moës, N., Dolbow, J., and Belytschko, T. (1999) A finite element method for crack growth without remeshing. *Int. J. Num. Methods Eng.*, **46** (1), 133–150.
65. Réthoré, J., Hild, F., and Roux, S. (2008) Extended digital image correlation with crack shape optimization. *Int. J. Num. Methods Eng.*, **73** (2), 248–272.
66. Réthoré, J., Roux, S., and Hild, F. (2009) An extended and integrated digital image correlation technique applied to the analysis fractured samples. *Eur. J. Comput. Mech.*, **18**, 285–306.

Chapter 6

1. Gabor, D. (1948) A new microscopic principle. *Nature*, **161**, 777–778.
2. Goodman, J. (1975) in *Laser Speckle and Related Phenomena*, vol. 9 (ed. J.C. Dainty), Springer, Berlin, Heidelberg, pp. 77–121.
3. Butters, J. and Leendertz, J. (1971) Speckle pattern and holographic techniques in engineering metrology. *Opt. Laser Technol.*, **3** (1), 26–30.
4. Macovski, A., Ramsey, S., and Schaefer, L.F. (1971) Time-laps interferometry and contouring using television systems. *Appl. Opt.*, **10** (12), 2722–2727.
5. Schwomma, O. (1972) Austrian Patent 298830.
6. Jones, R. and Wykes, C. (1983) *Holographic and Speckle Interferometry*, Cambridge University Press, London.
7. Løkberg, O. and Slettemoen, G. (1987) in *Applied Optics and Optical Engineering*, vol. 10 (eds J. Wyant and R. Shannon), Academic Press, New York, pp. 455–504.
8. Doval, A. (2000) A systematic approach to TV-holography. *Meas. Sci. Technol.*, **11**, R1–R36.
9. Rastogi, P.K. (2001) *Digital Speckle Pattern Interferometry and Related Techniques*, John Wiley & Sons, Inc.
10. Spooren, R. (1992) Double pulse subtraction TV-holography. *Opt. Eng.*, **31**, 1000–1007.
11. Slettemoen, G. (1979) General analysis of fringe contrast in electronic speckle pattern interferometry. *Opt. Acta*, **26** (3), 313–327.
12. Wykes, C. (1987) A theoretical approach to the optimisation of ESPI fringes with limited laser power. *J. Mod. Opt.*, **34** (4), 539–554.
13. Slettemoen, G. (1980) Electronic speckle pattern interferometric system based on a speckle reference beam. *Appl. Opt.*, **19** (4), 616–623.
14. Siebert, T., Splitthof, K., and Ettemeyer, A. (2004) A practical approach to the problem of the absolute phase in speckle interferometry. *J. Hologr. Speckle*, **1**, 32–38.
15. Gautier, B. (2005) Etudes et réalisation d'un interféromètre de speckle à mesure de forme intégrée. Thèse. Ecole des Mines de Paris, Alès.
16. Goudemand, N. (2005) 3D-3C speckle interferometry: optical device for measuring complex structures. ETHZ thesis no 15961. Swiss Federal Institute of Technology, Zürich.
17. Slettemoen, G. (1977) Optimal signal processing in electronic speckle pattern interferometry. *Opt. Commun.*, **23** (2), 213–216.
18. Lehmann, M. (1995) Optimisation of wavefield intensities in phase-shifting speckle interferometry. *Opt. Commun.*, **118**, 199–206.
19. Maack, T., Kowarschik, R., and Notni, G. (1998) Effect of reference beam in speckle interferometry. *Opt. Commun.*, **154**, 137–144.

20. Burke, J. (2001) Application and optimisation of the spatial phase shifting technique in digital speckle interferometry. PhD thesis. University of Oldenburg, Deparment of Physics, Oldenburg.
21. Owner-Petersen, M. (1991) Decorrelation and fringe visibility: on the limiting behaviour of various electronic speckle-pattern correlation interferometers. *J. Opt. Soc. Am. A*, **8** (7), 1082–1089.
22. Neiswander, P. and Slettemoen, G. (1981) Electronic speckle pattern interferometric measurements of the basilar membrane in the inner ear. *Appl. Opt.*, **20**, 4271–4276.
23. Gülker, G., Hinsch, K.D., and Kraft, A. (2001) Deformation monitoring on ancient terracotta warriors by microscopic TV-holography. *Opt. Lasers Eng.*, **36**, 501–512.
24. Gastinger, K. (2006) Low coherence speckle interferometry (LCSI) for the characterisation of adhesive bonded joints. PhD-thesis. Carl von Ossietzky University, Department of Physics, Oldenburg.
25. Gastinger, K. (2007) Low coherence speckle interferometry (LCSI) – when speckle interferometry goes sub-surface. 8th International Conference on Correlation Optics, SPIE Proceedings, Vol. 7008 (eds M. Kujawinska and O.V. Angelsky), p. 70081I.
26. Fujimoto, J. G., Drexler, W., Morgner, U., Kärtner, F. X., and Ippen, E. P. (2000) Optical coherence tomography: high resolution imaging using echoes of light. *Optics and Photonics News*, pp. **25**.
27. Fercher, A.F., Drexler, W., Hitzenberger, C., and Lasser, T. (2003) Optical coherence tomography – principles and applications. *Rep. Prog. Phys.*, **66**, 239–303.
28. Bouma, B.E. and Tearney, G. (2002) *Handbook of Optical Coherence Tomography*, 1st edn, Marcel Dekker, New York.
29. Mandel, L. and Wolf, E. (1995) *Optical Coherence and Quantum Optics*, 1st edn, Cambridge University Press, New York.
30. Fuji, T., Miyata, M., Kawato, S., Hattori, T., and Nakatsuka, H. (1997) Linear propagation of light investigated with a white-light Michelson interferometer. *J. Opt. Soc. Am. B*, **14**, 1074.
31. Gülker, G., Hinsch, K.D., and Kraft, A. (2003) Low-coherence ESPI in the investigation of ancient terracotta warriors. *Proc. Speckle Metrol.*, **4933**, 53–58.
32. Gastinger, K. and Winther, S. (2005) Optimisation of low-coherence speckle interferometry (LCSI) for characterisation of multi-layered materials. *Proc. SPIE*, **5858**, 157–168.
33. Gastinger, K., Hinsch, K.D., and Winther, S. (2006) Investigations of phase changes in semitransparent media using low coherence speckle interferometry (LCSI). Speckle06: Speckles, From Grains to Flowers, SPIE Volume 6341 Nimes, France (eds P. Slangen and C. Cerruti), p. 63410s.
34. Gastinger, K., Løvhaugen, P., Skotheim, O., and Hunderi, O. (2007) Multi-technique platform for dynamic and static MEMS characterisation. *Proc. SPIE*, **6616**, 6163K.
35. Leendertz, J.A. and Butters, J.N. (1973) An image-shearing speckle-pattern interferometer for measuring bending moments. *J. Phys. E*, **6**, 1107–1110.
36. Hung, Y.Y. (1982) Shearography, a new optical method for strain measurement and non-destructive testing. *Opt. Eng.*, **21** (3), 391–395.
37. Goodman, J.W. (2007) *Speckle Phenomena in Optics: Theory and Applications*, Roberts and Company Publishers, Greenwood Village, CO.
38. Steinchen, W. and Yang, L. (2003) *Digital Shearography: Theory and Applications of Digital Speckle Pattern Shearing Interferometry*, SPIE Press, Bellingham, WA.
39. Takeda, M. (1990) Spatial-carrier fringe-pattern analysis and its applications to precision interferometry and profilometry: an overview. *Ind. Metrol.*, **1**, 79–99.
40. Pedrini, G., Zou, Y.-L., and Tiziani, H.J. (1996) Quantitative evaluation of digital shearing interferogram using the spatial carrier method. *Pure Appl. Opt.*, **5** 313–321.
41. Huntley, J.M. and Saldner, H. (1993) Temporal phase-unwrapping algorithm for automated interferogram analysis. *Appl. Opt.*, **32** (17), 3047–3052.
42. Somers, P.A.A.M. and Bhattacharya, N. (2006) Three-bucket quadrature phase

stepping in a shearing speckle interferometer, Speckle06 Proceedings: Speckles, from Grains to Flowers, Proceedings of SPIE, Vol. 6341 (eds P. Slangen and C. Cerruti), pp. 63411K - 1–63411K-6.
43. Somers, P.A.A.M. and Bhattacharya, N. (2005) Maintaining sub-pixel alignment for a single camera two-bucket shearing speckle interferometer. *J. Opt. A: Pure Appl. Opt.*, **7**, S385–S391.
44. Kästle, R., Hack, E., and Sennhauser, U. (1998) Multiwavelength shearography for evaluation of in-plane strain distributions. *Proc. SPIE*, **3520**, 248–253.
45. Ettemeyer, A., Krupka, R., and Walz, T. (2004) Advanced inspection of helicopter structures using shearography. Conference Proceedings WCNDT 2004, http://www.ndt.net/. (accessed 2004–11).
46. Pezzoni, R. and Krupka, R. (2000) Laser-shearography for nondestructive testing of large area composite helicopter structures. Conference Proceedings WCNDT 2000, http://www.ndt.net/. (accessed 2000–11).
47. Andersson, J. and van den Bos, B. (2000) NDI on bonded sandwich structures with foam cores and stiff skins – shearography the answer? Conference Proceedings WCNDT 2000, http://www.ndt.net/. (accessed 2000–11).
48. Kalms, M. (2006) Mobile shearography for NDT of technical and artwork components. Conference Proceedings 12th A-PCNDT 2006 – Asia Pacific Conference on NDT, http://www.ndt.net/. (accessed 2007–11).
49. Somers, P.A.A.M. and Bhattacharya, N. (2010) Vibration phase-based ordering of vibration patterns acquired with a shearing speckle interferometer and pulsed illumination. *Strain*, **46** (3), 234–241 (online 2008). http://onlinelibrary.wiley.com/doi/10.1111/j.1475-1305.2008.00447.x/full
50. Leith, E. and Upatnieks, J. (1961) New technique in wavefront reconstruction. *J. Opt. Soc. Am.*, **51**, 1469.
51. Denisyuk, Y.N. (1962) Manifestation of optical properties of an object in wave field of radiation in scatters. *Dokl. Akad. Nauk. SSSR*, **144**, 1275–1278.
52. Kronrod, M.A., Merzlyakov, N.S., and Yaroslavskii, L.P. (1972) Reconstruction of a hologram with a computer. *Sov. Phys. Techn. Phys.*, **17**, 333–334.
53. Schnars, U. and Jüptner, W. (1994) Direct recording of holograms by a CCD target and numerical reconstruction. *Appl. Opt.*, **33**, 179–181.
54. Cuche, E., Bevilacqua, F., and Depeursinge, C. (1999) Digital holography for quantitative phase contrast imaging. *Opt. Lett.*, **24**, 291–293.
55. Javidi, B. and Tajahuerce, E. (2000) Three-dimensional object recognition by use of digital holography. *Opt. Lett.*, **25**, 610–612.
56. Nomura, T. and Javidi, B. (2007) Object recognition by use of polarimetric phase-shifting digital holography. *Opt. Lett.*, **32**, 2146–2148.
57. Yamaguchi, I., Kato, J., and Ohta, S. (2001) Surface shape measurement by phase shifting digital holography. *Opt. Rev.*, **8**, 85–89.
58. Picart, P., Diouf, B., Lolive, E., and Berthelot, J.-M. (2004) Investigation of fracture mechanisms in resin concrete using spatially multiplexed digital Fresnel holograms. *Opt. Eng.*, **43**, 1169–1176.
59. Pedrini, G., Schedin, S., and Tiziani, H.J. (2002) Pulsed digital holography combined with laser vibrometry for 3D measurements of vibrating objects. *Opt. Lasers Eng.*, **38**, 117–129.
60. Picart, P., Leval, J., Mounier, D., and Gougeon, S. (2003) Time averaged digital holography. *Opt. Lett.*, **28**, 1900–1902.
61. Picart, P., Leval, J., Grill, M., Boileau, J.-P., Pascal, J.C., Breteau, J.-M., Gautier, B., and Gillet, S. (2005) 2D full field vibration analysis with multiplexed digital holograms. *Opt. Express*, **13**, 8882–8892.
62. Kreis, T., Adams, M., and Jüptner, W. (1997) Methods of digital holography: a comparison. *Proc. SPIE*, **3098**, 224–233.
63. Wagner, C., Seebacher, S., Osten, W., and Jüptner, W. (1999) Digital recording and numerical reconstruction of lens less Fourier holograms in optical metrology. *Appl. Opt.*, **38**, 4812–4820.
64. Liebling, M. (2004) On fresnelets, interferences fringes, and digital holography. PhD thesis no. 2977. Ecole Polytechnique Fédérale de Lausanne, Switzerland.
65. Goodman, J.W. (1996) *Introduction to Fourier Optics*, McGraw-Hill, New York.

66. Picart, P. and Leval, J. (2008) General theoretical formulation of image formation in digital Fresnel holography. *J. Opt. Soc. Am. A*, **25**, 1744–1761.
67. Kriens, R.F.C. (1988) Quantitative holographic interferometry: measurement of solid objects deformations. PhD thesis no. 1613. Delft University of Technology.
68. Linet, V., Bohineust, X., and Dupuy, F. (1991) Three dimensional dynamic analysis of parts of automobile body by holographic interferometry. Proceedings of 3rd French-German Congress on Applications of Holography (ed. P. Smigielski), pp. 213–224.
69. Schedin, S., Pedrini, G., Tiziani, H.J., and Santoyo, F.M. (1999) Simultaneous three-dimensional dynamic deformation measurements with pulsed digital holography. *Appl. Opt.*, **38**, 7056–7062.
70. Picart, P., Mounier, D., and Desse, J.M. (2008) High resolution digital two-colour holographic metrology. *Opt. Lett.*, **33**, 276–278.
71. Tankam, P., Picart, P., Mounier, D., Desse, J.M., and Li, J.C. (2010) Method of digital holographic recording and reconstruction using a stacked colour image sensor. *Appl. Opt.*, **49**, 320–328.
72. Tankam, P., Song, Q., Karray, M., Li, J.C., Desse, J.M., and Picart, P. (2010) Real-time three-sensitivity measurements based on three-colour digital. Fresnel holographic interferometry. *Opt. Lett.*, **35**, 2055–2057.

Chapter 7

1. Buytaert, J.A.N. and Dirckx, J.J.J. (eds) (2010) *Optical Measurement Techniques for Structures and Systems*, Shaker Publishing, Maastricht, ISBN-978-90-423-0366-9.
2. Lee, H., Cho, H., and Kim, M. (2006) A new 3-D sensor system for mobile robots based on Moiré and Stereo vision technique. Proceedings of the International Conference on Intelligent Robots and Systems, pp. 1384–1389.
3. Gea, S.L.R., Decraemer, W.F., and Dirckx, J.J.J. (2005) Region of interest micro-CT of the middle ear: a practical approach. *J. X-Ray Sci. Technol.*, **13**, 137–147.
4. Dirckx, J.J.J. and Decraemer, W.F. (1997) Coating techniques in optical interferometric metrology. *Appl. Opt.*, **36**, 2776–2782.
5. Trucco, E. and Verri, A. (1998) *Introductory Techniques for 3-D Computer Vision*, Prentice Hall.
6. Chen, F., Brown, G.M., and Song, M. (2000) Overview of three-dimensional shape measurement using optical methods. *Opt. Eng.*, **39**, 10–22.
7. Lanman, D. and Taubin, G. (2009) Build your own 3-D scanner: 3-D photography for beginners. chapter in: ACM SIGGRAPH 2009 Courses, pp. 1–87.
8. Dorrio, B.V. and Fernandez, J.L. (1999) Phase-evaluation methods in whole-field optical measurement techniques. *Meas. Sci. Technol.*, **10**, R33–R55.
9. Takeda, M., Ina, H., and Koboyashi, S. (1982) Fourier-transform method of fringe-pattern analysis for computer-based topography and interferometry. *J. Opt. Soc. Am.*, **72**, 156–160.
10. Su, W.-H. and Liu, H. (2006) Calibration-based two-frequency projected fringe profilometry: a robust, accurate, and single-shot measurement for objects with large depth discontinuities. *Opt. Express*, **14**, 9178–9187.
11. Miao, H., Quan, C., Tay, C.J., and Fu, Y. (2007) Analysis of phase distortion in phase-shifted fringe projection. *Opt. Lasers Eng.*, **45**, 318–325.
12. Dirckx, J.J.J. and Decraemer, W.F. (1997) Optoelectronic moiré projector for real-time shape and deformation studies of the tympanic membrane. *J. Biomed. Opt.*, **2**, 176–185.
13. Creath, K. (1988) *Progress in Optics*, vol. **XXVI**, Elsevier, New York, pp. 357–373.
14. Osten, W. (2000) *Optical Methods in Experimental Solid Mechanics*, Springer-Verlag, pp. 308–363.
15. Yatagai, T., Idesawa, M., Yamaashi, Y., and Suzuki, M. (1982) Interactive fringe analysis system: applications to moiré contourogram and interferogram. *Opt. Eng.*, **21**, 901–906.
16. Mieth, U. and Osten, W. (1989) Three methods for the interpolation of phase

values between fringe pattern skeleton. chapter in: *Proc. SPIE*, **1163**, 151–154.

17. Benoit, P., Mathieu, E., Hormiere, J., and Thomas, A. (1975) Characterization and control of three-dimensional objects using fringe projection techniques. *Nouv. Rev. Optique*, **6**, 67–86.

18. Buytaert, J.A.N. and Dirckx, J.J.J. (2007) Design considerations in projection phase shift moiré topography, based on theoretical analysis of fringe formation. *J. Opt. Soc. Am. A*, **24**, 2003–2013.

19. Meadows, D., Johnson, W., and Allen, J. (1970) Generation of surface contours by moiré patterns. *Appl. Opt.*, **9**, 942–947.

20. Takasaki, H. (1970) Moiré topography. *Appl. Opt.*, **9**, 1467–1472.

21. Dirckx, J.J.J. and Decraemer, W.F. (1989) Phase shift moiré apparatus for automatic 3-D surface measurement. *Rev. Sci. Instrum.*, **60**, 3698–3701.

22. Dirckx, J.J.J. and Decraemer, W.F. (1990) Automatic calibration method for phase shift shadow moiré interferometry. *Appl. Opt.*, **29**, 1474–1476.

23. Amidror, I. (2000) *The Theory of the Moiré Phenomenon*, Periodic Layers, vol. I, Kluwer Academic Publishers, Dordrecht, ISBN-978-1-84882-180-4.

24. Coggrave, C.R. and Huntley, J.M. (1999) High-speed surface profilometer based on a spatial light modulator and pipeline image processor. *Opt. Eng.*, **38**, 1573–1581.

25. Quan, C., He, X.Y., Wang, C.F., Tay, C.J., and Shang, H.M. (2001) Shape measurement of small objects using LCD fringe projection with phase shifting. *Opt. Commun.*, **189**, 21–29.

26. Guo, H., He, H., and Chen, M. (2003) Gamma correction for digital fringe projection profilometry. *Appl. Opt.*, **43**, 2906–2914.

27. Takeda, M. and Mutoh, K. (1983) Fourier transform profilometry for the automatic measurement of 3-D object shapes. *Appl. Opt.*, **22**, 3977–3982.

28. Rajoub, B.A., Burton, D.R., and Lalor, M.J. (2005) A new phase-to-height model for measuring object shape using collimated projections of structured light. *J. Opt. A: Pure Appl. Opt.*, **7**, S368–S375.

29. Rajoub, B.A., Lalor, M.J., Burton, D.R., and Karout, S.A. (2007) A new model for measuring object shape using non-collimated fringe-pattern projections. *J. Opt. A: Pure Appl. Opt.*, **9**, S66–S75.

30. Maurel, A., Cobelli, P., Pagneux, V., and Petitjeans, P. (2009) Experimental and theoretical inspection of the phase-to-height relation in Fourier transform profilometry. *Appl. Opt.*, **48**, 380–392.

31. Zhang, X., Lin, Y., Zhao, M., Niu, X., and Huang, Y. (2005) Calibration of a fringe projection profilometry system using virtual phase calibrating model planes. *J. Opt. A: Pure Appl. Opt.*, **7**, 192–197.

32. Kemao, Q. (2004) Windowed Fourier transform for fringe pattern analysis. *Appl. Opt.*, **43**, 2695–2702.

33. Shulev, A., Van Paepegem, W., De Pauw, S., Stoykova, E., Degrieck, J., and Sainov, V. (2010) *Optical Measurement Techniques for Structures and Systems*, Shaker Publishing, pp. 369–378, ISBN-978-90-423-0366-9.

34. Yue, H.M., Su, X.Y., and Liu, Y.Z. (2007) Fourier transform profilometry based on composite structured light pattern. *Opt. Laser Technol.*, **39**, 1170–1175.

35. Fu, Y., Groves, R.M., Pedrini, G., and Osten, W. (2007) Kinematic and deformation parameter measurement by spatio temporal analysis of an interferogram sequence. *Appl. Opt.*, **46**, 8645–8655.

36. Gorthi, S.S. and Rastogi, P. (2010) Fringe projection techniques: whither we are? *Opt. Lasers Eng.*, **48**, 133–140.

37. Halioua, M., Krishnamurthy, R.S., Liu, H., and Chiang, F.-P. (1983) Projection moiré with moving gratings for automated 3-D topography. *Appl. Opt.*, **22**, 850–855.

38. Dirckx, J.J.J., Decraemer, W.F., and Eyckmans, M.M.K. (1990) Grating noise removal in moiré topography. *Optik*, **86**, 107–110.

39. Dirckx, J.J.J., Decraemer, W.F., and Dielis, G. (1988) Phase shift method based on object translation for full field automatic 3-D surface reconstruction from moiré topograms. *Appl. Opt.*, **27**, 1164–1169.

40. Andresen, K. (1986) Das phasenshift verfahren zum moiré-bildauswehrtung. *Optik*, **72**, 115–119.
41. Buytaert, J.A.N., Aernouts, J.E.F., and Dirckx, J.J.J. (2009) Indentation measurements on the eardrum with automated projection moiré profilometry. *Opt. Lasers Eng.*, **47**, 301–309.
42. Buytaert, J.A.N. and Dirckx, J.J.J. (2008) Moiré profilometry using liquid crystals for projection and demodulation. *Opt. Express*, **16**, 179–193.
43. Buytaert, J.A.N. and Dirckx, J.J.J. (2010) Phase-shifting moiré topography using optical demodulation on liquid crystal matrices. *Opt. Lasers Eng.*, **48** (2), 172–181.
44. Dirckx, J.J.J., Buytaert, J.A.N., and Vander Jeught, S.A.M. (2010) Implementation of phase-shifting moiré profilometry on a low-cost commercial data projector. *Opt. Lasers Eng.*, **48** (2), 244–250.
45. Ghighlia, D. (1998) *Two-Dimensional Phase Unwrapping: Theory, Algorithms, and Software*, John Wiley & Sons, Inc.
46. Buytaert, J.A.N. and Dirckx, J.J.J. (2012) Phase-stepping algorithms: overview and simulations, *Interferometry Principles and Applications*, NOVA Science Publishers, in press. ISBN 978-1-61209-347-5.
47. Carré, P. (1966) Installation et utilisation du comparateur photoélectriqueet interférentiel du Bureau International des Poids et Mesures. *Metrologia*, **2**, 13–23.
48. Schmit, J. and Creath, K. (1995) Extended averaging technique for derivation of error-compensating algorithms in phase-shifting interferometry. *Appl. Opt.*, **34**, 3610–3619.
49. Hariharan, P. and Oreb, B.F. (1987) Digital phase-shifting interferometry: a simple error-compensating phase calculation algorithm. *Appl. Opt.*, **26**, 2504–2505.
50. Zhang, S., Li, X., and Yau, S.-T. (2007) Multilevel quality-guided phase unwrapping algorithm for real-time three-dimensional shape reconstruction. *Appl. Opt.*, **46**, 50–56.
51. Aernouts, J.E.F., Soons, J.A.M., and Dirckx, J.J.J. (2010) Quantification of tympanic membrane elasticity parameters from in situ point indentation measurements: validation and preliminary study. *Hear. Res.*, **263**, 177–182.
52. Osten, W., Nadeborn, W., and Andrä, P. (1996) General hierarchical approach in absolute phase measurement. *Proc. SPIE*, **2860**, 2–13.
53. Zhao, H., Chen, W., and Tan, Y. (1994) Phase unwrapping algorithm for the measurement of three-dimensional object shapes. *Appl. Opt.*, **33**, 4497–4500.
54. Li, J.L., Su, H.J., and Su, X.Y. (1997) Two-frequency grating used in phase-measuring profilometry. *Appl. Opt.*, **36**, 277–280.
55. Ryu, W., Kang, Y., Baik, S.H., and Kang, S. (2008) A study on the 3-D measurement by using digital projection moiré method. *Optik*, **119**, 453–458.

Chapter 8

1. Thomson, W. (1855) On the thermoelastic properties of matter. *Q. J. Math.*, **I**, 57–77.
2. Jamieson, J.A., McFee, R.H., Plass, G.N., Grube, R.H., and Richards, R.G. (1963) *Infrared Physics and Engineering*, McGraw-Hill.
3. Nicholas, J.V. and White, D.R. (2001) *Traceable Temperatures*, 2nd edn, John Wiley & Sons, Ltd, Chichester.
4. Stanley, P. and Chan, W.K. (1984) Quantitative stress analysis by means of the thermoelastic effect. *J. Strain Anal. Eng.*, **20** (3), 129–137.
5. Stanley, P. and Chan, W.K. (1988) The application of thermoelastic stress analysis to composite materials. *J. Strain Anal. Eng.*, **23** (3), 137–142.
6. Potter, R.T. (1987) Stress analysis in laminated fibre composites by thermoelastic emission. Proceedings of the 2nd International Conference on Stress Analysis by Thermoelastic Techniques, SPIE Vol. 731, London.
7. Bakis, C.E. and Reifsnider, K.L. (1991) The adiabatic thermoelastic effect in laminated fiber composites. *J. Compos. Mater.*, **25**, 809–830.
8. Dulieu-Barton, J.M., Emery, T.R., Quinn, S., and Cunningham, P.R. (2006) A temperature correction methodology for

quantitative thermoelastic stress analysis and damage assessment. *Meas. Sci. Technol.*, **17** (6), 1627–1637.

9. Dulieu-Smith, J.M. (1995) Alternative calibration techniques for quantitative thermoelastic stress analysis. *Strain*, **31** (1), 9–16.

10. Whelan, M., Hack, E., Siebert, T., Burguete, R., Patterson, E., and Salem, Q. (2005) On the calibration of optical full-field strain measurement systems. *Appl. Mech. Mater.*, **3–4**, 397–402.

11. Dulieu-Smith, J.M., Quinn, S., Shenoi, R.A., Read, P.J.C.L., and Moy, S.S.J. (1997) Thermoelastic stress analysis of a GRP tee joint. *Appl. Compos. Mater.*, **4** (5), 283–303.

12. Dulieu-Barton, J.M., Earl, J.S., and Shenoi, R.A. (2001) Determination of the stress distribution in foam-cored sandwich construction composite tee joints. *J. Strain Anal. Eng.*, **36** (6), 545–560.

13. Boyd, S.W., Dulieu-Barton, J.M., and Rumsey, L. (2006) Stress analysis of finger joints in pultruded GRP material. *Int. J. Adhes. Adhes.*, **26** (7), 498–510.

14. Emery, T.R., Dulieu-Barton, J.M., Earl, J.S., and Cunningham, P.R. (2008) A generalised approach to the calibration of orthotropic materials for thermoelastic stress analysis. *Compos. Sci. Technol.*, **68**, 743–752.

15. El-Hajjar, R. and Haj-Ali, R. (2003) A quantitative thermoelastic stress analysis method for pultruded composites. *Compos. Sci. Technol.*, **63** (7), 967–978.

16. Pitarresi, G., Found, M.S., and Patterson, E.A. (2005) An investigation of the influence of macroscopic heterogeneity on the thermoelastic response of fibre reinforced plastics. *Compos. Sci. Technol.*, **65** (2), 269–280.

17. Wang, W.J., Dulieu-Barton, J.M., and Li, Q. (2010) Assessment of non-adiabatic behaviour in thermoelastic stress analysis of small scale components. *Exp. Mech.*, **50**, 449–461, doi:

18. Wong, A.K. (1991) A non-adiabatic thermoelastic theory for composite laminates. *J. Phys. Chem. Solids*, **52** (3), 483–494.

19. Lesniak, J.R. (1988) Internal stress measurement. *Proceedings of the 6th Congress on Experimental Mechanics, Portland*, SEM, Bethel, CT.

20. Sathon, N. and Dulieu-Barton, J.M. (2007) Evaluation of sub-surface stresses using thermoelastic stress analysis. *Appl. Mech. Mater.*, **7–8**, 153–158.

21. Mackenzie, A.K. (1989) Effects of surface coating on infra-red measurements. *Proc. SPIE*, **1084**, 59–71.

22. Quinn, S. and Dulieu-Barton, J.M. (2002) Identification of the sources of non-adiabatic behaviour for practical thermoelastic stress analysis. *J. Strain Anal.*, **37**, 59–72.

23. Welch, C.S. and Zickel, M.J. (1993) Thermal coating characterization using thermoelasticity. *Rev. Prog. Quant. Nondestructive Eval.*, **12**, 1923–1929.

24. Robinson, A.F., Dulieu-Barton, J.M., Quinn, S., and Burguete, R.L. (2010) Paint coating characterisation for thermoelastic stress analysis. *Meas. Sci. Technol.*, **21**, 085502.

25. Wong, A.K., Sparrow, J.G., and Dunn, S. (1988) On the revised theory of the thermoelastic effect. *J. Phys. Chem. Solids*, **49**, 395–400.

26. Eaton-Evans, J.M., Dulieu-Barton, J.M., Little, E.G., and Brown, I.A. (2006) Thermoelastic Studies on Nitinol Stents. *J. Strain Anal. Eng. Des.*, **41**, 481–495.

27. Dulieu-Barton, J.M., Eaton-Evans, J., Little, E., and Brown, I. (2008) Thermoelastic stress analysis of vascular devices. *Strain*, **44**, 102–118.

28. Wong, A.K., Dunn, S.A., and Sparrow, J.G. (1988) Residual stress measurement by means of the thermoelastic effect. *Nature*, **332**, 613–615.

29. Patterson, E.A. (2007) The potential for quantifying residual stress using thermoelastic stress analysis. Proceedings of SEM Conference, Springfield, MA.

30. Quinn, S., Dulieu-Barton, J.M., and Langlands, J.M. (2004) Progress in thermoelastic residual stress measurement. *Strain*, **40**, 127–133.

31. Harwood, N. and Cummings, W.M. (1991) *Thermoelastic Stress Analysis*, IOP Publishing Ltd, Bristol.

32. Dulieu-Barton, J.M. and Stanley, P. (1998) Development and applications of thermoelastic stress analysis. *J. Strain Anal. Eng. Des.*, **33**, 93–104.

33. Pitarresi, G. and Patterson, E.A. (2003) A review of the general theory of thermoelastic stress analysis. *J. Strain Anal. Eng.*, **38** (5), 405–417.
34. Dulieu-Smith, J.M. and Stanley, P. (1993) Progress in the thermoelastic evaluation of mixed-mode stress intensity factors. Proceedings of SEM Spring Conference on Experimental Mechanics, Dearborn, 1993, pp. 617–629.
35. Dulieu-Barton, J.M., Fulton, M.C., and Stanley, P. (1999) The analysis of thermoelastic isopachic data from crack tip stress fields. *Fatigue Fract. Eng. Mater. Struct.*, **23**, 301–313.
36. Tomlinson, R.A., Nurse, A.D., and Patterson, A. (1997) On determining the stress intensity factors for mixed mode cracks from thermoelastic data. *Fatigue Fract. Eng. Mater. Struct.*, **20** (2), 217–226.
37. Tomlinson, R.A. and Marsavina, L. (2004) Thermoelastic investigations for fatigue life assessment. *Exp. Mech.*, **44**, 487–494.
38. Dulieu-Barton, J.M. and Worden, K. (2002) Genetic identification of crack-tip parameters using thermoelastic isopachics. *Meas. Sci. Technol.*, **14**, 176–183.
39. Diaz, F.A., Patterson, E.A., Tomlinson, R.A., and Yates, J.R. (2004) Measuring stress intensity factors during fatigue crack growth using thermoelasticity. *Fatigue Fract. Eng. Mater. Struct.*, **27** (7), 571–584.
40. Zanganeh, M., Tomlinson, R.A., and Yates, J.R. (2008) T-stress determination using thermoelastic stress analysis. *J. Strain Anal.*, **43**, 529–537.
41. Stanley, P. and Dulieu-Smith, J.M. (1996) Devices for the experimental determination of individual stresses from thermoelastic data. *J. Strain Anal. Eng. Des.*, **31**, 53–63.
42. Barone, S. and Patterson, E.A. (1996) Full-field separation of stresses by combined thermo- and photoelasticity. *Exp. Mech.*, **36**, 318–324.
43. Barone, S. and Patterson, E.A. (1998) Polymer coating as a strain witness in thermoelastic stress analysis. *J. Strain Anal. Eng. Des.*, **33**, 223–232.
44. Barone, S. and Patterson, E.A. (1998) An alternative finite difference method for post-processing thermoelastic data using compatibility. *J. Strain Anal. Eng. Des.*, **33**, 437–447.
45. Lin, S.T., Miles, J.P., and Rowlands, R.E. (1997) Image enhancement and stress separation of thermoelastically measured data under random loading. *Exp. Mech.*, **37** (3).
46. Murakami, Y. and Yoshimura, M. (1997) Determination of all the stress components from measurements of the stress invariant by the thermoelastic stress method. *Int. J. Solids Struct.*, **34**, 4449–4461.
47. Crump, D.A., Dulieu-Barton, J.M., and Savage, J. (2010) Design and commission of an experimental test rig to apply a full-scale pressure load on composite sandwich panels representative of aircraft secondary structure. *Meas. Sci. Technol.*, **21** (16), 015108.
48. Freuhmann, R.K., Dulieu-Barton, J.M., and Quinn, S. (2010) Thermoelastic stress and damage analysis using transient loading. *Exp. Mech.*, **50**, 1075–1086.

Chapter 9

1. Dally, J.W. and Riley, W.F. (1991) *Experimental Stress Analysis*, McGraw-Hill, Inc., New York.
2. Theocaris, P.S. and Gdoutos, E.E. (1979) *Matrix Theory of Photoelasticity*, Springer-Verlag, Berlin.
3. Huard, S. (1997) *Polarisation of Light*, John Wiley & Sons, Ltd, Chichester.
4. Frocht, M.M. (1948) *Photoelasticity*, Vols. **I & II**, John Wiley & Sons, Inc., New York.
5. Durelli, A.J. and Shukla, A. (1983) Identification of Isochromatic Fringes. *Exp. Mech.*, **23** (1), 111–119.
6. Cloud, G.L. (1995) *Optical Methods of Engineering Analysis*, Cambridge University Press, New York.
7. Patterson, E.A. (2002) Digital photoelasticity: principles, practice and potential. *Strain*, **38**, 27–39.
8. Ramesh, K. (2000) *Digital Photoelasticity – Advanced Techniques and Applications*, Springer, Berlin.
9. Lesniak, J.R. and Zickel, M.J. (1998) Applications of automated grey-field polariscope. Proceedings of the SEM Spring

Conference on Experimental and Applied Mechanics, Houston, Texas, pp. 298–301.
10. Ramesh, K. and Ganapathy, V. (1996) Phase-shifting methodologies in photoelastic analysis: the application of Jones calculus. *J. Strain Anal.*, **31**, 423–432.
11. Ramji, M. and Ramesh, K. (2008) Whole field evaluation of stress components in digital photoelasticity -- issues, implementation and application. *Opt. Lasers Eng.*, **46** (3), 257–271.
12. Ghilgia, D.C. and Pritt, M.D. (1998) *Two-Dimensional Phase Unwrapping*, John Wiley & Sons, Inc.
13. Patterson, E.A., Brailly, P., and Taroni, M. (2006) High frequency quantitative photoelasticity applied to jet engine components. *Exp. Mech.*, **46** (6), 661–668.
14. Carazo-Alvarez, J., Haake, S.J., and Patterson, E.A. (1994) Completely automated photoelastic fringe analysis. *Opt. Lasers Eng.*, **21**, 133–149.
15. Ajovalasit, A., Barone, S., and Petrucci, G. (1995) Towards RGB photoelasticity: full-field automated photoelasticity in white light. *Exp. Mech.*, **35**, 193–200.
16. Pacey, M.N., James, M.N., and Patterson, E.A. (2005) A new photoelastic model for studying fatigue crack closure. *Exp. Mech.*, **45** (1), 42–52.
17. Kenny, B. and Patterson, E.A. (1985) Load and stress distributions in screw threads. *Exp. Mech.*, **25**, 205–213.
18. Quiroga, J.A., Pascual, E., and Villa-Hernandez, J. (2008) Robust isoclinic calculation for automatic analysis of photoelastic fringe pattern. *Proc. SPIE*, **7155**, 715530.
19. Zandman, F., Redner, A.S., and Dally, J.W. (1977) *Photoelastic coatings*, SESA Monographs No. 3, Iowa State University, Ames, IA.
20. Greene, R.J., Patterson, E.A., and Rowlands, R.E. (2008) Thermoelastic stress analysis, in *Handbook of Experimental Solid Mechanics* (ed. W.N. Sharpe Jr.), Springer, New York. pp. 743–767.
21. Gdoutos, E.E. (1985) in *Photoelasticity in Engineering Practice* (eds S.A. Paipetis and G.S. Holister), Elsevier Applied Science Publishers, London, pp. 181–204.
22. Smith, C.W. (1993) in *Experimental Techniques in Fracture* (ed. J.S. Epstein), VCH Publishers, Inc., New York, pp. 252–290.
23. Post, D. (1972) Photoelastic fringe multiplication -- tenfold increase in sensitivity. *Exp. Mech.*, **10**, 305–312.
24. Paipetis, S.A. (1985) in *Photoelasticity in Engineering Practice* (eds S.A. Paipetis and G.S. Holister), Elsevier Applied Science Publishers, London, pp. 205–224.
25. Fessler, H. (1992) An assessment of frozen stress photoelasticity. *J. Strain Anal.*, **27** (3), 123–126.
26. Aben, H. and Guillemet, C. (1993) *Photoelasticity of Glass*, Springer-Verlag, Berlin.

Abbreviations and Notations

Mathematical Notations

$*$	complex conjugation
$\star$	convolution
$\otimes$	correlation or convolution
$\approx$	approximately equal to
∇	gradient (nabla) operator
$\bullet'$	derivative of $\bullet$
a	bold lower case letters designate vectors
â	bold lower case letters with hat designate unit vectors
A	bold capital letters designate matrices
$\mathbf{A}^t$	transpose of matrix A
δ_{ij}	Kronecker delta symbol
FT	Fourier Transform
f^{-1}	inverse function
i	imaginary unit

Greek Symbols

α	phase shift angle
α	coefficient of linear thermal expansion
γ_{xy}	engineering shear strain
δ	separation
$\delta(\cdot)$	delta pulse
Δ	distance, difference or interval
$\Delta\varphi$	phase change or phase difference
ε	strain or strain component
$[\varepsilon_{ij}]$	strain tensor
$\varepsilon_{xx}, \varepsilon_{yy}, \varepsilon_{xy}$	tensorial components of strain in Cartesian co-ordinates
$\varepsilon_1, \varepsilon_2, \varepsilon_3$	principal strains
ε_0	dielectric permittivity
η	noise loss factor
θ, ϑ	tilt or incidence angle
κ	threshold value

Optical Methods for Solid Mechanics: A Full-Field Approach, First Edition. Edited by Pramod Rastogi and Erwin Hack.
© 2012 Wiley-VCH Verlag GmbH & Co. KGaA. Published 2012 by Wiley-VCH Verlag GmbH & Co. KGaA.

κ	Kolossov's constant
λ	wavelength or inter fringe plane distance
Λ	synthetic wavelength
λ and μ	Lamé constants
μ_0	magnetic permeability
$\mu\varepsilon$	microstrain, i.e. strain of 10^{-6}
υ	optical or carrier frequency
ν	Poisson's ratio
φ, ϕ	(wrapped) phase
Φ, ψ	unwrapped phase
ρ	mass density
ρ	lens insertion loss
σ	standard deviation
$\sigma_{xx}, \sigma_{yy}, \tau_{xy}$	stress components in Cartesian co-ordinates
$\sigma_1, \sigma_2, \sigma_3$	principal stresses
$[\tau_{ij}]$	stress tensor
ω	angular frequency
Ω	opening angle

Latin Symbols

a, A	amplitude
a	semi-major axis of elliptical hole, inner diameter of a thick cylinder
a_n, b_n	sample coefficients
b	Semi-minor axis of elliptical hole, outer diameter of a thick cylinder
b	global correlation vector
b	background
$b(\mathbf{x})$	field of contrast variation
B	Stefan Boltzmann constant
c	speed of light
c_n	complex sample coefficients ($c_n = a_n + i b_n$)
C	capacitance
C_i	stress-optic coefficient
C_ε	specific heat at constant strain
C_p	specific heat at constant pressure
$C(\cdot)$	cosine signal or coefficient
$C(\bullet, \clubsuit)$	covariance of variables $\bullet$ and $\clubsuit$
d	grating translation
D	inter grid distance
$D(\cdot)$	denominator term of a fraction
e	surface emissivity
E	Young's modulus
E_{ijkl}	elasticity tensor
E	electric field vector
f	focal length
f	frequency
F	force
f#	f number
G	shear modulus
h	Planck's constant

h	height or thickness
H	magnetic field vector
H_n	Hermite polynomial of order n
i, **j**, and **k**	unit vectors of Cartesian co-ordinates
I	intensity distribution or irradiance
J_n	Bessel function of order n
k	wavenumber
k	fringe order
k	thermal conductivity
K	Boltzmann constant
K	strain coefficient of photoelastic coating
K_I, K_{II}	stress intensity factors
K	relative permittivity
K	stiffness matrix
L	length
ℓ	element size
ℓ	grid to object plane distance along z-axis
ℓ'	grid to lens plane distance along z-axis
ℓ''	lens plane to object plane distance along z-axis
m	diffraction order, harmonic order
M	optical magnification
M_I	irradiance modulation
M	global correlation matrix
n	number of ...
n	refractive index
n_x, n_y, n_z	direction cosines of plane perpendicular to **n**
N	fringe order
N	number of ...
$N(\cdot)$	numerator term of a fraction
p	grid period or pixel pitch
p	pressure
$P(z)$	complex polynomial
P	projection matrix
Q	charge
R	reflectivity coefficient
R_f	reinforcement factor
$S(\cdot)$	sine signal or coefficient
S	Poynting vector
S	Stokes vector
t	transmission function
t	time
T	temperature
T	exposure time
T	threshold function
T	transformation matrix
T_x^n, T_y^n, T_z^n	Cartesian components of stress vector on plane perpendicular to **n**
u	displacement vector
V	voltage
x, y, z	Cartesian co-ordinates
x	2-D or 3-D co-ordinates
X, Y, Z	X-, Y-, Z-axis

Abbreviations

1-D	one-dimensional
2-D	two-dimensional
3-D	three-dimensional
A/D	Analogue-to-Digital
ADC	Analogue-to-Digital Conversion
AFM	Atomic Force Microscope
AOD	Average Optical Density
APS	Active Pixel Sensor
B/W	Black and White
C8-DVC	Digital Volume Correlation based on C8 FE shape functions
CCD	Charge-Coupled Device
CDS	Correlated Double Sampling
CFRP	Carbon-Fibre Reinforced Polymer
CMM	Coordinate Measuring Machine
CMOS	Complementary Metal-Oxide-Semiconductor
CTN	Compact Tension Notch
CW	Continuous Wave
DCDS	Digital Correlated Double Sampling
DHI	Digital Holographic Interferometry
DIC	Digital Image Correlation
DPSS	Diode Pumped Solid State
DVC	Digital Volume Correlation
EM	Electromagnetic radiation
ESPI	Electronic Speckle Pattern Interferometry
FE	Finite Element
FEI-DIC	Digital Image Correlation based on finite-element shape functions
FEM	Finite Element Modelling
FF	Fill Factor
FFT	Fast Fourier Transform
FOV	Field Of View
FPN	Fixed Pattern Noise
FSHS	Full-Scale Histogram Stretch
FTP	Fourier Transform Profilometry
FW	Full-Well
FWHM	Full-Width at Half-Maximum
GPU	Graphics Processing Unit
GRIN	Graded Index
I-DIC	Integrated Digital Image Correlation
IR	Infrared
LCM	Liquid Crystal Matrix
LCD	Liquid Crystal Display
LCP	Left Circularly Polarized wave
LCSI	Low-Coherence Speckle Interferometry
LSB	Least Significant Bit
LVDT	Linear Variable Displacement Transducer
μCT	micro Computed Tomography
MEMS	Micro-Electro-Mechanical System

MRI	Magnetic Resonance Imaging
NA	Numerical Aperture
NDT	Non-Destructive Testing
NIR	Near-Infrared spectral range
OCT	Optical Coherence Tomography
OPD	Optical Path length Difference
PBS	Polarizing Beam Splitter
PCB	Printed Circuit Board
PEM	Phase-Evaluation Method
PIV	Particle Image Velocimetry
pp	Pixel Pitch
PPD	Pinned photodiode
PSA	Phase-Stepping/shifting Algorithm
PZT	Piezoelectric material
Q4-DIC	Digital Image Correlation based on Q4 FE shape functions
QE	Quantum Efficiency
QWP	Quarter Wave Plate
RCP	Right Circularly Polarized wave
RGB	Red-Green-Blue
RMS	Root Mean Square
ROI	Region Of Interest
SCF	Stress Concentration Factor
SEM	Scanning Electron Microscope
SIF	Stress Intensity Factor
SLD	Super Luminescence Diode
SNR	Signal-to-Noise Ratio
TSA	Thermoelastic Stress Analysis
UV	Ultraviolet spectral range
VIS	Visible spectral range
WFTP	Windowed Fourier Transform Profilometry
X-DIC	eXtended Digital Image Correlation
XI-DIC	eXtended and Integrated Digital Image Correlation
XCMT	X-ray Computed Micro-Tomography
ZOI	Zone Of Interest

Index

a

aberration
– chromatic aberration   17, 55
– spherical aberration   17–18, 55–56
absorption   39, 47, 57, 61–62, 80, 193, 260–261, 304, 370, 380
acoustooptic modulators   104
adhesion   257–258
adiabatic conditions   346–347, 356–357, 361
Airy pattern   28–30, 32
analog-to-digital conversion   56, 58
analyzer   370–374, 376–377, 379, 381–382, 391
antireflective coating   20
aperture   17, 28–31, 33, 41, 49, 51, 53–54, 103–104, 107, 125, 185–186, 230, 232, 240–242, 252, 282, 299–300, 321, 333
astigmatism   17–18, 56
attenuation   5, 234, 241, 248, 256, 356, 363
automatic gain control   75

b

Babinet-Soleil compensator   376
bandwidth   25, 45, 66, 82, 111, 251, 255, 287
beam splitter   22–23, 103, 173–174, 236, 251–252, 255, 267, 285, 290
beat signal   83, 90, 100
Bessel function   28, 54, 237, 239, 293
bias   83, 85, 88–89, 101, 110, 122, 189, 194–195, 203, 207, 222
birefringence   44, 104, 154–155, 241, 367, 370, 374, 376, 378, 383
blackbody   348–349, 352, 354
Boltzmann constant   349–351
branch-cut method   132
Brewster angle   7–9

c

calibration   66, 103, 113–114, 124–126, 179, 216–217, 236, 267, 314, 321, 337, 352–355, 386
camera calibration   216
carrier frequency   87–88, 97, 107–112, 265
Cauchy's formula   143–144
caustics   160–161
CCD   57–59, 62, 65–66, 68, 82, 126, 128, 173–177, 183, 189, 199, 205, 240, 242, 244, 262, 277, 279, 284–285, 332–336, 343, 370–371
CFRP   273–274, 357
chain plate   167–172
characteristic length   384
characteristic polynomial   118, 120, 125–127, 139
circle of least confusion   18
cladding   42–43
CMOS   57–60, 62–63, 65–68, 81–82, 126, 128, 183, 279, 301
coating   20, 159, 162, 164, 168, 170–171, 304, 358–360, 363, 382–383
coefficient
– coefficient of linear expansion   359
– stress-optic coefficient   369
coherence
– coherence layer   247–257, 261
– coherence length   25–26, 40–41, 83, 89, 102, 106, 232–233, 244–245, 249, 251, 253–254, 256, 261–262
– coherence width   26–27, 45
– lateral coherence   26–27, 45
– spatial coherence   26, 38, 42, 90, 231–232, 234
– temporal coherence   41, 235, 252
coherent gradient sensor   161
color space   335

coma   17–18, 55–56
compact tension   241–242
contouring   135, 279
contrast   22–23, 25–27, 41, 51, 53, 64, 72–73, 75, 79, 81, 86, 104, 111, 122, 141, 155, 159, 188, 195, 231, 233, 236–237, 240–241, 247–249, 251, 256, 261, 268, 274, 279, 295, 299, 310, 328, 332–334
convolution   27, 77, 108, 110, 253, 282–283
coordinate measuring machine   303
core   42–43, 270, 272
correlation   64, 108–109, 158, 163, 173, 175, 177, 183–228, 237, 241, 264, 299, 305, 354
coupling efficiency   43–44
crack   153–154, 156, 160–161, 209–212, 222–224, 227, 268, 273–274, 364, 366, 387–390, 392
critical angle   7
curvature   12–13, 17–18, 33, 45, 56, 107, 157–158, 161–162, 193, 281, 305

d

damage   270, 273–274, 357, 365
dark current   56, 65–66, 80, 82, 127–128
dark-field   368, 370–372, 374–377, 382, 385, 391
decorrelation   241, 266, 274
defect   66, 258–259, 265, 270–273, 277–278, 299
delamination   258
demodulation   307, 318, 323, 325–326, 332–336, 343
depletion region   62–63
depth of field   49, 53, 80, 241
dielectric
– dielectric constant   2
– dielectric interface   7
– dielectric permittivity   1–2
diffraction
– diffraction grating   29–30, 104–105
– diffraction limit   54, 56, 332
– Fraunhofer diffraction   27–28
– Fresnel diffraction   27
diffuse reflection   90, 158, 175, 271, 304–305, 322
digital image correlation (DIC)
– beam-DIC   226
– FFT-based DIC   200
– global DIC   188, 213–214, 227
– integrated DIC   208–210, 228
– local DIC   187, 212
– Q4-DIC   209, 212–213, 220, 226–227
digital volume correlation   184, 220–221, 223

discontinuity   222, 341–342, 382
displacement
– in-plane displacement   201, 210, 241
– out-of-plane displacement   155, 158, 161, 173, 230, 236, 244, 263, 266, 269–270, 273, 287, 289, 300–301
– standard displacement uncertainty   194, 198–199, 204, 214–215, 221
distortion
– barrel distortion   19
– pincushion distortion   19
dynamic range   56, 64, 91, 126, 128, 183, 189–190, 195, 241, 256, 327, 333

e

elastic constant   146
elastic region   147
electric field   2–6, 8, 19, 33, 35–36, 43, 45, 57
electromagnetic wave   1, 3, 5, 7
emission
– spontaneous emission   39
– stimulated emission   39
emissivity   352, 354, 358–360, 363
error
– quantization error   71
– systematic error   107, 194, 203

f

f-number   24–25, 28, 31, 53, 232
Fabry-Perot   20–22, 40, 45
failure plane   142–143
fatigue   153, 163, 166–167, 171, 222, 270, 364, 367, 384, 388
feature   70, 78–79, 100, 112, 165, 262, 299, 304–305, 331, 342
feature extraction   70, 79
fiber
– multimode fiber   11, 43
– optical fiber   42, 162, 236, 242
– single mode fiber   42–43, 45
field of view   49, 53, 113, 131, 165, 258, 260–261, 272, 285, 299, 305, 370, 374, 379–380
field stop   53–54, 58
fill factor   56, 60–61, 111, 335
filtering   44, 93, 108, 111, 132–134, 173, 189, 206, 252, 254, 299, 306–307, 310, 313–315, 317, 326, 341
finesse   21
finite element analysis (FEA)   356
flux-transfer   59
focal length   13–16, 18, 24, 28, 31, 45, 50–54, 58, 81, 173, 232, 271
focal plane   15, 17, 28–29, 55

focal point   28, 51–53, 216, 218
four-point bending   355, 386
Fourier holography   279
Fourier spectrum   111, 316
Fourier transform   23, 28, 97, 107, 110–111, 115, 187, 252–253, 283, 306–307, 311, 313–317, 337, 379–380
Fourier transform profilometry   307, 311, 313–315, 317
Fourier transform spectrometry   23
fracture mechanics   153–154, 160, 387
frame rate   56, 66, 69, 76, 241, 336, 343
Fraunhofer diffraction   27–28
Fresnel holography   230, 279, 292, 296, 301
Fresnel transform   280, 282–283, 297, 301
fringe
– correlation fringe   175
– fractional fringe   376–377
– fringe constant   370, 372, 383–384, 386, 388
– fringe order   84, 130, 134, 136, 163, 170–171, 341, 369, 374–376, 379–380, 383, 385–391
– fringe plane   309, 318, 321, 329–332, 337, 342
– fringe tracing   329
– isochromatic fringe   163, 368, 374–375, 380
– isoclinic fringe   376
fringe projection   84, 113–114, 129, 134–135, 303–306, 308, 310, 312, 314, 316, 318, 320, 322, 324, 326, 328, 330, 332, 334, 336, 338, 340–343
full well   63

g
Gaussian beam   32–33, 45
geometrical optics   10–11, 28, 47, 49–50, 304
graphics processing unit (GPU)   336
Gray code   341–342
gray level   72–73, 78, 128, 185–186, 190–191, 193, 196–199, 212–213, 221
grid   77–78, 112, 114, 156, 164, 305–326, 328–338, 341–343, 387

h
half wave plate   35
heat transfer   346, 356–357
high-speed camera   67, 183
histogram   72–76, 79, 125, 231–233
hole   49, 57, 60, 62, 151–154, 169–171, 178, 357–358

holography   1, 19, 24, 41, 44, 155, 157, 163, 173, 175, 178, 229–230, 239, 247, 278–279, 281, 283–285, 287, 289, 291–293, 295–297, 299, 301, 304
honeycomb   269–273
Hooke's law   146, 355, 383
HRR field   160

i
image correlation   158, 163, 183–184, 186, 188, 190, 192, 194, 196, 198–220, 222, 224, 226–228, 299, 305
image formation   1, 10, 44, 47, 49, 51, 53, 55, 58
image plane   15–16, 19, 48–51, 53–54, 58, 217, 263
image processing   48–49, 70–73, 75, 77, 79–80, 108, 156, 183, 279, 299
imager formats   68
infrared   38, 47, 50, 62, 159, 165, 186, 229, 345, 348–349, 351–354, 363, 387
intensity   3, 5–6, 9, 19–22, 24, 26, 28, 32, 45, 59, 70, 78, 83–85, 89, 93, 97, 101, 105, 114–115, 154–158, 173, 184–185, 231–234, 236–238, 241–242, 254–256, 262–264, 269, 281, 299, 306, 309–312, 314, 322, 326, 334–335, 338, 349, 364, 371–373, 377–381, 388, 391
interference
– geometric interference   316, 322, 325
– interference pattern   19–20, 25–27, 45, 231, 235–236, 238–239, 262–266, 318, 334
interferometer
– Fizeau interferometer   102–103, 105, 126
– Mach-Zehnder interferometer   103–104, 251, 285
– Michelson interferometer   22–23, 102, 109, 236
– Twyman-Green interferometer   102–103
interferometry
– heterodyne interferometry   23, 88, 134, 304
– holographic interferometry   24, 230, 235, 299
– low-coherence speckle interferometry   246–247, 249, 251, 253, 255, 257, 259, 261
– moiré interferometry   164
– speckle pattern interferometry   229, 235, 237, 239, 241, 243, 245, 298, 304
– speckle pattern shearing interferometry   230, 262–263, 265, 267, 269, 271, 273, 275, 277, 298

interpolation   77–78, 189–194, 196, 199, 201–203, 207, 209, 214–215, 221–222, 307, 329, 341
irradiance   3, 83, 85–88, 91–97, 101, 110, 115–116, 118, 122, 125, 127–129
isochromatic fringe   163, 368, 374–375, 380
isoclinic fringe   376
isopachics   155, 364
isothetics   157

j
Jones
 – Jones calculus   35, 37–38, 370
 – Jones matrix   35, 37–38
 – Jones vector   4, 35–37

l
Lamé constant   346
Lamé's problem   171, 178
Laplace transform   115
laser   1, 5–6, 22–24, 26, 31–32, 38–45, 86, 89, 102–105, 107, 113, 127, 158, 173–174, 176, 229–231, 236–237, 241, 244–245, 275–277, 284–285, 289–290, 295, 297–298, 300–302, 304
lens
 – concave lens   15
 – convex lens   18
 – thick lens   13–14, 45
 – thin lens   13–14, 16, 50–52
lens combination   14, 16
lens design   19
liquid crystal   334, 336, 342
Lissajous figure   93, 99, 139

m
magnetic field   1–2, 4
magnetic permeability   1–3
magnification   10, 15, 18, 24, 53, 77, 81, 112, 173, 205, 211, 232, 313, 319–320, 322, 330–333, 343
Malus' law   6
material
 – birefringent material   367, 369
 – isotropic material   146–147
 – orthotropic material   146, 156, 345, 353, 355
Maxwell's equations   1, 49
measurement uncertainty   127
media
 – absorbing media   5
 – anisotropic media   4, 7, 33–34
 – biaxial media   34
 – isotropic media   33–34
 – uniaxial media   34
membrane   172, 177, 257, 259–260, 295, 337, 340
MEMS   141, 155, 158, 161, 163, 165, 172–173, 175, 177–179, 257, 259
microinterferometry   178
miscalibration   124, 126
mode
 – longitudinal mode   40–41, 45
 – transversal mode   40–41, 43
modulation   83, 85–88, 91, 93, 97, 101, 104–105, 107, 110–112, 115–116, 127, 187–188, 194, 201, 203, 237–238, 264, 269, 274–275, 295, 310–312, 318, 332–334, 380
Mohr's circle   163, 387
moiré
 – geometric moiré   164
 – moiré interferometry   164
 – projection moiré   307–308
 – shadow moiré   157–158, 308–309, 316, 318–319, 321
morphological operation   77, 79
motion detection   78–79
Mueller matrix   371

n
noise
 – 1/f-noise   65–67
 – digitization noise   128–129
 – fixed-pattern noise   56
 – grid noise   306–307, 309, 318–319, 321–326, 335, 337–338, 342–343
 – kTC-noise   64, 67–68, 80
 – random noise   76, 101, 127, 316, 327
 – readout noise   56, 127–128
 – shot noise   65–66, 76, 80, 127–128
 – thermal noise   65–67
nondestructive evaluation   391
nonlinearity   122, 126, 329–330, 335
numerical aperture   31, 107, 125, 282

o
off-axis   18, 107, 279, 281
optical coherence tomography   247
optical flow   185–186
optical path length difference   233
optical system   12, 14–19, 30, 49–50, 53–54, 58–59, 236, 256, 300, 370

p
paraxial approximation   12, 17, 33, 47, 50
phase
 – phase jump   130

- phase map   83–84, 87–89, 111, 129–130, 132–133, 137, 242, 244–245, 250–251, 258, 279, 300, 313, 316, 339, 342
- unwrapped phase   84, 270, 277, 299, 314
- wrapped phase   84, 87–89, 111, 129, 137, 244, 265, 270, 287, 291, 299, 329

phase shifting
- spatial phase shifting   88, 107
- temporal phase shifting   86, 102

phase stepping   107, 237–238, 265, 333
phase unwrapping   84, 129–131, 133–135, 137, 139, 265, 270, 307, 313, 315, 317, 321, 328, 341–342
photodetector   57, 60, 63, 65, 351
photodiode   57, 61–65, 82

photoelasticity
- reflection photoelasticity   170, 368, 381–383, 387–388

photoelectric effect   47–48
photomechanics   229
photometry   50
piezoelectric   102, 236, 252, 267, 343
pinhole camera   49, 216
pixel pitch   58–59, 61, 112, 297
pixel size   216, 265
Planck's constant   60, 349
Planck's law   348–349, 354
plastic region   147–148
point source   6, 10, 44, 322
point-by-point   154, 159, 162, 181
Poisson's ratio   171, 192, 210, 228, 230, 242, 346, 355
polariscope   167–170, 368–382, 385–386, 389, 391–392
polarization   3–4, 7–9, 34–38, 44–45, 101, 104–105, 241, 267, 279, 334, 370–371, 374
polarized light   3–4, 6, 44, 231, 334, 367, 377, 382
polarizer   6, 36–37, 334, 370–377, 391
Poynting vector   3
pressure sensor   165, 172–173, 175, 177–179, 257, 259
principal plane   16
principal point   14
printed circuit board   285
profilometry   303–308, 310–338, 340, 342–343
projector   113–114, 136, 138, 316, 331–332, 334–336, 342–343

q

quantum efficiency   56, 60
quarter wave plate   35

r

radiant flux   352
radiation   21–22, 39–40, 47, 49–50, 54, 165, 304, 348–349, 352
radiometry   49–50
radius of curvature   12–13, 33, 45, 107

ray
- chief ray   53
- marginal ray   53
- ray-tracing   50–52

reflection
- reflection coefficient   8–9, 81, 248
- total internal reflection   7, 9, 42

reflectivity   22, 126, 252–253, 280, 311–312, 322, 352
refraction   7–9, 11–13, 51–52, 367, 369
refractive index   2–3, 5, 7–8, 10–11, 20–21, 31, 40, 42–43, 45, 55, 80, 104, 106–107, 155, 248, 254, 257, 261–262, 367, 369, 379
regularization   186, 188, 213–214
residual   44, 78, 99, 156, 165, 181, 184, 186, 189, 191, 195–196, 212, 222, 363, 365, 367–368, 391

resolution
- angular resolution   31–32
- spatial resolution   31–32, 56, 81, 89, 101, 109, 111, 204, 206, 222, 224, 229–230, 252, 268, 282, 287, 289, 297–299, 301, 306, 310, 315, 318, 326, 357

resonator   39–41, 45
retardation   367, 369–372, 379, 381–382, 391–392
RGB   290–292, 335, 380, 390–391
rigid body movement   263
Ronchi ruling   333

s

sawtooth image   84, 89, 96, 109, 132–133
sensitivity   58–60, 66, 84, 113, 120–122, 129, 134–137, 175, 184, 195, 197, 221–222, 230, 236, 239–241, 244–245, 258, 263, 266–269, 278, 285, 287, 289–290, 295, 297–300, 302, 327, 331, 349, 354, 357, 360, 363, 365, 383
sensitivity vector   84, 175, 240, 244, 263, 266–267, 297, 302
shading   331, 338–339, 341
Shannon theorem   279, 281, 297
shape reconstruction   216, 218, 340
shearing distance   262–264, 266, 268–269, 271, 273–274, 277, 298, 301
shearography   158, 163, 173, 175, 177–179, 230, 262–271, 273–275, 277–278, 298, 300–301
shutter   64, 67–69, 127, 241

signal-to-noise ratio   50, 91, 99, 328
Snell's law   7–9, 12
specific heat   346–348
speckle
– objective speckle   231
– speckle averaging   239, 266, 271, 273, 301
– speckle correlation   158, 175
– speckle field   131, 236, 265, 269, 280, 299
– speckle pattern   24, 43, 111, 130, 158, 229–235, 237, 239, 241, 243, 245, 256, 262–263, 265–267, 269, 271, 273, 275, 277, 297–298, 304
– speckle phase   130, 132, 237–238, 263
– speckle size   24, 232, 235, 266, 299
– subjective speckle   230
spectral range   48–49, 60, 72
spectral width   25, 41
St. Venant's principle   384
Stefan-Boltzmann   349–353
stereo vision   216–218
Stokes vector   371–373, 377–378, 381
strain
– linear strain   149
– plane strain   209, 228, 367
– principal strain   149, 156, 160, 170–171, 180, 369
– shear strain   145, 215
– strain invariant   346
– strain witness   355, 359, 382–383
strain gauge   159–160, 162, 164, 181, 204–206, 348, 354, 365
strain-optic law   170
streamline fillets   163
stress
– assembly stress   164, 168
– normal stress   143, 161, 163, 180
– plane stress   146, 171, 180, 209–210, 367, 382, 387
– principal stress   148–150, 154–155, 159, 163, 180, 369, 372, 385–387
– residual stress   363, 365
– shear stress   143, 145, 150, 163, 180, 389
– stress concentration   141, 151, 153, 156, 160–162, 171, 391
– stress concentration factor   151, 171
– stress tensor   143–145, 149, 151, 178, 180, 346, 390
– yield stress   150, 179, 348, 362
stress intensity factor   388
stress-freezing   378, 384, 386–387
stress-optic coefficient   369
stress-optic law   369–370, 372, 374, 383, 390
stroboscopic illumination   239, 295
structured light   304–305, 310–312, 338, 342

super luminescence diode   251
surface geometry   305, 318
synthetic wavelength   101, 114, 134, 304

t

Tardy compensation   376–377, 380
tensile test   205–206, 228, 241, 244, 299, 348
thermal conductivity   346, 356
thermoelastic constant   347, 354–355, 359, 361, 363
thermoelastic effect   345, 347, 359, 363
thermoelastic stress analysis   159, 161, 163, 345–346, 348, 350, 352, 354, 356, 358, 360, 362, 364, 366, 387
thermography   1, 345, 348–349, 351, 356
thresholding   78–79, 314
time-average   157, 178
time-edge effect   383–384
topography   257, 303, 316, 318, 337–340, 342
transmission   7, 21–23, 93, 102–104, 114, 125, 156, 160, 165, 167–169, 178, 192, 257, 310, 312, 322, 328, 334–335, 376, 390
triangulation   304–305
TSA   159, 164, 345–346, 352, 354–358, 361–365
TV-holography   173–175
tympanic membrane   337, 340

v

V-number   43, 45
van Cittert-Zernike theorem   231
vibration   19, 24, 89, 104, 115, 127, 155, 157, 230, 235, 237–239, 241, 269–270, 274–277, 279, 292–293, 295–296, 298, 301
video   48, 58, 66, 69–71, 75, 79, 89–90, 109, 134, 235–236, 241

w

wafer   58, 61, 165, 172, 238, 259, 301
wave
– circularly polarized wave   4, 35
– electromagnetic wave   1, 3, 5, 7
– elliptically polarized wave   35–36
– evanescent wave   9
– extraordinary wave   34
– linearly polarized wave   2, 5, 35–37
– monochromatic wave   2, 280
– ordinary wave   34
– plane wave   2–7, 10, 28–29, 34–36, 214, 269
– reference wave   24, 83, 86, 91, 100, 102, 104, 110, 173, 177, 229, 236, 241, 252, 254, 263, 269, 279–282, 285, 287, 297
– spherical wave   6, 10, 33, 269

wave equation   1
wavefront   10, 26, 32, 83–85, 88, 91, 101, 109–110, 130
wavelength   2–3, 9, 11, 17, 20–21, 24–25, 30–31, 35, 41, 43, 45, 47, 49–50, 55, 59–60, 62, 83–84, 101–102, 105–107, 113–114, 134–136, 165, 171, 195, 206, 229, 247, 251, 256, 260–261, 282–283, 285–286, 289, 297, 299, 302, 304, 334, 348–354, 359, 369–370, 372, 374, 380, 391
wavelet   316
white-light   81, 106, 112–114, 374

Wien's displacement   350
Wiener-Khintchine theorem   253

y
yield criterion   150, 180
Young's double-slit experiment   19–20, 26
Young's modulus   171, 230, 242, 346, 348, 355, 361, 383–384

z
Z-transform   115–118, 123